Translational Systems Sciences

Volume 45

In 1956, Kenneth Boulding explained the concept of General Systems Theory as a skeleton of science. He describes that it hopes to develop something like a "spectrum" of theories—a system of systems which may perform the function of a "gestalt" in theoretical construction. Such "gestalts" in special fields have been of great value in directing research towards the gaps which they reveal. There were, at that time, other important conceptual frameworks and theories, such as cybernetics. Additional theories and applications developed later, including synergetics, cognitive science, complex adaptive systems, and many others. Some focused on principles within specific domains of knowledge and others crossed areas of knowledge and practice, along the spectrum described by Boulding. Also in 1956, the Society for General Systems Research (now the International Society for the Systems Sciences) was founded. One of the concerns of the founders, even then, was the state of the human condition, and what science could do about it. The present Translational Systems Sciences book series aims at cultivating a new frontier of systems sciences for contributing to the need for practical applications that benefit people. The concept of translational research originally comes from medical science for enhancing human health and well-being. Translational medical research is often labeled as "Bench to Bedside." It places emphasis on translating the findings in basic research (at bench) more quickly and efficiently into medical practice (at bedside). At the same time, needs and demands from practice drive the development of new and innovative ideas and concepts. In this tightly coupled process it is essential to remove barriers to multi-disciplinary collaboration. The present series attempts to bridge and integrate basic research founded in systems concepts, logic, theories and models with systems practices and methodologies, into a process of systems research. Since both bench and bedside involve diverse stakeholder groups, including researchers, practitioners and users, translational systems science works to create common platforms for language to activate the "bench to bedside" cycle. In order to create a resilient and sustainable society in the twenty-first century, we unquestionably need open social innovation through which we create new social values, and realize them in society by connecting diverse ideas and developing new solutions. We assume three types of social values, namely: (1) values relevant to social infrastructure such as safety, security, and amenity; (2) values created by innovation in business, economics, and management practices; and, (3) values necessary for community sustainability brought about by conflict resolution and consensus building. The series will first approach these social values from a systems science perspective by drawing on a range of disciplines in trans-disciplinary and cross-cultural ways. They may include social systems theory, sociology, business administration, management information science, organization science, computational mathematical organization theory, economics, evolutionary economics, international political science, jurisprudence, policy science, socio-information studies, cognitive science, artificial intelligence, complex adaptive systems theory, philosophy of science, and other related disciplines. In addition, this series will promote translational systems science as a means of scientific research that facilitates the translation of findings from basic science to practical applications, and vice versa. We believe that this book series should advance a new frontier in systems sciences by presenting theoretical and conceptual frameworks, as well as theories for design and application, for twenty-first-century socioeconomic systems in a translational and trans-disciplinary context.

Masahiro Kuroda
Editor

Intangible Capital Investment and Economic Growth

Science for Science, Technology and Innovation Policy

Springer

Editor
Masahiro Kuroda
Minato-ku, Tokyo, Japan

ISSN 2197-8832 ISSN 2197-8840 (electronic)
Translational Systems Sciences
ISBN 978-981-96-7674-3 ISBN 978-981-96-7675-0 (eBook)
https://doi.org/10.1007/978-981-96-7675-0

This Springer imprint is published by the registered company Springer Nature Singapore Pte Ltd.
The registered company address is: 152 Beach Road, #21-01/04 Gateway East, Singapore 189721, Singapore

Preface

After the Renaissance in the fifteenth and sixteenth centuries, Western countries moved away from the medieval power structure dominated by religious authority centered on "God as Creator."

This book embarks on an exploration of the intricate relationships between science, intangible capital, and the trajectory of economic growth. It investigates how advances in scientific knowledge and technological innovation have continuously reshaped the structures of modern economies, not only fueled periods of rapid expansion but also generated profound societal shifts. By tracing the evolution of science and technology policy, the study seeks to uncover the underlying mechanisms that link intangible investments—such as research, education, and human capital—to the transformation of economic systems and the emergence of new social challenges. Through this lens, the narrative sets the stage for a deeper inquiry into how these unseen drivers of growth have come to influence not just markets and productivity, but the very fabric of contemporary society.

In the nineteenth century, scientific knowledge of material structures based on quantum physics is deepened and enhanced our understanding of life sciences grounded in evolutionary theory. Moving into the twentieth century, the material and energy revolution led to the discovery of semiconductor materials, further advancing biological sciences and bringing unprecedented prosperity to humanity.

However, such progress simultaneously brought to light new global social challenges, such as the depletion of natural resources and environmental destruction. These developments not only intensified conflicts between nations and ethnic groups but also triggered social phenomena within countries—exacerbating income disparities and class consciousness. By the twenty-first century, the growing inequality in distribution and heightened awareness of social disparity has generated tension among social classes, ethnic groups, and nations, increasingly constraining the freedom of thought and action of individuals.

The background of current issues includes structural changes in the relationship between science, economy, and society caused by the technological innovations resulting from the scientific revolution and the excessive pursuit of economic growth based on them. It is also said that this has led to the emergence of issues that science alone cannot resolve—the so-called trans-science problems. How can we accurately assess the merits and drawbacks of the scientific revolution and solve the social problems it has caused? If these issues have structurally affected the foundations of social norms established since the nineteenth century—namely, democracy, the principles of market competition, and the logic of rational scientific thinking—then what can science and technology policy achieve in the trans-science era to correct the structural basis of science and socioeconomics? Tackling this challenge must begin with deepening our understanding of the nature of modern science and technology through scientifically grounded analysis.

We are now in an era that demands an accurate assessment of both the benefits and drawbacks of the scientific revolution, as well as a thoughtful response to the social problems it has brought about. These issues have exerted structural influences on the very foundations of social norms—such as democracy, the principles of market competition, and the logic of rational scientific thinking—that have been established since the nineteenth century.

In the trans-science era, what can science and technology policy do to correct the structural foundations of science and the socioeconomic system? To address this challenge, we must begin by gaining a profound understanding of the essence of modern science and technology through scientifically grounded analysis.

Minato-ku, Tokyo, Japan Masahiro Kuroda

Acknowledgments

I would like to express my heartfelt gratitude to Professor Dale W. Jorgenson of Harvard University for providing me with invaluable experiences through my research. From 1978, I had the opportunity to spend more than two years as the visiting researcher at Harvard University, supported by the Fulbright Foundation. The discussions with him and our subsequent joint research were invaluable experiences for mastering empirical economics, focusing on U.S.-Japan productivity comparisons. It was a wonderful opportunity to rigorously understand the theoretical structure of neoclassical economics and its empirical research methods, building on the empirical research achievements accumulated at Keio University. During my stay in Cambridge, Boston, I brought my wife with my four children. During our stay, Professor Jorgenson and his wife Linda gave their heartfelt care to us; we would like to say our heartfelt thanks to Dale and Linda.

Unfortunately, Professor Jorgenson passed away in the spring of 2022 at the age of 87. His sincere dedication to academia remains unforgettable.

I dedicate this book which was born from our joint research with him and sincerely pray for his eternal peace.

During my stay at Cambridge, I had great support from the Japan Institute of Harvard University, which was directed by Professor Reischauer. If we had not had that support, we could not have stayed there for 2 years. I would like to extend our deepest gratitude to Professor Reischauer and his institute.

The achievements of this book stem from the author, Masahiro Kuroda's, academic journey, which began with participating in Professor Ichiro Okuma's seminar at the Faculty of Economics, Keio University, Japan, and continued through the Graduate School of Keio University and the Keio Economic Observatory (KEO). The spirit of the Keio Economic Observatory, founded by Professors Kotaro Tsujimura, Keiichiro Obi, and Iwao Ozaki, was inherently aimed at empirically elucidating the economic structures within society by revealing the data generating mechanisms of economic phenomena, which are challenging to observe through controlled laboratory experiments.

Professor Tsujimura focused on analyzing household saving and consumption behavior. Professor Obi and Professor Ozaki focused on the labor supply analysis of households and the analysis of input-output technical structures based on industrial linkage analysis, respectively. Professor Tsujimura empirically demonstrated the variability of consumer preference fields through the formation of consumption habits, while Professor Obi investigated the preference fields of household income and leisure, particularly the labor supply behavior of household heads and non-head adult members. Professor Ozaki conducted studies on economies of scale in industrial technology selection and the interdependence of input-output. In these distinguished and invaluable research environments, the author received rigorous training through weekly discussions with other esteemed professors, alongside the three founding professors. This setting provided an extraordinary and rare opportunity for the author's intellectual and professional development.

Although my research is not yet complete, I reached this point due to years in a rich research environment and opportunities for discussions. I would like to express my heartfelt gratitude to the professors at the Keio University Industrial Research Institute, including Professors Tsujimura, Obi, and Ozaki, and other senior professors, as well as the current active faculty members like Koji Nomura, for their long-term support. I could have many life friends through the activity in my university life. Th friendly relation-ship is continuing until now. It was a great experience in my life, and I could obtain many suggestions from my activity with my friends.

Since 2008, my work with the Center for Research and Development Strategy (CRDS) at the Japan Science and Technology Agency (JST) has broadened my views on science and society. Collaborating with natural scientists expanded my understanding beyond economics and provided insights into current science, technology, and policies. My interest in science and technology policy's impact on society grew from my work on the 2011 MEXT 4th Basic Plan for Science and Technology committee. Later, I joined the MEXT project "Science for Redesigning Science, Technology, and Innovation Policy (SciREX project)." The SciREX project, which continues to this day, endeavors to promote science and technology policy "scientifically."

The research outcomes presented in this book are attributed to the SciREX project initiated at the National Graduate Institute for Policy Studies (GRIPS). I owe a great deal to Project Leader Professor Tateo Arimoto and project members Professor Kenta Ikeuchi, Yasushi Hara, Michael C. Huang, Tatsuro Sasaki, as well as Hiroaki Harada from JST.CRDS. Primarily, Michael C. Huang played an essential role in the publication of this book. He participated in discussions until the very end, assisted with the writing, and even helped with proofreading. His substantial contribution makes him a co-author of this book. He is a professional in the science and technology system in his country. I asked him to introduce the recent situations and policy thoughts in his country. His message and the introduction are introduced in Chaps. 6 and 7 briefly.

I would like to thank other members, including Mr. Satoshi Otagaki from Mitsubishi Research Institute and Mr. Masatoshi Yokohashi from Applied Research Institute, who engaged in the creation of input-output tables for evaluating science and technology policies and the construction of the empirical general equilibrium model for our policy simulation. I would like to take this opportunity to express my deep gratitude to all these individuals for their support.

In such a research environment, my focus was on empirically examining the general interdependent structure that describes the mechanisms by which various economic agents, such as producers, consumers, and the government as policy entity, both domestically and internationally, achieve supply-demand equilibrium in product and factor markets through their participation in the market. While this field is known as general equilibrium theory in theoretical economics, my aim was to empirically capture the behavioral characteristics of consumers, producers, governments, and foreign entities participating in the market. I aimed to analyze how these behaviors interact through the market to achieve supply-demand equilibrium and to link these findings to the evaluation of policies and their implications, thereby contributing to the resolution of policy issues.

This book was published by Springer Nature Singapore Pte Ltd., and I would like to express my gratitude to Mr. Yutaka Hirachi and especially to the editor, Ms. Manonmani Pradeepkumar, for their tremendous support during the publication process.

Lastly, although this book is the result of a project conducted as part of the SciREX initiative, it owes much to the Japanese-language volume authored by me and published by Keio University Press, titled "*Science, Technology and Japan's Economic Growth (Intangible Capital Investment and Economic Growth: Science, Technology and Innovation*)." For the English publication, Keio University Press kindly granted permission for translation and publication of the original Japanese edition. I would like to express my deepest gratitude for their generous cooperation. I would like to once again extend my deepest thanks to all the individuals who supported the compilation of this book. If there are any errors in the volume, they are entirely responsible.

I wholeheartedly welcome constructive criticism from readers.

December, 2025 Masahiro Kuroda

Competing Interests The author has no competing interests to declare that are relevant to the content of this manuscript.

Contents

Contributor and Editor

About the Editor and Author

Masahiro Kuroda was born on January 17, 1941. While enrolled at Keio University, he studied abroad at Harvard University in the United States for 2 years from 1979 to 1981, with support from the Fulbright Foundation. At Harvard, he received numerous forms of support while serving as a Visiting Professor in the Department of Economics and as a Visiting Scholar at the Harvard University Japan Institute. His collaborative research with Professor Dale W. Jorgenson, a leading economist at Harvard, has continued since that time, and this book is the culmination of that research. During his stay, he also received generous support from Dr. Edwin O. Reischauer, then Director of the Harvard University Japan Institute. Masahiro Kuroda is a Principal Fellow, Center for Research and Development Strategy, Japan Science and Technology Agency (JST). He is a Professor Emeritus of Keio University. He has a Ph.D. in Business and Commerce. He was promoted to Professor in 1982. Through the Director of the Keio Economic Observatory and Vice President (provost) of Keio University during June 2001–May 2005, he moved on to President of Economic and Social Research Institute and Cabinet Office during June 2005–May 2008. In June 2008, he got affiliated with the current position. He served as President of the International Input-Output Association (IIOA) during 1997–1999 and as Chairman of the Advisory Committee for Natural Resources and Energy, Energy and Supply Sub-committee in Japan from 1998 to 2010. Dr. Kuroda has received the Economist Award and the Japan Statistical Society Prize. His publications include the following books: *General Equilibrium Analysis of Japanese Economy* (co-author, 1974); *Introduction to Positive Economics* (1984), and *Quantitative Analysis to General Equilibrium* (1989).

About the Contributor

Michael C. Huang is a Senior Research Fellow at the Ocean Policy Research Institute (OPRI) of the Sasakawa Peace Foundation, a Visiting Researcher at the SciREX Center of the National Graduate Institute for Policy Studies (GRIPS), and an adjunct Lecturer in charge of Special Economic Seminar in the Faculty of Economics at Toyo University in Tokyo, Japan. Dr. Huang's research expertise includes general equilibrium analysis, GTAP modeling for policy simulations, and blue impact evaluation. At OPRI, he is responsible for economic analysis for projects related to the blue economy, blue finance, science, technology, innovation policy assessment, and disaster risk management. Dr. Huang aims to connect the ocean to the social system on interdisciplinary topics, including renewable energy, deep sea water industry, global supply chain resilience, eco-friendly shipping, and hydrogen society. Prior to joining OPRI, he served as a Research Associate in the Asian Development Bank Institute and a resource speaker for capacity building and training. He holds a Ph.D. in Public Economics from GRIPS and is a member of the Japan Ocean Policy Society, Japan Economic Association, and the Japan Castle Foundation. Dr. Huang has been awarded the best presenter in the International Input-Output Association (IIOA) in 2014 and 2016. He has edited several books, including *Innovative Social Practice and Community-Based Approach to the Ocean* (2025, Springer-Nature, co-eds.), *Achieving Ocean Equity* (2026, Edward Elgar Publishing, co-eds.), *Moving toward Net-Zero Carbon Society* (2023, Springer-Nature, co-eds.), *Blue Economy and Blue Finance* (2022, ADBI press, co-eds.), and *Climate Change and Disaster Management Law and Policy* (2020, New Sharing Publishing, co-eds.).

List of Figures

List of Tables

Chapter 1
Toward Science for Science and Innovation Policy

Masahiro Kuroda

Abstract The Meiji Restoration's reforms were pivotal in transforming Japan into a competitive industrialized state, albeit at the cost of rising militarism in the early twentieth century. Japan's economic challenges, from post–World War I to the Great Depression, eventually led to its participation in World War II. After Japan's defeat in 1945, the nation's political and economic systems underwent further restructuring, guided by democratic principles under the new Constitution of Japan. The reforms in agrarian and industrial sectors, notably the dissolution of zaibatsu and the implementation of the priority production system, played crucial roles in Japan's postwar recovery. These policy experiments in restoring market mechanisms and expanding production capacity heavily contributed to the Japanese postwar economic growth and industrial modernization.

While scientific and technological advancements have fueled economic expansion, they have also contributed to widening societal disparities, creating tensions across nations and social strata. The chapter calls for a rethinking of the balance between growth, wealth distribution, and societal well-being, advocating for a new paradigm that harmonizes science, technology, and equitable progress. Through an analysis of Japan's post–World War II economic trajectory, this chapter examines how science and technology influenced its recovery and eventual stagnation in the twenty-first century. The chapter aims to empirically investigate the interrelationship between technological evolution and economic growth, assessing its dual role as both a driver and outcome of development. Finally, it offers recommendations for future science and technology policies, industrial strategies, and the need for a new scientific philosophy to achieve sustainable, balanced societal advancement.

Keywords Economic growth · Science and technology policy · Japanese economy · Wealth of Nations · Trans-Science Age

M. Kuroda (✉)
Professor Emeritus, Keio University, Tokyo, Japan

M. Kuroda (ed.), *Intangible Capital Investment and Economic Growth*, Translational Systems Sciences 45,
https://doi.org/10.1007/978-981-96-7675-0_1

1.1 Toward a "Prosperous and Equal Society Through Economic Growth"

The quest for an "affluent and equitable society" through economic growth has been a long-standing human aspiration. Since the early twentieth century, economic growth has shifted from a means to achieve prosperity and equality to an end. This shift has resulted in a stark reality: scientific and technological advancements have fueled economic growth but have also created substantial societal disparities, even amidst the wealth achieved. These disparities have heightened tensions among nations and social strata, fundamentally altering the original human aim of balancing wealth and equality through economic growth. The unrelenting drive for economic growth, while initially yielding positive results, has also spawned adverse effects such as inequality and conflict. Thus, we must reassess our goals and methods to establish a new paradigm that harmonizes science and society.

This calls for the creation of a new scientific philosophy and corresponding policy instruments that can govern the nexus of science and society. To accomplish this, we need to scrutinize how the evolution of science and technology in the twentieth century contributed to the pursuit of economic growth and what outcomes these advancements yielded in the context of economic growth. These inquiries necessitate empirical substantiation.

This book aims to trace the trajectory of the Japanese economy after World War II, charting its path toward recovery and sustained growth. It also seeks to analyze the structural shifts that have occurred in the Japanese economy, especially as it entered a period of extended stagnation at the dawn of the twenty-first century, thereby altering its growth path. Our analysis will focus on the empirical examination of the relationship between the evolution of science and technology and the achievement of economic growth, viewing both as means and ends.

Drawing on this thorough analysis, we plan to suggest strategies for future science and technology policies, as well as industrial policies. These recommendations will be grounded in our empirical investigation to foster a more harmonious integration of science, technology, and societal progress.

1.2 Adam Smith's Principle of Market Competition and Neo-Classical Economics

In the eighteenth century, Adam Smith focused on elucidating the actual economy's structure and addressing institutional deficiencies. He challenged the belief that markets could operate spontaneously without intervention, arguing against the laissez-faire policy's ability to create "optimal and efficient markets" in "harmony with natural phenomena." Smith identified competitive disparities within the real market, such as the hierarchical relationship between experts and apprentices, which obstructed the market mechanism's optimal function. He advocated policy intervention to ensure equality among market participants, enabling the price mechanism to operate effectively.[1]

In the tangible economy and society following the eighteenth and nineteenth century's Industrial Revolution, the market seemed to be able to guide economic growth ideally and seem to be able to realize the ideal market without requiring any policy modifications. Rapid globalization and economic expansion created an illusion of a thriving real economy and society under laissez-faire principles. However, this era's reality involved both increased globalization and sustainable economic growth seemingly guided the market not to be idealistic. This success was illusory, overlooking the underlying complexities and disparities within the economy, which necessitated policy interventions for sustainable and equitable progress. Smith identified competitive disparities within the real market, but however, neoclassical economists, in this context, sought to construct a comprehensive equilibrium model to systematically explain market mechanisms.

Leon Walras, in his work *Outline of Pure Economics*, defined economics' scope as the examination of "social facts" regarding policy and morality.[2] He introduced a "perfectly organized competitive market" to analyze this scope. His approach involved logically deducing normative market functions from a mathematical model devoid of any "frictions," then proposing adjustments required in real market functions based on these normative propositions. While Walras' method introduced a scientific "theory and demonstration" approach from natural sciences to elucidate economic phenomena, it contributed to clarifying economics' logical framework.

Walras acknowledged that his foundational premise of a "frictionless world" did not align with economic society's reality. He recognized the need for policies to bridge this gap and rectify it in the real world. The nineteenth-century economists assessed the real economy without questioning their theoretical assumptions. Neoclassical economists often saw the real economy as frictionless and supported laissez-faire, opposing government intervention. This stance diverged significantly from Smith's advocacy of "market principles."

1.3 Critiques of Neo-Classical Economics by Leontief

In his 27th J. W. Gibbs Memorial Lecture, "Economics and Mathematics" delivered to the American Mathematical Society in Baltimore, USA, Wassily Leontief, an advocate of input-output analysis, expressed sharp criticism against neoclassical "general equilibrium theories." Leontief noted, "Rarely has modern empirical science built such a complex theory on limited and contradictory facts."[3] There has been scant effort to refine a theory by empirically estimating the values of parameters

[1] Smith, Adam (1776) and (1759).

[2] Walras, Leon (1874).

[3] Leontief, Wassily (1966).

within a "pure" theory. This traditional approach persists among today's economists, regardless of their mathematical inclination. A cursory examination reveals that the empirical assumptions supporting these theories are primarily qualitative, marked by ambiguity and generality. Therefore, pure theory yields only a limited number of practical propositions.

1.4 Empirical General Equilibrium Theory: Input-Output Analysis

Leontief's input-output analysis, a pioneering work in economics, provided an empirical framework for the general equilibrium theory. This analysis, conducted using the Harvard Mark I computer at Harvard University in the 1930s, resulted in an input-output table for the US economy, which was subsequently used for policy evaluation. The analysis utilized the input-output structure of commodity production to describe the mechanism that determines transaction quantities and market prices in equilibrium with supply and demand. This tool allowed for the examination of how the supply-demand balance alters the structure of a commodity's supply and demand when market demand composition changes, assuming technological characteristics remain constant. As science advances, changes in production techniques will adjust the use of inputs for each commodity. This, in turn, can shift the market's supply-demand balance. Increasing observations can enhance the model's autonomy, but this does not negate the potential for further autonomy through additional observations. The focus, rather, is on determining the direction of future economic and social trends based on current technological characteristics. If problems arise, the challenge is to guide the current technological characteristics to change the economy and society in the desired direction. This involves providing policy guidelines on how to steer these technological characteristics. This approach to policy analysis was characteristic of Leontief's work" (Kuroda & Nomura 2004).

1.5 Science and Society in the Trance-Science Age

In his paper "Science and Trans-Science," German physicist Alvin M. Weinberg introduces the concept of "Transcendental Science" (Weinberg 1972).[4] This new paradigm transcends traditional modern science and addresses the challenges posed by the current scientific paradigm. Weinberg discusses the relationship between science, technology, and society, noting that it has created various challenges. While some of these challenges are being addressed through the evolution of science and technology, many cannot be resolved by science alone, necessitating a new scientific paradigm. This new paradigm acknowledges the need for transformation. It recognizes that the cost of problem-solving is high, the problems are complex, and the scientific paradigm must evolve. These issues extend beyond the realm of science, encompassing ethics, and artistic sensibilities. Weinberg suggests that humanity faces challenges not encompassed by the conventional scientific paradigm. He advocates for the creation of a "trans-science" that includes the humanities and social sciences, disciplines traditionally excluded from the scientific paradigm. This perspective aligns with Leontief's critique of economics.

In addressing this task, a re-examination and reorganization of existing findings reveal several key perspectives to consider when contemplating the state of trans-science in contemporary science and society. While it may not be possible to provide answers to all these perspectives, it is crucial to consider the nature of "science, technology, and society" considering these issues.

1.6 Science, Technology, and Innovation Policy in the Trans-Science Age

Since the advent of the modern scientific paradigm in the seventeenth century, there has been a significant advancement in knowledge within natural science fields such as physics, chemistry, and biology. These fields have primarily focused on elucidating natural phenomena, adopting a mechanical view of nature. Many humanities and social science subjects, linked to social values, do not fit well within this paradigm due to their indirect relationship with natural phenomena. However, economics stands as an exception within the humanities and social sciences, having partially incorporated the paradigm of the natural sciences into its methodology. **In the current era of trans-science, characterized by a complex relationship between science and society, it is key to construct an economic model that reevaluates the relationship between science and economics and the nature of economic policy, considering the accomplishments and shortcomings of economics thus far.**

How has the natural sciences paradigm influenced economics? Since the emergence of neoclassical economics in the nineteenth century, economics has relied more on normative theories than on the observation and analysis of real markets to elucidate the structure of the "market," a central issue in economic phenomena. The approach involved making assumptions, deriving propositions from the structure, and modifying the real market to align with the normative theory if the propositions were desirable. This approach was influenced by the real-world experiences of developed countries like the UK, which experienced stable growth. It suggested that a market economy thrives on competition in a laissez-faire environment and should be adjusted to support such a market. This implies that the challenge in the real economy and society is to modify the real market by removing various constraints and shifting toward a laissez-faire market. However, if the competitive base of workers and entrepreneurs participating in the market competition is uneven in the real market, a laissez-faire policy without policy intervention would not achieve equality in the competitive base of these participants, and the market would not align with the vision of neoclassical economists. The theories could only be propositions derived from premises without examining the empirical validity of the premises and could no longer be empirical propositions reconcilable with real phenomena. This is partly due to the difficulty of verifying premises in real market structures where experimentation is challenging.

To construct a "trans-science" that addresses issues unsolvable by modern science, it is first necessary to understand the structure of the economy and society from which observed facts emerge, without making unnecessary assumptions. It is crucial to reproduce, as much as possible, the structure of the market under the given institutional conditions and legal constraints of the observed economy and society, given the behavior of the various economic agents participating in that market. It is crucial that the market trends from theoretical models reflect the real market and that the effects of new policies are identifiable, allowing us to reduce changes in economic agents' behavior. The policy must be clear, with an analytical framework for implementation. **In other words, if the observation environment cannot be fixed in advance, as in a laboratory experimental environment, it is necessary to demonstrate, as stably as possible, from what market environment the observed values are created and from what Data Generating Mechanism the observed values from that environment are derived. Understanding the structure of these observables is the primary goal of experimental design.**

As noted, the twentieth century marked an era of unprecedented sustained economic growth for humankind. Scientific advancements since the late nineteenth century have greatly enhanced our understanding of energy and material structures. This understanding has revealed the structure of new materials and paved the way for their creation. Furthermore, the development of semiconductor materials, computer technology, and information and communication technology has fundamentally transformed the structure of global society and the relationship between science and society.

In addition, the accumulation of empirical observations of natural phenomena, beginning with Darwin in the nineteenth century, led to the emergence of biological evolution and the elucidation of the origins of life and bioscience. In the twentieth century, this extended to the elucidation of life structures and genetics, and brain science and mental structures, leading to rapid advances in medicine, medical science, psychology, and spiritual science. These scientific and technological advances have spurred progress in thermochemistry, electronic science, materials science, and bioscience, described as an energy revolution. **Concurrently, the pace of technological innovation has accelerated through the utilization of the intangible knowledge accumulated in these fields, and the speed of information transmission has increased through the evolution of information processing technology**. This has expanded the economy of knowledge on a temporal and spatial scale. In today's discussions on the relationship between science and society, it is crucial to evaluate this intangible fixed asset of knowledge assets, particularly its economies of scale in time and space.

The scientific and technological advances of the twentieth and twenty-first centuries have confronted humanity with previously unknown challenges and made us aware of the limits of global-scale development. We now understand that the scale of fossil fuels required by Earth is not inexhaustible, there are limits to the scale of supply, and the emission and accumulation of greenhouse gases will cause excessive pollution of the global environment, affecting the ecosystem and even threatening human survival. While human activities have led to an increase in income and wealth on average, they have also resulted in significant disparities in wealth distribution. These growing disparities are creating friction between classes in different countries and manifesting as international disharmony and conflict. The rapid progress of science in recent years has accelerated its impact on society, creating various forms of disharmony and posing further challenges to society. These challenges are not unrelated to the economic effects of the temporal and spatial scale of knowledge assets, a characteristic of modern-era science and technology.

1.7 Overview of the Book

This book examines the complex interactions between modern science, technology, and society, which have resulted in various contemporary issues. By investigating the post–World War II evolution of the Japanese economy, it seeks to identify key issues, empirically understand the interplay between science and technology in the modern era, and discern its structure and associated problems. The goal is to contemplate future science and technology policies, as well as industrial strategies. The book concludes by assessing the current state of science and proposing future technology and industrial policies.

Chapter 2 begins by examining Japan's socio-economic conditions at the end of World War II and during the US occupation, followed by an evaluation of reconstruction and rehabilitation policies. Postwar Japan faced severe challenges: the GDP per capita in 1946 was 55% lower than the prewar peak of 1934–1936, the manufacturing production index was a mere fifth of the 1944 level, and rice production, the country's staple food, hit its lowest since 1903. The populace was forced to survive on minimal living costs. Compounding these hardships, the wholesale price index surged by 300% in 1946 and another 100% in the subsequent year, leading to hyperinflation. Over five million people, including military personnel and civilians, were repatriated from abroad, resulting in widespread unemployment. In response to these challenges, the Japanese government embarked on a radical reform of the prewar market structure and a comprehensive overhaul of the socio-economic fabric. This included agrarian reform, zaibatsu dissolution, and labor law revisions, all implemented under occupation. These reforms represented a significant experiment in dismantling prewar privileges of various hierarchies and organizations, and in artificially creating an open market where all citizens could compete equally. This rare policy induction of Adam Smith's "free competition principle"—leveling the competitive playing field in a free market—provided invaluable experience in designing the system that would eventually drive Japan's high economic growth.

Chapters 3 and 4 outline the evolution of the Japanese economy from the 1960s to the 1990s, building on the postwar reforms discussed in Chapter 3. These chapters highlight the key characteristics of this economic progression. Chapter 4 introduces the concept of Total Factor Productivity (TFP), a metric used in subsequent chapters to gauge economic growth. This chapter scrutinizes the economic expansion of the Japanese economy from its postwar recovery to the high-growth period of the 1970s, contrasting it with the US economy. Comparing the TFP trends reveals the macro and sector-specific international competitiveness of both countries, highlighting Japan's high economic growth outcomes. Chapter 5 addresses the growth trajectory of the Japanese economy post-1970s. It reveals a shift in investment patterns following period, transitioning from tangible fixed asset accu-mulation to intangible fixed asset accumulation, particularly in research and development (R&D) and information processing. **This chapter presents a novel approach to empirical science by exploring the measurement of intangible assets such as R&D and information processing and their services.** Chapter 5 outlines the framework for measuring intangible fixed capital investment, essential for R&D and information processing. It explains how the results of these activities are measured by accumulating this capital. The chapter explains how services from intangible capital investments fit into the wider economic framework. **We propose a method for understanding the incorporation of such intangible capital investments into the overall framework and their economic impact.** The first issue addressed is the concept and measurement of the quantity of these intangible capital services and their prices. For measuring the impact of these capital services on economic accounting, we suggest a method to capture the structure of the production of these intangible capital services through an inter-industry analysis. This might be covered in Chap. 5.

Chapter 6 employs a trans-log price function to measure the cost structure of each industry. This approach aids in understanding the technological attributes of R&D and information processing activities (ICT), as well as their impact on the main production activities of each industry. This analysis is based on the time-series data from the input-output table, which encapsulates the R&D and the ICT information processing activities proposed in Chap. 5. Chapter 7 probes the economic implications of the shift in investment activities toward R&D and information processing. This examination is conducted in the context of the recent Japanese economy, which is perceived to have transitioned from a high-growth phase to a low-growth phase. This transition marks a deviation in the economic growth pattern that the Japanese economy has pursued since the postwar period, particularly evident in the 1990s. The chapter contemplates the significance of the economic effects resulting from this shift in investment toward R&D and information processing activities.

In Chapter 7, we propose to develop a "policy simulation" model to assess the interdependence and effects of industrial policy and science, technology, and innovation policy. This model should be based on our analysis of the shift in investment structure from tangible to intangible fixed capital. Our aim is to determine the future direction of science, technology, and innovation policy in alignment with industrial policy. **We explore how the enhancement of intellectual assets, through intangible capital investments such as R&D and ICT, can contribute to economic**

and societal development. Ultimately, we propose a vision for science and society in a "trans-scientific" era, emphasizing the crucial role of intangible capital investments in information processing.

The Japanese economic structure experienced a significant shift following the burst of the bubble economy post the Plaza Accord in the mid-1980s. Subsequently, from the 1990s, **the Japanese economy transitioned from a mass production and consumption model, which had propelled its growth from the postwar reconstruction era to the early 1970s, to a model where capital investment pivoted from tangible to intangible assets, particularly in R&D and information processing technology**. This evolution in information processing technology instigated a shift from a heavy industrial structure to an era dominated by intangible fixed capital in the form of information, driving societal structural changes. The economy is now based on information processing technology and advanced computers and software that analyze this information. This shift has also led to an expansion of intangible fixed assets in the form of intellectual assets related to R&D investment, significantly altering the structure of market value. The division of labor, where knowledge development in the form of information and R&D as intangible fixed capital enhances industrial productivity and transforms the industrial structure, is shaping a new economic society. This trend is permeating not only domestic but also international division of labor structures and innovations in the international value chain. These developments in information and knowledge are transforming and evolving the social structures of the twenty-first century. **The question then arises: how will the expansion of knowledge assets transform and expand the interdependent relationships of economic systems, creating a new industrial structure based on the characteristics of contemporary technological structures?** This book answers these questions by using a policy simulator to explore how combining science and technology innovation policy with industrial policy can create a new economic structure.

References

Kuroda, M., & Nomura, K. (2004). Technological change and accumulated capital: A dynamic decomposition of Japan's growth. In E. Dietzenbacher & M. L. Lahr (Eds.), *Wassily Leontied and input-output economics*. Cambridge University Press.

Leontief, W. W. (1953). *Studies in the structure of the American economy*. Oxford University Press.

Leontief, W. W. (1966). *Essays in economics, theories and theorizing*. Oxford University Press. (Japanese translation: Kazuhiko Tokuyama).

Smith, A. (1759). *The theory of moral sentiments*. A. Millar, A. Kincaid and J. Belims. (Japanese translation: Mizuta Y (1990) The Theory of Moral Sentiments. Chikuma Shobo).

Smith, A. (1776). In E. Cannan (Ed.), *An inquiry into nature and causes of the wealth of nations* (pp. 1–2). Methuen. (Japanese translation: Ouchi H, Matsukawa S (1969) The Wealth of Nations I, II. Iwanami Shoten).

Walras, L. (1874). *Elements of pure economics or the theory of social wealth*. Porcupine Press. (Japanese Translation: Hisatake Masao (1983), Iwanami Shoten).

Weinburg, Alvin M. (1972). Science and Trans-Science" Minerva, vol.10.No.2.

Chapter 2
Departure from Collapse: The Rebirth and Development of the Japanese Economy in the Post World War II

Masahiro Kuroda

Abstract While scientific and technological advancements have fueled economic expansion, they have also contributed to widening societal disparities, creating tensions across nations and social strata. This book calls for a rethinking of the balance between growth, wealth distribution, and societal well-being, advocating for a new paradigm that harmonizes science, technology, and equitable progress. Through an analysis of Japan's post–World War II economic trajectory, this book examines how science and technology influenced its recovery and eventual stagnation in the twenty-first century. **The chapter aims to empirically investigate the interrelationship between technological evolution and economic growth, assessing its dual role as both a driver and outcome of development. Finally, I would like to offer my recommendations for future science and technology policies, industrial strategies, and the need for a new scientific philosophy to achieve sustainable, balanced societal advancement**.

In the mid-nineteenth century in Japan, the leaders of Meiji period recognized the need for modernization after observing the advancements in Western societies, particularly in political systems, market economies, and the evolution of science and technology. Japan, previously isolated under the Edo shogunate, underwent rapid reforms to build a constitutional monarchy, establish a parliamentary system, and dismantle its feudal societal structure. These changes facilitated the country's modernization and engagement with global economies. The Meiji Restoration's reforms were pivotal in transforming Japan into a competitive industrialized state, albeit at the cost of rising militarism in the early twentieth century. Japan's economic challenges, from post–World War I to the Great Depression, eventually led to its participation in World War II. After Japan's defeat in 1945, the nation's political and economic systems underwent further restructuring, guided by democratic principles under the new Constitution of Japan. The reforms in agrarian and industrial sectors, notably the dissolution of zaibatsu and the implementation of the priority production system, played crucial roles in Japan's postwar recovery. These policy

---

M. Kuroda (✉)
Professor Emeritus, Keio University, Tokyo, Japan

M. Kuroda (ed.), *Intangible Capital Investment and Economic Growth*, Translational Systems Sciences 45,
https://doi.org/10.1007/978-981-96-7675-0_2

experiments in restoring market mechanisms and expanding production capacity have laid the foundation for Japan's rapid postwar economic growth and industrial modernization.

While scientific and technological advancements have fueled economic expansion, they have also contributed to widening societal disparities, creating tensions across nations and social strata. The chapter calls for a rethinking of the balance between growth, wealth distribution, and societal well-being, advocating for a new paradigm that harmonizes science, technology, and equitable progress. Through an analysis of Japan's post–World War II economic trajectory, this chapter examines how science and technology influenced its recovery and eventual stagnation in the twenty-first century. The chapter aims to empirically investigate the interrelationship between technological evolution and economic growth, assessing its dual role as both a driver and outcome of development. **Finally, it offers recommendations for future science and technology policies, industrial strategies, and the need for a new scientific philosophy to achieve sustainable, balanced societal advancement**.

Keywords Meiji Restoration · Constitutional monarchy · Zaibatsu dissolution · Economic reforms · Postwar recovery, Science, Technology, and Innovation Policy · Trans-Science Age

2.1 Japanese History of the Modernization After Edo Era

In the mid-nineteenth century, Japan's Meiji leaders visited the US and European countries including Britain and France. **They observed societies advanced by three key developments: the establishment of a legitimate state based on democratic principles, the construction of a market economy, and the evolution of science and technology**. These elements, underpinned by scientific logical thinking, supported the legitimacy of the state in the early modern period. This exposure underscored the urgency of modernizing Japan, then a semi-civilized society in East Asia.

The political system choices—monarchy, aristocracy, or republic—required the establishment of a parliamentary system for support. This system needed to define the relationship between political sovereignty and parliament, particularly in the context of member elections. The enactment of a constitution as a legal norm to support this political system represented a rare and challenging effort to revolutionize Japan's feudal society and change the national mentality. This was particularly difficult given the challenges of opening the country after long isolation and harmonizing it with the international environment.

The Edo shogunate's isolation system, which lasted for over 200 years, had been stable under the decentralized feudal system of samurai clans. The Tokugawa Shogun led this system, and the status system of samurai, farmers, merchants, and Confucian ethical thought formed a significant foundation of Japanese society. **This system represented the establishment of Japan's spiritual and material civilization. However, the modern era's Western society's development demanded global**

socio-economic development and cooperation that transcended geographical conditions. Japan had to seek new national legitimacy in this global context.

The Meiji Restoration's systemic reforms began with the Grand Council of State and led to rapid social reforms. These included the establishment of a constitutional monarchy, the abolition of feudal domains, the reform of the samurai, farmer, artisan, and merchant status system, and the enactment of the Meiji Constitution. The establishment of a parliamentary system opened Japan's borders to the outside world. The Meiji period also improved international amity and trade conditions, including addressing unequal trade treaties with the United States and other Western nations. It also established autonomous tariff decisions and extraterritorial jurisdiction, issues that the Japanese people encountered for the first time in their history in the modern international environment. **Domestically, the enactment of the Meiji Constitution, which established a constitutional monarchy, and the shift to a parliamentary cabinet political system, represented a rapid, bloodless revolution. This revolution sought to rewrite the hierarchical order of feudal society and change the people's sense of rights to equality of all four classes, including the liberal civil rights movement**.

The Meiji and Taisho periods in Japan saw rapid societal development, spurred by the industrial revolution in the textile industry and other technological reforms. This era also witnessed a growing desire to secure overseas markets and resources. Events such as the Sino-Japanese and Russo-Japanese wars, an escalating arms race with Western powers, and pressure to expand influence on the Asian continent transformed Japan's political system into a militaristic, unified totalitarian state.

Japan entered World War I in 1914, aligning with the Allied side under the Anglo-Japanese Alliance. The war, instigated by the Sarajevo Incident involving the German Austrian Turkish allies and the British French Russian coalition, lasted over 4 years and ended with an Allied victory. The Treaty of Versailles in 1919 marked this victory but imposed harsh terms on the defeated Germany, including reparation claims and territorial cessions. These conditions gave rise to Nazism in Germany, complicated European state relations, and led to the establishment of a totalitarian state.

Postwar economic recovery and intensifying international competition to expand the global market led to an imbalance in global supply and demand. This imbalance increased market risks, causing stock market instability, which triggered a price collapse in the US stock market and its subsequent spread worldwide, leading to the Great Depression of 1929. Global oversupply fears lowered agricultural commodity prices, prompting many countries to adopt protectionist trade policies. These policies triggered high tariffs, shrunk world markets, and created unemployment, resulting in a vicious cycle in the global economy. The effects spread to Germany, the war's defeated nation, causing a financial depression, increased unemployment, and stagnation in the payment of World War I reparations. These effects further spread to other European economies. In Germany, this led to the birth of Hitler's regime, while in the Soviet Union, the expansion of socialist ideas led to the formation of Stalin's regime in 1936. The German invasion of Poland in 1939 sparked World War II. In 1940, Italy joined the war, and in 1941, following the attack on Pearl Harbor,

Japan extended its war front across the Asian continent. Japan's defeat in 1945 marked its entry into the war against the United States. More than a quarter-century after opening to the outside world and after two world wars, Japan entered a period of momentous change, echoing the Meiji Restoration.

2.2 Departure from Collapse

In August 1945, upon accepting the Potsdam Declaration, the Japanese government sought to maintain the emperor's position under the Imperial Japanese Constitution (Meiji Constitution) and to "uphold the national polity." However, the Potsdam Declaration called for a fundamental reform of the existing constitution, advocating for "the establishment of a responsible government with peaceful tendencies," "a revival and strengthening of democratic tendencies," and "respect for fundamental human rights." In April 1946, a general election for the House of Representatives was held under new election laws, which granted women the right to vote. In May, the 90th Imperial Diet convened, and on June 20, the "Imperial Constitution Revision Bill" was introduced. The House of Councilors passed it on October 7, and it was then promulgated.

The newly enacted provisions that would shape the "democratic normative orthodoxy" of future Japanese society. These provisions included "the status of the emperor and sovereignty of the people," "renunciation of war," "duties and rights of the people," "the right to select and dismiss public officials, the essence of public service," "guarantee of universal suffrage and voting secrecy," "academic freedom," and "guarantee of fundamental human rights." The "Constitution of Japan" was unprecedented globally. Although enacted under GHQ control, it indicated a new period of legitimacy aimed at guiding Japan towards democracy. Labor Standards Law was enacted in 1947 under Article 27(2) of the Constitution. It established minimum labor conditions, including labor contracts, wages, working hours, holidays and annual paid leave, accident compensation, and work regulations. It was the law with the aim of human rights, particularly workers' right to survival. The Labor Standards Act, along with the Trade Union Act and the Labor Relations Adjustment Act, collectively known as the three labor laws, ensured workers' rights and equal relations with employers.

Following Japan's unconditional surrender in August 1945, efforts to rebuild the postwar economy commenced. However, the real GDP per capita in 1946 was less than 55% of the prewar peak (1934–1935), and the manufacturing production index was only one-fifth of the 1944 level. Compounded by poor rice production since 1903 and the return of over five million people from abroad, the country faced a severe food shortage; life was barely sustainable. **The Economic Planning Agency reported that by 1952, real GDP had returned to the prewar levels of 1934–1936. Real private investment and consumption expenditure recovered by 1951, the manufacturing production index by 1952, and GDP per capita by 1955. It was not until 1956 when the Economic White Paper declared that the Japanese economy was "no longer postwar," that the struggle for economic recovery continued for more than a decade.**

The inaugural "General Commentary" of the first postwar "Economic White Paper: Report on Economic Facts" (July 4, 1947), compiled by the Economic Stability Headquarters under the Japanese government, expressed the government's intent to engage the public in addressing the nation's economic challenges. The government sought to use all available data and statistics to understand and solve these problems collectively with the people, despite the urgency of the situation. At that time, the economy was in a state of deficit across all sectors, including the government, private industry, and households. The government's expenditure exceeded its revenue, resulting in a deficit in both its general and special accounts. Households also faced significant income deficits, and the private sector struggled with substantial technological gaps in supply capacity. There were notable disparities in household income distribution and enterprise balance of payments. **A key objective of economic recovery was to eliminate these deficits and strive for an economic structure where legitimate labor could reap its rewards. The White Paper concluded that even after the war, the rulers' tendency to obscure unpleasant realities, distort facts, and misrepresent reality brought immense suffering to the people. It asserted that the path to a democratic Japan lay in establishing a government of the people, by the people, and for the people. The government appealed to the Japanese people to brace themselves for the hardships of reconstruction and expressed its commitment to collaborate with the people in rebuilding the nation**.

In 1946, the government's general and special account deficits were 40% more than the expenditure, with revenue of JPY 115.6 billion compared to expenditure of JPY 192 billion. Major industries such as coal and steel also ran deficits, with total expenditure exceeding sales at official prices. Households, described as living a **"bamboo shoot lifestyle,"** continued to deplete their savings and sell their possessions to afford life's necessities. The war had decimated the domestic industry's production capacity, and the supply of raw materials was insufficient. The influx of soldiers and other returnees from abroad exacerbated household overcrowding and unemployment. Concurrently, the severe shortage of goods made it impossible for people to purchase even necessities, leading to hyperinflation. The economy continued to struggle, relying on foreign aid and debt to sustain itself.

The struggle with inflation, unemployment, and supply shocks persisted. Hyperinflation, as reflected in effective prices, surged more than tenfold from 1946 to 1949 in both consumer and wholesale prices. However, it rapidly declined in 1949, with the disparity between the illegal market and official prices gradually diminishing and the black market's role contracting. Beginning in 1946, official prices were subject to multiple revisions. A rationing system was implemented, the old yen was replaced by new yen, and households experienced limitations on bank withdrawals. These measures aim to curtail the circulation of new yen and suppress market demand by limiting household bank withdrawals.

According to an earlier citation from the Fiscal History Office of the Ministry of Finance, the average planned monthly household consumption expenditure in March 1946 was 526 yen, distributed among food (50%), housing (10%), clothing (6%), utilities (6%), and miscellaneous expenses (28%). However, an actual household survey revealed a total living cost of approximately 738 yen, allocated across food (73%), housing (4%), clothing (5%), utilities (4%), and miscellaneous expenses (12%). This not only exceeded the planned total, but even the food allocation was inadequate, indicating severe impoverishment. Given that the total observed cost of living in 1937 was 72.8 yen, and prices escalated from a 1.0 yen increase in 1935 to 48.7 in 1946, a 50-fold increase would necessitate a total monthly consumption expenditure of about 3500 yen to sustain the prewar standard of living. Consequently, the Japanese populace endured significant poverty relative to their prewar status, living either below survival cost or barely scraping by.[1]

War damage reduced capital equipment capacity, and many unemployed people moved into agriculture due to an underperforming labor market. This situation arose from the repatriation of five million people from overseas at war's end. The manufacturing sector's bottleneck was raw materials, compounded by disrupted inter-industry dependencies, leading to a collapse in the supply structure for intermediate goods such as raw materials and energy. Despite the war's impact, it is estimated that 80% of prewar capital facilities could be operational. The government's lower pricing of intermediate materials for key commodities led to a demand surge that outstripped rationing at official prices. This imbalance in market supply and demand accelerated the rise in raw material and energy prices, subsequently driving up manufactured goods prices.

2.3 Structural Reform in the Agricultural Sector

In this context, the policy adopted by the Japanese government's Economic Stability Headquarters, in collaboration with the General Headquarters of the Supreme Commander of the Allied Powers (GHP), **aimed at restoring the market economic mechanism. This policy, rooted in Adam Smith's assertion that the principle of competition in a market economy lies not in government laissez-faire but in developing the market structure through government policy intervention, was highly rational.** The practical application of Smith's argument included pivotal policy decisions that guided the subsequent development of the Japanese economy.

Market reforms initially targeted structural reforms in the agricultural sector. Specifically, **the idea of improving peasants' status under the landownership system**, where peasant fees accounted for half of the harvest, and creating tenant farmers were long-standing aspirations of the prewar agricultural and forestry bureaucracy. In the second Yoshida Cabinet, the second agrarian reform plan won GHQ's trust, fulfilling **the Ministry of Agriculture and Forestry's prewar goal to create tenant farmers and improve peasants' status**.

The reform involved forcibly acquiring 1.77 million hectares of land from 2.52 million landowners, equivalent to 35% of all farmlands (including 75% of all small

[1] For the living conditions immediately after the war, see Kuroda M (1993) Price and goods control in the Japanese postwar inflationary period. In: Teranishi J, Kosai Y (eds) The Japanese Experience of Economic Reforms. Macmillan Press LTD.

farmland), and selling 1.94 million hectares, including farmland paid for with property tax, to 4.2 million farm households. The government refrained from introducing an inflation-induced price indexing clause, and the purchase price remained unchanged until the reforms' conclusion in 1950, despite dramatic postwar inflation depreciating money's value. **This agrarian reform, the postwar period's largest, eliminated inequality between landowners and peasants, restored social stability, and expanded food production**. This measure was introduced to help restore and rebuild the Japanese economy, along with the implementation of a sloping production system to address the shortage of agricultural fertilizers. The Basic Agricultural Law of 1961 was enacted to correct the income gap between farmers and industrialists and alleviate food shortages by expanding agriculture's scale, improving productivity, and diversifying agricultural products to meet diverse demand growth. **It also fostered the agricultural sector as a labor source from the traditional sector to meet the manufacturing industry's growing labor demand in the 1960s**.

2.4 Industrial Policy with "Dismantling Zaibatsu"

The market system underwent a significant reform known as the "dismantling of zaibatsu." This policy aimed to break down the large industrial and financial conglomerates that dominated most of Japan's commerce and production. This reform was a provision of the Japanese economic restructuring stipulated in the GHQ's "Initial US Policy toward Japan after the mantling the zaibatsu. Surrender." issued in September 1945. The policy's objective was to eliminate the adverse effects of Japan's industrial structure, which was controlled by a few capitalist group, known as Zaibatsu.

On October 16, Raymond C.Kramer, Director of the Economic and Scientific Bureau at GHQ, stated that they anticipated Japan' voluntary action in dismantling the zaibatsu. Following this, the Japanese Government engaged in discussion of family members of the four zaibatsu from the director and auditor positions of the holding ions with four zaibatsu including Mitsubishi and Sumitomo, and GHQ. They submitted a draft plan for zaibatsu dismantling to GHQ. The plan proposed that all securities and other assets agencies and froze. The payment for these transferred assets would be made in government bonds, with a prohibition on transfer and conversion for 10 years. The plan also included the expulsion of family members of the four zaibatsu from the committee director and auditor positions of the holding companies and imposed strict regulations on the holding companies' control over their subsidiaries.

The agrarian reform and zaibatsu dissolution were significant postwar reforms that safeguarded citizens' human rights, as guaranteed by the Constitution and the Labor Standards Law.

These reforms systematically ensured market position parity among various classes. **As discussed in Chap. 1, the "establishment of national legitimacy based on democracy" was a crucial norm for building the foundation of democratic states in modern Britain, the US, France, and other Western countries. This norm guided the development of political and legal systems to guarantee this legitimacy and bold institutional reforms to ensure equal rights among citizens. The structure of Japanese society and economy postwar was largely supported by these reforms.**

2.5 Introduction of "Gradient Production System"

From the perspective of establishing a market economy, the implementation of an industrial policy known as "Gradient production system" was fundamental to the subsequent economic structure. Adam Smith posited that the expansion of industries' productive capacity, which would be brought about by an increase in production efficiency through the expansion of the division of labor, was the basis for the realization of "national wealth." This expansion would be facilitated by two factors: the evolution of technology promoting the division of labor within each industry and changes in the inter-industry division of labor structure. While advances in production technology in each industry significantly contributed to the establishment of the division of labor within the industry, the inter-industry division of labor heavily relied on the efficiency of coordinating input-output structures among industries, or the efficient construction of industrial structures.

After the war, many unemployed people, including repatriates, flooded the job market, leading to an abundant labor supply. Conversely, the capacity of capital equipment had declined because of war damage. However, a postwar survey indicated that the manufacturing industry's capital equipment was operating at approximately 80% of its prewar level. The bottleneck in production expansion was not labor, capital, or other production factors, but rather the lack of intermediate raw materials, including energy. The economy initially struggled due to limited energy resources like coal. Aging facilities led to a significant decline in coal production, making the energy supply to the manufacturing sector a major bottleneck. The decline in energy production pressured production in major manufacturing industries, leading to shortages and price hikes in intermediate goods production and causing inflation to rise, creating a vicious cycle. **This cycle was adjusted by an industrial policy called the "priority production system."**

Assuming the coal, steel, and chemical fertilizer production industries are the primary bottlenecks of supply shortages, these sectors receive priority in the allocation of intermediate raw materials and financial support through government subsidies. This prioritization aims to expand production, with increased coal production allocated primarily to the steel industry and increased steel production allocated to facility expansion in the coal and chemical fertilizer industries. **This approach, designed to smooth the overall industrial structure's circulation and eliminate**

bottlenecks, considers the interdependence of the economy's industrial structure. The elimination of industrial supply capacity bottlenecks through a gradient production system served as an empirical test of economic theory's general equilibrium system and a policy experiment to realize the market mechanism. This strategy is known for revitalizing Japan's economy since the 1960s and boosting market mechanisms. Political, legal, and institutional reforms helped these efforts by facilitating the market mechanism. This strategy revitalizes growth by focusing on economic norms, democratic legitimacy, and technological advancements. In subsequent chapters, I will trace the path of Japan's postwar economic development to elucidate this policy experiment's significance and outcomes.

This policy experiment, initiated in the 1960s, is widely recognized for its pioneering role in rejuvenating the Japanese economy and enhancing the market mechanism. The outcomes of these endeavours were supported by a comprehensive range of political, legal, and institutional reforms, which collectively improved the operation of the market mechanism. This approach to revitalization, grounded in the three norms of modern economic growth, considered the establishment of democratic legitimacy and the attributes of science and technology. In the subsequent chapter, I will delineate the trajectory of Japan's postwar economic development to elucidate the significance and outcomes of this policy experiment.

The economy was initially limited by a shortage of energy resources, particularly coal. The aging of facilities led to a significant drop in coal production, creating a major bottleneck in energy supply for the manufacturing sector. The interplay among the energy, manufacturing, and agriculture industries aimed to streamline the overall industrial structure and alleviate bottlenecks. The removal of the industrial supply capacity bottleneck through the gradient production system served as a practical test of the general equilibrium system of economic theory. It was also a policy experiment to actualize the market mechanism, in conjunction with the interdependence of the economy's industrial structure. **This initiative is a policy experiment to revitalize Japan's economy and market since the 1960s. Political, legal, and institutional reforms enhanced market mechanisms. Recovery was based on democratic legitimacy and modern economic growth norms, considering scientific and technological features**.

This policy experiment, which aimed to revitalize the Japanese economy and the market mechanism, has been highly regarded since the 1960s. The outcomes were supported by a range of political, legal, and institutional reforms that enhanced the operation of the market mechanism. This revitalization path, based on the three norms of modern economic growth, contemplated the establishment of democratic legitimacy and the attributes of science and technology. In the subsequent chapter, I will delineate the trajectory of Japan's postwar economic development to elucidate the significance and outcomes of this policy experiment.

References

Economic White Paper: Report on Economic Facts. (4 July 1947), compiled by the Economic Stability Headquarters under the Japanese Government.

Kuroda, M. (1993). Price and goods control in the Japanese postwar inflationary periods. In J. Teranishi & U. Kosai (Eds.), *The Japanese experience of economic reforms*. Macmillan Press and St. Martin's Press.

Chapter 3
High Growth After "No More Postwar" in Japan

Masahiro Kuroda

Abstract The evolution of this grand national design reform in the Japanese economy since the 1956 White Paper on the Economy, which stated, "It is no longer postwar," requires evaluation from various perspectives. This chapter aims to trace the subsequent transition of the Japanese economy. When comparing the international competitiveness of an industry by comparing the average price of its products, the challenge lies in calculating the differences in the industry's production technical efficiency and the international differences in the prices of the input factors of production as comparable indicators. Production factors include labor, capital, raw materials, and other elements. The structure of each factor input is determined by technology with technological advancements. Consequently, given each production factor price, it is necessary to aggregate the productivity of each production factor. Total Factor Productivity (TFP) is a productivity measurement concept that evaluates the overall efficiency of all input factors. This chapter will evaluate changes in the international competitiveness of industries between Japan and the US from 1960 to 1990.

Keywords Market Mechanism · Agrarian Reform · Gradient production system · Dismantling of conglomerates · Revision of labor laws · Total Factor Productivity (TFP) · Purchasing Parity Index

3.1 Recovery of the Japanese Economy After the World War II

According to "Long-Term Economic Statistics" by Hitotsubashi University, Japan's real gross domestic product (in 1990 fixed prices) had an annual growth rate of 2.90% from the Meiji Restoration to the end of World War II in 1945. This rate increased to 8.36%

M. Kuroda (✉)
Professor Emeritus, Keio University, Tokyo, Japan

M. Kuroda (ed.), *Intangible Capital Investment and Economic Growth*, Translational Systems Sciences 45,
https://doi.org/10.1007/978-981-96-7675-0_3

during the postwar period of 1945–1950, and further to 9.40% during 1950–1973. However, it decreased to 3.47% for 1973–1987, before rising again to 4.73% during 1987–1992. The growth rates from 1945 to the "oil shock" in 1973 were notably high. The annual growth rate of 2.90% from the Meiji Restoration to 1945 was high globally.

Post–world war economic recovery is due to new market design institutions ensuring efficient market competition and constitutional reforms establishing a democratic state. **The equalization of bargaining power among capitalists, peasants, and urban workers through agrarian reform, the dismantling of conglomerates, and the revision of labor laws was based on the fundamental principle of market design to realize market competition's rationality and efficiency.** The industrial policy, known as the "Gradient production system," aimed at realizing a division of labor structure based on inter-industry interdependence. This policy was facilitated by scientific and technological advancements that increased production efficiency and improved market division of labor efficiency. These market and industrial structures were based on a constitution designed to achieve a democratic state with equal citizenship. **The evolution of this grand national design reform in the Japanese economy since the 1956 White Paper on the Economy, which stated, "It is no longer post-war," requires evaluation from various perspectives. This chapter aims to trace the subsequent transition of the Japanese economy and evaluate the results of realizing the national orthodoxy and the merits and demerits of economic policies from an economic perspective.**

In this chapter, we aim to trace the subsequent transitions in the Japanese economy and evaluate the realization of national legitimacy and the merits and demerits of economic policies from the perspective of changes in international competitiveness of industries and the productivity that supports them. To achieve this, we will explain the theoretical framework for measuring industrial productivity and attempt to capture the changes in the competitiveness of Japanese and US industries based on the historical changes in productivity by industry in both countries.

Postwar economic reconstruction was supported by the establishment of new market design systems that enabled efficient market competition and by constitutional reforms that laid the foundation for a democratic state. The equalization of bargaining power among capitalists, farmers, and urban workers—achieved through agricultural land reforms, the dissolution of the zaibatsu conglomerates, and revisions to labor laws—was based on the fundamental principle of market design aimed at realizing the rationality and efficiency of market competition.

Furthermore, the industrial policy known as the "priority production system" sought to establish a division of labor based on the interdependent structure among industries. **This policy was underpinned by advances in science and technology that improved production efficiency and the efficiency of intra-market division of labor**. These market and industrial structures were founded on a constitutional framework designed to achieve a democratic state with equal citizenship.

Since the 1956 Economic White Paper declared that "Japan is no longer in the postwar era," the progress of this grand national design reform in the Japanese economy must be evaluated from multiple perspectives. This chapter aims to trace the subsequent transitions of the Japanese economy and to evaluate, from an economic standpoint, the realization of national policy goals and the achievements and drawbacks of economic policies.

Reconstructing the industrial structure and enhancing the international competitiveness of industries were key tasks in revitalizing Japan's postwar economy. The international competitiveness of a commodity, including the qualitative evaluation of its products, can be assessed by comparing the prices of specific commodities produced by an industry on an international scale. This comparison allows for an evaluation of price competitiveness among countries in the global market. However, when assessing the international competitiveness of an industry, the industry's competitiveness is evaluated by considering the industry's output as a composite of various commodities it produces and calculating and comparing the prices of that composite commodity in the international market.

In this scenario, it is assumed that the type of commodity produced by industry varies from country to country. The price of the industry's output is based on the average price of the output each country in the industry produces, allowing for a comparison of the industry's price competitiveness. The price of a product depends on the price of each input factor of production and the technical efficiency, or productivity, of each input factor. When comparing the international competitiveness of an industry by comparing the average price of its products, the challenge lies in calculating the differences in the industry's production technical efficiency and the international differences in the prices of the input factors of production as comparable indicators.

Production factors can be classified into labor, capital, intermediate raw materials, and other factors. The structure of each factor input is determined by technology, with technological advancements reflected in improvements in factor input efficiency. Consequently, given each production factor price, it is necessary to aggregate the productivity of each production factor, such as labor productivity, capital productivity, and productivity of intermediate raw materials, including energy, to comprehensively evaluate overall production efficiency. The impact of technological progress on the input structure of labor, capital, intermediate raw materials, and others is expected to vary from one factor to another. **The productivity index for each factor will also vary depending on the production efficiency of each production factor, making it challenging to evaluate the superiority of overall efficiency by comparing the productivity variation of any single factor. In many instances, comparing productivity variations of any one factor makes it difficult to evaluate the superiority or inferiority of overall efficiency**.

Total Factor Productivity (TFP) is a productivity measurement concept that evaluates the overall efficiency of all input factors. It does this by weighing and averaging the productivity gains of each production factor due to technological progress. The extent to which technological progress has altered an industry's productivity during a country's economic development is a crucial indicator. It aids in understanding the promotion of basic research through science and technology policy and the evaluation of industrial technology development policy based on that research. In terms of an industry's international competitiveness, it is an important question to ask whether one country's productivity change outperforms or underperforms another is in an international comparison. This leads to the question of whether the productivity level in one country and industry is more efficient relative to other countries. Evaluating the relative differences in output prices and each production factor across countries is crucial. Developing comparable indicators that accurately reflect these differences in the relative productivity of all industry factors across countries is necessary. One method to develop comparable indices for an industry's output prices and the prices of labor, capital, and intermediate raw materials across international borders is to compare their international comparisons in terms of the relative currency value across international borders. This is measured in terms of the relative ratio of exchange rates at each point in time, in the sense of a comparison of average monetary value. This can also serve as an indicator to evaluate the benefits of Adam Smith's division of labor. This is because changes in the technological efficiency of individual industries and the evolution of the division of labor structure among industries enhance a country's productivity.

Regarding an industry's international competitiveness, it is crucial to determine whether one country's productivity changes to outperform or underperform in an international comparison. The initial step involves assessing whether the productivity level in one country and industry is more efficient compared to other countries. It is also vital to evaluate the relative differences in output prices and individual production factor prices across countries to develop comparable indicators that accurately reflect these differences in the relative productivity of all production factors in the industry across countries. A method to develop comparable indices for an industry's output prices and the prices of labor, capital, and intermediate raw materials across international borders is to use their international comparisons in the relative currency value across international borders. This is done using the relative ratio of exchange rates at each point in time, in the sense of a comparison of average monetary values.

However, while exchange rates are a relative measure of average international currency values, they are a macro-level indicator of international competitiveness. This includes the influence of financial markets as well as real markets and does not necessarily indicate only relative values that reflect the international competitiveness of individual products. For the relative comparisons of the production efficiency of individual industries and the price competitiveness of their products, it is necessary to have information on the relative international prices of the products of individual industries and the input factors such as labor, capital, and intermediate materials that produce those products. This also requires an international comparison of TFP.

This chapter will assess how the international competitiveness of industries between Japan and the US changed from 1960, when the Japanese economy recovered to a level that was no longer postwar and entered a period of high economic growth, to 1990, which is also regarded as the peak of growth. We will evaluate the post–World War II development of Japan's economy by confirming how it has changed as an observable fact. During this period, the Japanese economy experienced a period of high economic growth from 1960 to the early 1970s, a period of change in the international economic environment due to the transition from a fixed to a floating exchange rate system in the early 1970s, a period affected by energy price fluctuations known as the oil shock, and the Plaza Accord in 1985. The Japanese economy has responded to and adjusted to these changes with the evolution of science and technology and changes in the political environment during this period.[1]

[1]This study is the result of the joint research between the Keio Economic Observatory of the Research Institute of Industry at Keio University and Professor Dale W. Jorgenson of Harvard University in the US, including Dr. Mieko Nishimizu, former vice president of the World Bank. On the Japanese side, in addition to the author, many researchers at Keio's Institute of Industrial Research have contributed to this project.

3.2 Theoretical Framework for the Total Factor Productivity

The concept of TFP is briefly described as follows.[2] The balance between the inputs and outputs of a specific production sector j can be expressed using the prices q^j and quantities z^j of the output, and the prices, and prices p_i^j and quantities x_i^j of the i production factors.

$$q_j z_j = \Sigma_{(j=1,\ldots n)} p_j^{\,i} x_j^{\,i} \tag{3.1}$$

These inputs are divided into n production factors, such as labor, capital, energy, and intermediate raw materials. By differentiating both sides of Eq. (3.1) over time, we obtain:

$$\dot{q}_j / q_j + \dot{z}_j / z_j = \Sigma v_j^{\,i} \left(\dot{p}_j^{\,i} / p_j^{\,i} + \dot{x}_j^{\,i} / x_j^{\,i} \right) \tag{3.2}$$

where v_i^j is the distribution rate of i production factors.

The growth rates of the aggregated input for the quantity and price, which are aggregated *in inputs* for quantity and price, are defined by the Divisia index as follows:

$$\dot{X}_j / X_j = \sum_{(i=1,\ldots n)} v_j^{\,i} \left(\dot{x}_j^{\,i} / x_j^{\,i} \right), \tag{3.3}$$

$$\dot{P}_j / P_j = \sum_{(i=1,\ldots n)} v_j^{\,i} \left(\dot{p}_j^{\,i} / p_j^{\,i} \right). \tag{3.4}$$

By substituting Eqs. (3.3) and (3.4) into Eq. (3.2), we derive:

$$\begin{aligned} \dot{z}^j / z^j - \dot{X}^j / X^j &= \dot{z}_j / z_j - \sum_{(i=1,\ldots n)} v_j^{\,i} \left(\dot{x}_i^{\,i} / x_j^{\,i} \right) \\ &= \sum_{(i=1,\ldots n)} v_j^{\,i} \left(\dot{p}_j^{\,i} / p_j^{\,i} \right) - \dot{q}_j / q_j \\ &= \dot{P}_j / P_j - \dot{q}_j / q_j . \end{aligned} \tag{3.5}$$

For output z_j and aggregate input X_j in sector j, define productivity as:

$$\varphi_j = z_j / X_j . \tag{3.6}$$

Differentiating it over time and defining it in terms of growth rate, we get:

$$\dot{\varphi}_j / \varphi_j = \dot{z}_j / z_j - \dot{X}_j / X_j = \dot{P}^j / P_j - \dot{q}_j / q_j . \tag{3.7}$$

This allows us to define the growth rate of TFP in sector j.

---

[2] *Masahiro Kuroda* (1989). Quantitative Analysis of General Equilibrium (*Japanese*), *Iwanami Co.*

The growth rate of TFP, as defined in Eq. (3.7), can be estimated from both the quantity and price perspectives. The productivity growth rate in sector j can be estimated by subtracting the aggregate input growth rate from the output growth rate. Alternatively, from the price perspective, the price of an input can estimate by the difference between the aggregate growth rate of the price of the input factors of production $\left(\dot{P}^j / P^j\right)$ and the growth rate of the price of that output $\left(\dot{q}^j / q^j\right)$ over a given period. In essence, the rate of technological progress during a period can be consistently determined by the difference between the rate of change in the quantity of all inputs and outputs of all production factors aggregated, and the difference between the rate of change in the price of inputs as an aggregate of the input prices of all input factors and the price of outputs. This implies that technological progress can decompose and understand in terms of changes in the output and input quantities of any production factor or in the output and input prices of any production factor.

The concept of total factor productivity (TFP) was developed as an indicator to evaluate the production efficiency of industries in each country. It is a time-series indicator that captures the transition of technological efficiency in various industries within each country. This indicator can also use for the cross-sectionally assess the differences in productivity between industries or products at a specific point in time, such as between Japan and the United States.

In this context, it is necessary to determine the relative prices of the output of the j industry in both countries at a given point in time and the prices of the inputs of each factor of production used in its production. Once the relative prices of output and factors of production between the two countries have been estimated and the price differentials between the two countries have been estimated for output and factors of production, we can estimate the difference in technical efficiency in the production of industry j in the two countries according to the definition of TFP above. The time-series changes in this disparity will provide an indicator of the international competitiveness of the industries of the two countries. The following results are referred from Jorgenson-Kuroda's (1992) comparison of US-Japan international competitiveness for the period 1960–1985 and the Kuroda-Nomura paper's measurement results for the period 1960–1992. Although there are some differences in the basic data, including the period of observation, the two papers use the same method of measurement in principle. The former uses 1970 as the base year for measurement, while the latter uses 1990 as the base year.

In the estimation of output prices and factors prices including labor, capital, and intermediate goods for the specific commodity, the methodologies for the 1970-based and 1990-based estimates are similar. The estimation method for the relative specific prices of output prices at the base time involves the following steps. First, we try to observe the price per single unit for goods that are homogeneous across countries at the base time. These are the purchaser prices for Japan and the US for each specific commodity. Next, these prices must transform into a common currency unit at the current exchange rate to estimate their relative ratio. Furthermore, the commodity corresponds to the US-Japan International Input-Output Table across 163 sectors. The commodity margin ratio is derived from the US-Japan Margin Matrix, and the producer price-based output price is calculated by subtracting the margin from the purchaser's price. By mapping the 163 sectors in the input-output table to the 31 industrial sectors, the Purchasing Power Parities (PPP) Index for Japan and the US of the product prices at the base time (1970 or 1990) is obtained based on the 31 sectors.

The relative US-Japan prices of labor input service prices are estimated by considering labor input in each industrial sector in Japan and the US. This is done by considering gender, age, education, and employment type (employed, self-employed, or family worker) in terms of hours worked, number of persons, and wage per hour (wage rate). The proportion of each labor type's contribution to the total labor cost is recorded as a weight. By using Divisia aggregation, the time series of labor price indexes for total labor input reflecting changes in each type of employment is estimated. Additionally, by converting the wage rates by employment type in Japan and the US at the base time (1970 and 1990) and comparing the wage rates by employment type at the base time between the two countries, the relative labor price of total labor input in Japan relative to the labor price of total labor input in the US at the base time (1.0) can be obtained. By comparing this to the aggregate labor price index (base year 1970 or 1990), we can trace changes in labor production efficiency due to industry-specific labor patterns. Time series of these information propose are useful to estimate industry-specific labor price indexes for PPP in Japan and the United States.

The relative prices of capital services between Japan and the US are determined by calculating the user cost of capital services at a base time and comparing it to the relative ratio of the user cost of capital services at that same base time. Chapter 4 will explain the connection between capital services and their user costs. The user cost of capital services is the sum of the discounted value of future services generated by the capital stock at any given time. This calculation uses the capital return rate as the discount rate. The present value can be equated to the current price of the capital service of the capital stock. Given this assumption, the price of the capital service produced by the current capital stock can be derived as a function of the price of the investment good, the rate of return on capital (discount rate), and the tax rate on time. The relative differences in the rates of return on capital and tax rates are calculated based on the capital stock amount, the cost of capital, and the prices of capital goods in Japan and the US at the base time. This allows for the estimation of the capital service prices at the base time for both countries. Relative prices of intermediate input goods, including energy, at the base time between Japan and the US are estimated by using the intermediate transaction in the Japan-US International Input-Output Table that we developed.

3.3 Japan-US Purchasing Power Parity Index and the Productivity Comparison

Consider the changing international competitiveness by industry between Japan and US from 1960 to 1985, as estimated by TFP indices across various industries.

Table 3.1 presents the PPP price indexes for 29 industries in Japan, when the US index in 1970 assumes to be 1.0 in the base year of 1970, in which the prices of products, capital, labor, energy, and intermediate raw materials in the US are normalized to 1.0 respectively. In six industries in Japan, the output price was higher than the US including (1) Agriculture, Forestry and Fishery, (3) Construction, (4) Foods Products (12) Petroleum and Coal Products (13) Rubber Products and (25) Electricity, Gas, and Water supply. In terms of factors, the PPP price of capital services in Japan is lower across all industries, ranging from 0.8 to 0.95 compared to 1.0 in the US. For labor services, the PPP price ranges from 0.1 to 0.3, indicating that Japanese labor prices are less than 30% of US prices in purchasing power terms. Energy prices range from 1.2 to 1.5, suggesting that Japan's prices are 20–50% higher than those of the US. Conversely, prices of intermediate raw materials excluding energy range from 0.6 to 0.9, indicating that Japan is cheaper than the US In 1970, the exchange rate was fixed at 360 yen to the dollar. However, the balance of supply and demand in the Japanese labor market and the yen's appreciation against the US dollar began to shift in the 1970s. This shift, particularly the yen's appreciation, became a significant factor in the Japanese economy and had a substantial impact on the relative parity of labor services.

Table 3.2 provides an index of time-series changes in macro-based output, labor, and capital services prices. This index is based on the base-time industry purchasing power index, adjusted to the base-time basis for output, labor, and capital services prices across industries in the time series. This base is then used to create a Divisia-aggregated industry price index to track time-series changes in output, labor, and capital services prices on a macro basis. The exchange rate used is that of 1971. Since the transition to a floating exchange rate system in 1971, the yen has been gradually appreciating from its previous level of 360 yen to the dollar.

In 1960, the price of capital services in Japan (0.62449) was approximately 78% of that in the US (0.79723) in the same year. As the yen appreciated, Japan's competitiveness gradually increased until 1985, when it fell to about 45% of the US level. Conversely, in 1960, the price of labor services in Japan was about 10% of that in the US, significantly lower than that of capital services. However, by 1985, it had risen to 35% of the US price, indicating a more rapid increase in the price of labor services in Japan than in the US.

These changes in factor prices were affected to the Japan-US relative price ratio.

As concerns the gross output prices in PPP terms, the Japanese output price was about 60% of US gross output price in 1960 and rose to approximately 79% in 1973, partly due to changes in exchange rates. By 1985, the price of labor and capital services as factors of production had declined to about 50%. These changes vary depending on labor and capital market conditions in both countries. In the 1960s, the Japanese labor market was an unrestricted labor supply market, with labor mobility from traditional sectors such as agriculture, forestry, and fisheries to the manufacturing sector. It resulted in a reduction in the cost of labor services. However, such situation was significantly changing in the 1970s.

Table 3.1 Purchasing power parity index of outputs and inputs by industry in Japan in the year 1970, in comparison with the US level in the year 1970 = 1.0

Industry category	Product price	Capital service price	Labor service price	Energy price	Intermediate goods price
1. Agriculture, forestry, and fisheries	1.04556	0.90835	0.21352	1.48236	0.92204
2. Mining	0.72155	0.88095	0.21263	1.44013	0.70573
3. Construction	1.03487	0.92127	0.18607	1.42641	0.72203
4. Food manufacturing	1.03569	0.92191	0.21894	1.26564	0.88483
5. Textile manufacturing	0.77898	0.90871	0.24099	1.18329	0.75976
6. Apparel manufacturing	0.76962	0.86037	0.18975	1.24298	0.74821
7. Wood and wood products manufacturing	0.79154	0.84363	0.22806	1.22681	0.90163
8. Furniture manufacturing	0.67945	0.84214	0.22952	1.22178	0.74429
9. Paper and pulp manufacturing	0.58858	0.89567	0.22171	1.18606	0.65664
10. Printing	0.78107	0.86742	0.21251	1.13482	0.65975
11. Chemical	0.66211	0.91711	0.25039	1.33631	0.71201
12. Petroleum and coal products	1.59952	0.89688	0.21846	1.31298	0.88291
13.Rubber products manufacturing	1.06186	0.86013	0.24042	1.22499	0.76731
14. Leather products manufacturing	0.71273	0.82076	0.23569	1.31561	0.81086
15. Ceramics, stone, and clay products manufacturing	0.69603	0.89998	0.23083	1.31627	0.72567
16. Iron and steel	0.81706	0.91205	0.25201	1.37079	0.80318
17. Metal products manufacturing	0.81514	0.90205	0.21072	1.32346	0.77507
18. Machinery	0.61327	0.92021	0.22546	1.28346	0.72093
19. Electrical machinery	0.68127	0.92036	0.22308	1.24327	0.71054
20. Automobile manufacturing	0.78627	0.91647	0.18581	1.17291	0.76428
21. Transport equipment manufacturing	0.94794	0.87722	0.21944	1.24063	0.76549
22. Precision machinery manufacturing	0.71912	0.86402	0.23151	1.22607	0.71774
23. Other manufacturing	0.69473	0.88034	0.22549	1.27395	0.71238
24. Transportation and communication	0.47247	0.91027	0.22713	1.43624	0.68624
25. Electricity, gas, and water utilities	1.02936	0.90389	0.26605	1.49491	0.78528
26. Wholesale and retail trade	0.66155	0.93094	0.25889	1.35118	0.73683

(continued)

Table 3.1 (continued)

Industry category	Product price	Capital service price	Labor service price	Energy price	Intermediate goods price
27. Finance	0.86176	0.83301	0.30796	1.14491	0.77297
28. Other services	0.56751	0.91719	0.25592	1.22718	0.73724
29. Public administration	0.30797	n.a.	0.19482	1.35489	0.68436

Such market conditions are reflected in the prices. In contrast, the stabilization of investment goods prices since the 1960s, centered on the heavy and chemical industries, has suppressed the increase in capital service prices over time.

Based on the above theoretical framework of TFP, we can estimate how these differences in factor prices in Japan and the US affect changes in productivity in industries in the two countries. However, there are differences in the time-series patterns of change reflecting the conditions in the respective factor markets.

Changes in TFP growth rates by industry in Japan and the US help classify shifts in international competitiveness. The technology disparity classification in Table 3.3 is defined as follows:

Type 1: Industries where the US surpassed Japan in the level of technical efficiency as measured by TFP in 1980, a gap that is expected to persist.

Type 2: In 1980, US technological efficiency was higher than Japan's. In the 1970s, Japan's efficiency gains slightly surpassed those of the US, but the US is likely to retain its efficiency for a while.

Type 3: In 1980, the US was more technologically efficient than Japan, but Japan's improvement in technological efficiency is strong, and the industry is expected to catch up with Japan in the near future.

Type 4: In 1980, the US was mand US technological superiorty is expected to persist.

Type 5: Industries where Japan's technological efficiency surpassed the US in 1980 and is expected to continue leading.

Estimates of the international competitiveness of Japanese and US industries, based on 1970 data, provide insightful observations that trace the trajectory of the Japanese economy during its recovery from the Great War.

1. In the 1960s, Japan's economy seemed highly price-competitive, with the US-Japan ratio of macro-level output prices (gross value added) exceeding 60%. This was primarily due to the extremely low prices of production factors, with labor prices at over 10% of the US level and capital prices at about 80% of the US level. In the decade leading up to 1970, when the exchange rate was fixed at 360 yen to the dollar, labor prices were about 20% of the US price, capital prices were about 90%, and macro-level output prices were 75% of the US price. The price of labor as a production factor increased about 3.5 times from 1960 to 1970 in Japan, compared to 1.6 times increase in the US. The price of capital rose 1.2 times in the US, compared to 1.4 times in Japan. The primary goal of postwar economic growth was to improve living standards by raising labor prices, enabling them to escape from

Table 3.2 US–Japan time-series comparisons of prices of value-added, labor, and capital service: at macro level during the years 1960–1985

		Capital service price			Labor service price			Value-added price		
Year	Exchange rate	Capital service price evaluated by PPP	UAS capital service price	Relative purchasing parity price: Japan/USA	Labor service price evaluated by PPP	UAS labor service price	Relative purchasing parity price: Japan/USA	Value-added evaluated by PPP	UAS value-added labor service price	Relative purchasing parity price: Japan/USA
1960	360	0.62499	0.79723	0.783952	0.06759	0.60926	0.11094	0.49401	0.78454	0.62968
1961	360	0.70011	0.80034	0.874766	0.07795	0.64391	0.12106	0.53183	0.79409	0.66974
1962	360	0.64268	0.87577	0.733846	0.08871	0.65408	0.13563	0.55298	0.80279	0.68882
1963	360	0.62544	0.91311	0.684956	0.10203	0.66726	0.15291	0.57685	0.80639	0.71538
1964	360	0.68795	0.96814	0.710589	0.10864	0.68739	0.15805	0.59492	0.81563	0.7294
1965	360	0.68865	1.05671	0.651692	0.12425	0.70308	0.17672	0.62978	0.83047	0.7463
1966	360	0.71741	1.08764	0.659602	0.13732	0.74533	0.18424	0.64779	0.86171	0.75172
1967	360	0.78291	1.06235	0.736961	0.15215	0.79066	0.19243	0.67604	0.88078	0.76755
1968	360	0.86281	1.07711	0.801042	0.17714	0.8549	0.20721	0.69657	0.91007	0.7654
1969	360	0.89842	1.09371	0.821443	0.20104	0.90917	0.22112	0.72318	0.95491	0.75733
1970	360	0.89842	1	0.898421	0.23211	1	0.23211	0.75878	1	0.75878
1971	348	0.81956	1.07581	0.761807	0.26643	1.07431	0.24801	0.77834	1.0476	0.74297
1972	303.1	0.83773	1.16855	0.726897	0.30113	1.14898	0.26208	0.80947	1.09325	0.74043
1973	271.1	0.9224	1.22005	0.756035	0.38076	1.24142	0.30671	0.92428	1.16623	0.79254
1974	293.1	0.99464	1.12504	0.884093	0.46834	1.36978	0.34191	0.8719	1.29731	0.67208
1975	269.8	0.9234	1.29908	0.710811	0.55019	1.49983	0.36883	0.99093	1.42734	0.69425
1976	296.5	0.94393	1.42287	0.663399	0.59518	1.62713	0.36579	1.05665	1.49954	0.70456
1977	268.3	0.96151	1.63368	0.588555	0.6492	1.73529	0.37412	1.10367	1.60448	0.68787

(continued)

Table 3.2 (continued)

		Capital service price			Labor service price			Value-added price		
Year	Exchange rate	Capital service price evaluated by PPP	UAS capital service price	Relative purchasing parity price: Japan/USA	Labor service price evaluated by PPP	UAS labor service price	Relative purchasing parity price: Japan/USA	Value-added evaluated by PPP	UAS value-added labor service price	Relative purchasing parity price: Japan/USA
1978	210.1	1.15219	1.78198	0.646579	0.67337	184,918	0.36451	1.18892	1.73642	0.6847
1979	219.5	1.21611	1.82541	0.666212	0.70365	2.00071	0.35171	1.21565	1.89859	0.64029
1980	203	1.00809	1.84044	0.547741	0.75423	2.21758	0.34011	1.27198	2.09651	0.60671
1981	219.9	0.98245	2.00438	0.490152	0.79732	2.39774	0.33257	1.30588	2.20653	0.56863
1982	235	1.04394	1.96229	0.532001	0.8339	2.54319	0.3779	1.34193	2.43295	0.55156
1983	232.2	1.06156	2.13698	0.496757	0.83456	2.64133	0.31596	1.36365	2.51156	0.54295
1984	251.1	1.10386	2.43909	0.452571	0.85129	273,005	0.31182	1.37795	2.56711	0.53665
1985	224.1	1.1502	2.46379	0.466842	0.89202	2.86401	0.31146	1.38862	2.66754	0.52056

Table 3.3 Industrial classification based on the international competitiveness: annual growth rate of productivity by industry

	Industry classification	Technology difference in 1980	1960–1970		1970–1980		1980–1985		Technology difference in 1985
			Japan	USA	Japan	USA	Japan	USA	
Type 1	1. Agriculture, forestry and	U > J	0.452	1.178	−1.641	0.673	−0.274	4.431	U > J
	2. Construction	U > J	0.854	0.228	0.717	−1.07	−1.707	0.516	U > J
	4. Food products	U > J	−0.155	0.556	0.37	0.208	−0.917	0.801	U > J
	5. Textile manufacturing	U > J	0.526	1.437	1.22	0.187	0.188	0.309	U > J
	10. Printing industry	U > J	0.858	0.647	−1.469	0.218	0.02	0.979	U > J
	12.Petrorium and coal prod	U > J	−1.358	1.616	−3.889	4.56	−1.29	3.422	U > J
	17. Metal products	U > J	2.668	0.293	0.837	0.616	0.009	0.376	U > J
Type 2	8. Woods products	U > J	1.405	0.03	1.364	0.792	1.02	1.475	U > J
	11. Rubber products	U > J	1.499	0.868	0.55	0.981	2.623	3.502	U > J
	15. Ceramic products	U > J	2.794	0.339	−1.248	0.555	0.414	2.443	U > J
	20. Motor vehicle manufac	J > U	0.086	0.155	0.512	0.282	−1.286	2.553	U > J
	21. Transportation machin	U > J	6.649	1.395	0.706	−4.26	2.107	3.436	U > J
Type 3	6. Clothing manufacturing	U > J	2.294	0.625	1.414	1.16	0.42	0.203	U > J
	28. Other service	U > J	1.378	0.701	3.033	0.018	0.502	−1.179	U > J
	27. Financial service	U > J	1.811	0.535	0.15	0.181	3.311	−1.179	U > J
Type 4	25. Electricity, gas, water	U > J	3.222	2.111	2.991	−1.497	0.603	−1.668	U > J

Table 3.3 (continued)

	Industry classification	Technology difference in 1980	1960–1970		1970–1980		1980–1985		Technology difference in 1985
			Japan	USA	Japan	USA	Japan	USA	
Type 5	2. Mining	J > U	1.662	1.084	1.722	−5.584	0.301	0.045	J > U
	7. Woods products	J > U	2.781	0.965	2.032	0.738	3.522	1.211	J > U
	9. Paper and pulp products	J > U	1.616	0.338	0.505	0.233	1.982	1.207	J > U
	11. Chemical products	J > U	3.343	1.501	0.731	−1.517	2.671	1.63	J > U
	14. Leather products	J > U	0.926	0.452	0.713	1.066	0.552	−4.352	J > U
	16. Iron and steel manufac	J > U	0.915	0.088	0.781	0.534	0.624	−2.294	J > U
	18. General machinery ind	J > U	2.212	0.801	0.377	0.693	−1.073	0.785	J > U
	19. Electric machinery pro	J > U	3.304	0.093	3.663	0.693	3.222	0.5	J > U
	22. Precision machinery pr	J > U	1.943	0.729	3.626	0.131	1.513	3.105	J > U
	23. Other manufacturing	J > U	1.741	0.647	1257	0.795	0.252	0.23	J > U
	24. Transportation and comr	J > U	3.056	1.085	0.49	0.995	1.186	0.251	J > U
	26. Trade	J > U	2.507	0.077	0.838	0.316	0.607	2.6	J > U

bare-minimum survival. Conversely, the policy of "heavy industrialization" through increased capital investment aimed to enhance production efficiency, but it also led to a relative increase in capital costs.

2. After the shift to a floating exchange rate in 1971 and the subsequent appreciation of the yen, the ratio of labor prices to those of the US changed from 0.11 in 1960 to 0.23 in 1970, and then to 0.31 in 1985. The ratio of capital prices to those of the US also changed from 0.78 in 1960 to 0.89 in 1970, and then to 0.46 in 1985. This indicates that the factor price of capital in Japan became more competitive than in the US after the shift to a floating exchange rate.
3. Japan's industrial policy, which focused on heavy and chemical industries, significantly contributed to stabilizing capital costs through the expansion of capital goods and to increasing the rate of economic growth. The rate of increase in the price of capital factors at the macro level was suppressed compared to the US, leading to a corresponding increase in the price of labor factors and a relative increase in the price of labor to capital (labor price/capital price). This resulted in a greater substitution of labor for capital.

Based on the 1970-standard technology gap between Japan and the US, industries are classified into several types as shown in Table 3.3. We extended the observation period to 1990 to verify the validity of the industry-type classification based on the 1970 standard by estimating the Japan-US technology gap by industry.

In the 1990 standards, compared to the 1970 standards, mining was divided into coal mining and other mining, rubber products and leather products were combined, transportation and communications were divided, electricity and gas were separated, and other services and public services were combined, resulting in 30 industrial sectors. Construction and retail trade are not classified as non-tradable goods, and it is challenging to determine the difference in purchasing power between Japan and the US. Therefore, it is assumed that there is no difference between Japan and the US at the base year of 1.0 in 1990. The estimation method remains unchanged from the previous estimates up to 1985.

Table 3.4 presents the estimated Japan-US relative prices for output and input (capital and labor) prices by industry sector for Japan and the US at the 1990 base time.

In 1990, the relative product prices of Japanese and US industries revealed that Japan's output prices in 4 out of 30 sectors—textile manufacturing, the automobile industry, other transportation machinery, and finance and real estate—were lower than those in the US. The oil shock led to higher energy prices in Japan compared to the US. In 1990, the exchange rate was 144.81 yen per dollar, nearly 200 yen higher than the 1970 rate of 348 yen, reducing Japan's price competitiveness.

When evaluating labor prices at 1990 exchange rates, all industrial sectors in Japan had lower relative US-Japan prices. The sectors with the lowest prices were agriculture, forestry, and fisheries, while the highest were the oil and coal production sector, telecommunications, and electricity and gas. This evaluation includes hourly wages for employees, self-employed individuals, and family workers. Industries with many family workers and self-employed individuals, such as agriculture, forestry, fisheries, and transportation, had relatively low wages. In contrast,

Table 3.4 Japan-US Relative prices for outputs and inputs by industry between Japan and US in year 1990

Industry	Product price	Capital service price	Labor service price
1. Agriculture, forestry, and fisheries	1.51661	1.27271	0.43785
2. Coal mining	1.17058	0.75949	9.91293
3. Other mining	1.65424	3.83152	0.47146
4. Construction	1	3.77015	0.65335
5. Food manufacturing	1.83264	1.28003	0.64843
6. Textile manufacturing	0.9645	0.44781	0.72985
7. Apparel manufacturing	1.07435	2.9897	0.62524
8. Wood and wood products manufacturing	1.39376	1.35129	0.5929
9. Furniture manufacturing	1.13526	2.46739	0.75049
10. Paper and pulp manufacturing	1.53708	1.12114	0.64306
11. Printing	1.1349	1.4416	0.75952
12. Chemical	1.00803	1.0807	0.69326
13. Petroleum and coal products	1.98749	1.83697	0.9213
14. Leather manufacturing	1.35516	5.5848	0.77581
15. Ceramics, stone, and clay products manufacturing	1.61002	1.29264	0.62769
16. Iron and steel	1.06876	1.6924	0.70353
17. Nonferrous metal manufacturing	2.02036	1.26962	0.61216
18. General machinery	1.65571	2.31605	0.66507
19. Electrical machinery manufacturing	1.32681	2.17578	0.65934
19. Automobile manufacturing	0.75483	1.11477	0.68122
20. Other transport equipment	0.95379	1.50084	0.82485
21. Precision machinery	1.06941	1.11466	0.61423
22. Other manufacturing	1.36975	1.94207	0.71387
23. Transportation	1.05021	0.99413	0.57036
24. Telecommunication	1.33553	0.70838	0.9077
25. Electric power	1.81397	1.1973	0.87677
26. Gas	2.52396	1.19072	0.76664
27. Wholesale and retail trade	1	1.56086	0.83684
28. Finance, insurance, and real estate	0.8782	2.14436	0.62192
29. Other services	1.37671	0.89986	0.74343

industries with fewer self-employed and family workers, such as oil and coal, electricity, and gas had higher wages.

On the other hand, differences in capital returns and taxation systems led to variations in capital service prices among industries. However, these differences significantly deviated from the trends observed in the relative ratio price estimates in 1970. This deviation is largely due to the yen's appreciation affecting the capital return rate, which resulted in particularly high capital service prices in less capital-intensive industries such as construction, clothing manufacturing, and leather goods. The previous estimate showed that the expansion of heavy and chemical industries since the

1960s led to a decrease in capital service prices due to increased capital investment. The capital return rate decline and yen appreciation since the 1970s have raised the cost of capital services in Japan.

By using these relative US-Japan ratio prices of output and factor prices, along with the distribution of production factors in both countries, we estimated the 1990-based disparity in TFP by industry for Japan and the US. Table 3.5 presents the results.

In 1990, a comparison of TFP between Japan and the US revealed that Japan surpassed the US in only eight sectors: pulp and paper manufacturing; ceramic, soil, and stone manufacturing; iron and steel production; general machinery

Table 3.5 1990-Based rate of estimated disparity of total factor productivity by industry between Japan and the US

	1960	1985	1970	1975	1980	1985	1990
1. Agriculture/forestry/fishery	1.199	1.218	1.086	1.201	1.106	0.865	0.844
2. Coal mining	0.154	0.207	0.267	0.313	0.287	0.252	0.184
3. Other mining	0.195	0.254	0.415	0.421	0.799	0.802	0.703
4. Construction	0.635	0.564	0.681	0.761	0.758	0.718	0.823
5. Food products	0.747	0.799	0.873	0.882	0.837	0.754	0.696
6. Textile manufacturing	1.025	0.988	0.997	0.886	0.842	0.838	0.842
7. Closing manufacturing	1.071	1.053	1.132	0.918	0.893	0.903	0.845
8. Woods products	0.887	0.903	0.966	1.067	0.964	1.027	0.933
9. Furniture products	0.587	0.586	0.671	0.632	0.589	0.661	0.619
10. Paper and pulp manufacturing	0.691	0.749	0.881	0.902	0.918	0.941	1.021
11. Printing	0.868	0.745	0.661	0.584	0.583	0.625	0.679
12. Chemical products	0.546	0.602	0.726	0.746	0.821	0.894	0.961
13. Oil and coal products	0.654	0.665	0.598	0.661	0.769	0.726	0.816
14. Leisure products	0.586	0.642	0.787	0.651	0.637	0.749	0.816
15. Ceramic products	0.821	0.921	1.118	1.078	1.108	1.191	1.073
16. Iron and steel manufacturing	0.701	0.735	0.895	0.975	1.046	1.047	1.006
17. Metal products	0.571	0.561	0.708	0.661	0.703	0.789	0.801
18. Machinery manufacturing	0.736	0.802	1.095	1.033	1.143	1.233	1.136
19. Motor vehicle manufacturing	0.681	0.753	0.985	1.076	1.319	1.375	1.396
20. Other transp. machinery	0.657	0.955	0.995	0.826	0.918	1.178	1.181
21. Precision machinery products	0.518	0.555	0.708	0.684	0.804	0.836	0.858
22. Other manufacturing	0.657	0.742	0.891	0.804	0.878	0.861	0.855
23. Transportation service	0.996	0.963	1.107	0.925	0.993	0.931	0.917
24. Communication service	0.512	0.604	0.624	0.869	0.735	0.934	1.042
25. Electricity/gas/water service		0.491	0.578	0.458	0.427	0.531	0.587
26. Trade	0.334	0.355	0.378	0.437	0.458	0.553	0.619
27. Wholesale and retail	0.385	0.476	0.634	0.671	0.771	0.737	0.829
28. Financial service	0.912	1.118	1.169	1.114	1.094	1.064	1.081
29. Other service	1.218	0.976	1.061	0.975	0.894	0.907	0.869
Total	0.487	0.548	0.743	0.728	0.728	0.793	0.836

Table 3.6 Contribution of sources of economic growth in Japan in the year 1960–1992

	Output	Labor input		Capital input		Annual growth rate
Year	Real GDP growth rate	Growth rate	Contribution ratio (%)	Growth rate	Contribution ratio (%)	Total factor productivity (%)
1960–1965	10.126	3.343		6.409		5.402
	100%		18%		29%	53%
1965–1970	11.791	3.661		7.611		6.223
	100%		17%		30%	53%
1970–1975	5.009	1.305		9.944		−0.085
	100%		14%		88%	−2.00%
1975–1980	4.277	4.278		4.175		0.896
	100%		42%		47%	21%
1980–1985	3.795	1.859		4.087		1.071
	100%		30%		42%	28%
1985–1990	4.626	2.225		4.899		1.306
	100%		28%		44%	28%
1960–1972	10.425	3.372		7.687		5.031
	100%		18%		34%	48%
1972–1992	3.887	1.737		5.185		0.741
	100%		27%		54%	19%
1990–1992	6.339	2.351		6.123		2.349
	100%		21%		42%	37%

manufacturing; automobile manufacturing; other transportation machinery manufacturing; telecommunications; and finance, insurance, and real estate.

Table 3.6 provides a macro-level summary of the factors that contributed to Japan's economic growth from 1960 to 1992, using 1990 as the base year.

During this time, labor contributed 18%, capital 29%, and technological progress 53% to growth in the 1960–1965 period. In 1960, only a few labor-intensive industries in Japan, like agriculture and textiles, had higher productivity than in the US. By 1965, this was even rarer. Japan's lower labor costs likely contributed to their competitiveness. Conversely, during the rapid economic growth period from 1960 to 1970, the rate of technological progress (or TFP growth rate) was notably high, at 5.402% per year in the first half and 6.223% in the second half. The policy of promoting heavy chemical industrialization increased the growth rate of capital input in industries such as chemicals, steel, general machinery, and automobile manufacturing, a trend that persisted until the early 1970s.

However, the global economic changes of the 1970s, the appreciation of the yen due to the introduction of a floating exchange rate system, and the oil price hikes caused by the oil shocks, had a significant impact on the Japanese economy. From 1960 to 1972, the average annual growth rate of real GDP exceeded 10%, but from 1972 to 1992, it was only 3.887%. In the first half, growth was fueled by capital expansion and technological progress. In the second half, reliance on labor and capital input increased as the impact of technological progress declined. Postwar heavy and chemical industrialization led to the stabilization of capital service prices, and the increase in factor prices relative to labor service prices increased capital intensity and promoted the introduction of capital-intensive technologies. Although the level of labor service prices was still low compared to the US, the rate of increase was high, and this, coupled with the expansion of capital investment, led to an increase in effective demand such as consumer spending during the period of rapid economic growth.

References

Jorgenson, D. W., Kuroda, M., & Nishimizu, M. (1992). Japan-U.S. industry-level productivity comparison, 1960–1979. *Journal of the Japanese and International Economies, 1*(March), 1–30.

Kuroda, M. (1989). Quantitative analysis of general equilibrium. In I. Shoten, K. Okawa, & R. Minami (Eds.), *Modern Japan's economic development: analysis based on long-term economic statistics*. Toyo Keizai Shinposha.

Chapter 4
Measuring the Spillover Effect of Total Factor Productivity by R&D and ICT Activities: Theoretical Framework of Capitalization

Masahiro Kuroda

Abstract Technology in a commodity production activity is interdependent with all related activities including the transactions of the intermediate raw material inputs and the intrinsic factors of capital and labor of production. The technology spillovers from interdependence are influenced by commodity production technologies and the market structure of capital and labor inputs. Technological spillovers affect not only current production, but also past technologies embedded in factor inputs from accumulated experiences. In the paper "Kuroda and Nomura" (2004), we proposed the analytical framework using input-output table, in which we can evaluate the spillover effects of the technological property among industry both statically and dynamically. **The potential for technological spillovers is evaluated by dividing the total factor productivity (TFP) into static and dynamic unit TFP**.

We observed that recent technological innovations are driving rapid structural changes through the accumulation of knowledge stock, which are intangible fixed assets created by investments in Research and Development (R&D) and Information and Communication Technology (ICT). The accumulation of these intellectual assets, such as R&D and ICT investments, influences the production efficiency of the related production sector. **We elucidate the relationship between the accumulation of intangible fixed assets and the knowledge capital services derived from these assets because of R&D investment and ICT activities, using the structure of the input-output table. We propose a theoretical framework that measures the impact of knowledge service stock accumulation through R&D and ICT investments on static and dynamic unit TFP**.

The previous chapter explained that growth was mainly due to the accumulation of tangible fixed assets like facilities and structures. Since the latter half of the twentieth century, there has been a shift in the Japanese economy from investment in hard productive facilities to soft utilities like R&D and ICT as intangible assets.

M. Kuroda (✉)
Professor Emeritus, Keio University, Tokyo, Japan

M. Kuroda (ed.), *Intangible Capital Investment and Economic Growth*, Translational Systems Sciences 45,
https://doi.org/10.1007/978-981-96-7675-0_4

Simultaneously, there is another consensus in the society that the Japanese economy might enter a prolonged period of stagnation in the 1990s. Therefore, the impact of this scientific and technological evolution has not been as positive in Japan as expected, especially when compared to other countries. Postwar expansion of tangible fixed capital investment is gradually changing.

We must consider the recent meaning and the effect of the knowledge stock by the R&D and ICT investment deeply. We aim to measure the quantity and quality of intellectual services provided by the knowledge stock through the accumulation of intellectual assets. We also propose a methodology to measure and evaluate its impact on industries. This approach will help us understand the role of these new technological innovations in economic development and the structural changes they bring about.

We propose a framework and measurement method for understanding and measuring intangible fixed capital, the stock of knowledge resulting from R&D and ICT investment. This framework allows for a systematic analysis of the structure in which R&D and ICT investment alters the productivity of an industry.

Keywords Knowledge stock as intangible capital assets · Static unit TFP · Dynamic unit TFP · Capital service price · Rate of return on capital · Gross capital cost

4.1 Increasing Technical Efficiency and the Impacts of Knowledge Stock as Intangible Capital Assets

Kuroda and Nomura (2004) introduced a method to measure the spillover effects of technological progress. The method is based on the structure of static and dynamic technological linkages in commodity production within the framework of input-output analysis. The technology in a commodity production activity is interdependent with all related activities in the market. Interdependence is because of the allocation of intermediate raw material inputs and the intrinsic factors of production, namely capital and labor. Therefore, changes in technological efficiency in the production of any commodity will directly or indirectly affect the efficiency of all related production activities. This interdependence requires systematic evaluation to accurately assess the impact of these changes. Spillovers from interdependence are affected by both current and historical commodity production technologies in capital service inputs. These capital services come from the accumulated capital stock. It implies that the capital stock depends on the past technology embodied in the production of the investment good at each point in the past. Technological spillovers also impact future commodity production methods, not just the current ones. It also includes the impact of the past technology embodied in capital through capital accumulation from past investments, referred to as dynamic spillovers. The evaluation of these dynamic spillovers is also crucial.

The evaluation of technological spillovers involves dividing total factor productivity (TFP) into static unit TFP and dynamic unit TFP. As shown in our 2004 publication, these metrics can be derived from Leontief's input-output framework (Kuroda & Nomura, 2004). The traditional TFP measure by industry assesses the efficiency of a specific commodity's production activity within a growth accounting framework, based on the commodity's own input structure at present. Conversely, the static unit TFP evaluates the efficiency of a commodity's production activity considering the interdependence of intermediate inputs, and capital and labor factors in the associated production activity. This measure is thoroughly evaluated in terms of the interdependence of intermediate inputs in related production activities and the demand-supply structure of capital and labor inputs in their respective markets.

Furthermore, dynamic unit TFP assesses the efficiency of a specific commodity's production at a given time. It considers the capital service input required to produce that commodity as the capital stock, which is formed by the accumulation of capital through past investment. This past investment is dependent on the technology available at that time. Therefore, the process of capital accumulation is retrospectively assessed with a dynamic perspective on the current capital service's technology. The production of a certain commodity at present involves the input of intermediate raw materials, labor services, capital services, and other factors of production. The capital services used in this process are the accumulation of past capital investment, and their technology embodies the technology at the time of the investment. According to Peterson (1979), the productivity index of a "dynamically and vertically integrated sector" evaluates production efficiency by reducing all factors invested in the commodity to all past labor input. Aulin-Ahmavara positioned it as corresponding to an integrated TFP index (a "fully effective" rate of TFP), which evaluates the input of capital factors in the past using dynamic input-output analysis. Aulin-Ahmavara calculates this indicator as a solution to the equilibrium growth path in a closed dynamic input-output model that expresses the technological structure using intermediate input coefficients and capital coefficients as structural parameters. In contrast, we attempted to define and measure the index "dynamic unit TFP" as an indicator to evaluate the efficiency of the current technology by looking back in time using historical time-series input-output table data.

Thus, our focus was on the dynamic effect of capital accumulation, particularly through investment in tangible fixed assets. We also acknowledge that recent technological innovations are driving rapid structural changes through the accumulation of knowledge stock, which are intangible fixed assets created by investments in Research and Development (R&D) and Information and Communication Technology (ICT). The accumulation of these intellectual assets, such as R&D and ICT investments, influences the production efficiency of the related production sector. Therefore, a logical framework to enhance production efficiency is necessary. This framework should incorporate the structure of accumulated knowledge stock produced by R&D and ICT activities, into the input-output structure of commodity production. It should also have the theoretical framework for measuring its technological spillover effects. For R&D investment, we differentiate the measurements between public R&D investment by the government and private non-profit organizations, R&D investment by the R&D sector as a private industry, and R&D investment within companies in each industrial sector. These are integrated into the input-output framework of relevant industries, mirroring production activities' structure. Moreover, it is crucial to understand the impact of accumulating ICT-related intangible fixed assets, such as software and system development, through ICT activities. This technology extends to the activities of the so-called information industry, including the ICT industry, the Internet, and other ICT and communication-related industries that promote such activities. It is also important to comprehend the impact of R&D investment in information-related intangible fixed assets, such as ICT activities in each industrial sector. Each aspect is integrated into the input-output structure of the relevant industry.

In addition to the ICT industry, which includes the Internet and other ICT and communication-related industries, it is crucial to understand the impact of ICT activities within companies across all industrial sectors. These activities contribute to the accumulation of ICT-related intangible fixed assets, such as software and system development.

In Sect. 4.2 we outline our theoretical approach to the measurement of the accumulation of each type of knowledge capital stock and the resulting knowledge service flows. Specifically, we elucidate the relationship between the accumulation of intangible fixed assets and the knowledge capital services derived from these assets because of R&D investment and ICT activities, by using the structure of the input-output table. We then introduce the theoretical structure of capital service quantity and its measurement method. Sections 4.3 and 4.4 focus on the estimation results of capital service quantity and price, based on the connected input-output table data from 1995 to 2011. Sections 4.5–4.7 present a total factor productivity system that incorporates the input structure of knowledge services from R&D and ICT investments into the concept of static and dynamic unit TFP structure. Here, we propose a theoretical framework that measures the impact of knowledge service stock accumulation through R&D and ICT investments on static and dynamic unit TFP. By evaluating the impact of knowledge stocks as intangible fixed assets on production efficiency, similarly to the assessment of the contribution of ordinary tangible fixed capital investment to TFP, we posit that the influence of knowledge expansion on the market economy can be understood in terms of production efficiency enhancement.

In late 2016, our System of National Accounts (SNA) began capitalizing R&D expenditures. We have updated the input-output table to include these changes. References for a summary of issues related to the capitalization of R&D investment and the estimation of an input-output table with the capitalization of R&D and ICT expenditures bases upon the time-series data of the Japanese economy's input-output table from 1995 to 2011. The concluding section of this chapter, Sect. 4.5, applies the static and dynamic TFP measurement frameworks presented in Sects. 4.3 and 4.4 to measure the contribution of intangible fixed assets, such as R&D and ICT investment, to Japan's productivity growth from 1995 to 2011. It also assesses the investment efficiency of R&D and ICT activities in a deflationary economy over the same period. We empirically evaluate the investment efficiency of ICT activities to comprehend the reality of Japan's economic stagnation during this period. The structural analysis of the results will be discussed in later chapters.

4.2 Capitalization of Knowledge Stock to R&D and ICT Investment and the Measurement Method of the Capital Service

4.2.1 Contribution of Knowledge Stock to Economic Growth and the Presentation of the Intangible Fixed Assets in the Input-Output Table

In the previous chapter, we discussed how Japan, since the end of World War II, has experienced significant economic growth. This growth was fueled by accumulating tangible assets like facilities and structures, leading to technological advances and higher industrial productivity. However, since the latter half of the twentieth century, there has clearly observed a shift in the Japanese economy from the investment as hard productive facilities to the soft utilities like R&D and ICT as intangible assets. This shift, fueled by rapid advances in information technology and other knowledge-based areas, has surely impacted on the efficiency of the market structure. It has also led to changes in societal consciousness and the hierarchy in society.

Indeed, the evolution of scientific knowledge in the twentieth century has stimulated research and development (R&D) activities in all scientific fields, promoted the accumulation of knowledge, and contributed to progress the industrial productivity and market efficiency. However, the development pattern during the high economic growth period, where scientific and technological progress contributed to enhanced economic growth, seems to change significantly since we entered the twenty-first century. However, there is a growing social consensus around the new view that the Japanese economy has entered a period of prolonged stagnation. Indeed, compared to other countries, the impact of the evolution of science and technology on the Japanese economy in the twenty-first century has not been as significant as expected, and someone are suggesting that Japan is entering an era of long-term stagnation.

It is a highly important task to scientifically re-examine how these new technological innovations affect economic development, and by clarifying the mechanisms of structural changes caused by these innovations, it is essential to build a market structure that allows for sustainable development, distinct from the high economic growth period of Japan. Indeed, in the structure of the twenty-first-century Japanese economy, a different relationship between science-technology and economics is recognized in comparison with the pattern of economic growth driven by the expansion of tangible capital investment. To confirm this, it is necessary to establish methodologies that empirically grasp the impact of changes in the quantity and quality of intellectual services provided by the knowledge stock accumulated through intellectual assets in the twenty-first century on industries. Through this, there is a growing need to empirically understand the role that new technological innovations play in economic development and the resulting structural changes.

The evolution of scientific knowledge has stimulated and refined R&D activities across all scientific fields. This has promoted the accumulation of knowledge stock, contributing to industrial productivity and market efficiency. This has resulted in a major structural change in the cycle between science and society. On the other hand, there are another consensus in the new society that the Japanese economy began a prolonged period of stagnation in the 1990s. The impact of this scientific and technological evolution has not been as positive in Japan as expected, especially when compared to other countries. It is crucial to evaluate the nature of these new technological innovations in economic development, and it is highly important to clarify the mechanism of structural change in their impact, and to formulate a sustainable market structure that diverges from the one promoted in Japan during the period of high economic growth. The postwar structure was marked by economic growth driven by the increase in tangible fixed capital investment. However, we can recognize the new structure between science and technology.

Specifically, we first need to develop a new market structure that is intellectually sound. We aim to measure the quantity and quality of intellectual services provided by the knowledge stock through the accumulation of intellectual assets. We must propose a methodology to measure and evaluate its impact on industries. This approach will facilitate our understanding of the impact of new technological innovations on economic development and the structural changes they are likely to induce.

It was often suggested that the Japanese economy might enter the period of stagnation in the 1990s, because the impact of technological evolution was not as positive as anticipated, especially when compared to other countries. This necessitates an evaluation of these new technological innovations in economic development, by the clarification of the mechanism of structural change in their impact, and the formulation of a new, sustainable market structure. This new structure should diverge from the one promoted in Japan during the period of high economic growth, which was characterized by economic growth driven by the expansion of tangible fixed capital investment. First, we aim to clarify a new market structure and consider the meaning that the new structure is intellectually robust. We attempt to measure the volume of intellectual services provided by the stock of knowledge through the accumulation of intellectual assets. We also propose a methodology to measure and evaluate its impact on industries and society.

We propose a framework for understanding the concept of the intangible fixed capital and the methodology for measuring impacts of the intangible fixed capital, as the stock of knowledge resulting from R&D and ICT investments. The framework might give us a systematic analysis of the structure in which R&D and ICT investments might have impacts on the level of the productivity of an industry. This confirms the theoretical consistency between industries' input-output structures and TFP measures. Furthermore, we propose a framework for analyzing the structure of changes in industrial productivity caused by intellectual services arising from the accumulation of knowledge stock as intangible assets. We also might propose a measurement method for this structure. To achieve this, we first recombine the current input-output table to capture the input structure of R&D activities (development of knowledge through R&D) and ICT activities (development of software and systems for ICT activities) as intangible asset production activities.

In our input-output system, the R&D and ICT activities within an industry are designed to differentiate from the main production activities by the independent sectors like the university and the specific academic institute. This system explicitly captures the creation of intellectual intangible fixed assets (knowledge stock) through R&D and ICT activities within a company. It tries to evaluate the impact of such knowledge creation activities as intra-firm activity, which support the main production activities of goods and services.

In addition to its main production activities, in each production sector, there also captures the input structures of intermediate goods, labor services, and capital services with respect to R&D and ICT activities within the firm. Both tangible and intangible capital stocks are identified as the accumulation of investment activities in tangible and intangible capital goods in that industry. We consider that tangible and intangible capital services, generated from the stocks at the beginning of each period formed by investment actions up to the previous period, are input into the main production activities and intra-firm R&D and ICT activities in the current period. Therefore, we theoretically distinguish between the amount of tangible and intangible capital stock and the amount of capital services generated by the stock. We would like to estimate each of them in a structurally consistent manner.

Regarding the accumulation of knowledge stock, apart from R&D and ICT activities within companies that support the main production activities within each industry, there are independent industries such as R&D sectors (private think tank, research institute in university, etc.) and ICT Institute for software and system development. These sectors build knowledge stock and offer intellectual services to other industries as market goods. These sectors not only support the main production activities within each industry but also contribute to the extension of the overall knowledge in the economy and society.

Knowledge assets, which are created and accumulated through research activities at public and private universities and research institutes, are the product of public R&D activities by the official subsidies. These subsidies, both tangible and intangible, support researchers and their R&D activities, leading to the accumulation of knowledge stock as intangible fixed assets. The intellectual services from this knowledge stock are public R&D knowledge services, provided as a common social good. These services differ from knowledge service provided by intra-firm or industrial activities dedicated to R&D and ICT. Except for patent rights and transactions, the knowledge stock is provided as a costless, non-competitive public good that influences the production of market goods and the output of related industrial sectors. Government-supported public goods, as intangible assets, influence market behavior and drive economic and social progress.

Understanding quantitatively how these knowledge assets generate sophistication and interdependence among industries and individuals, and how they lead to economic development and solutions to social issues, is crucial for evaluating the impacts of science and technology policies. The accumulation of knowledge created by R&D activities affects production efficiency and demand structures such as consumption and investment in the market. Government-created intellectual assets funded publicly are public goods and lack marketability as private goods once disclosed. Identifying the industry impacted by knowledge will non-marketably affect its cost function's production efficiency, influencing other sectors through price changes.

If the firm succeeded in its R&D activities and obtained patent rights before the knowledge is published, the rights will have marketability and generate patent fees or other compensation in transactions. In the current Japanese input-output table, patent rights are treated as generating the acquisition of patent fees in the sector that created the technology and the firms which used the technology in the production process must pay as payment of patent fees, which is recorded in the negative. In this case, the buying and selling of patents between domestic sectors would be offset macroscopically, and only the result of buying and selling with foreign countries would be recorded as the transfer of property income.

Patents from abroad are recorded as research sector revenues or retained earnings and affect exports or imports. Patents to abroad are treated similarly but in reverse. However, we do not explicitly represent patent transfers between domestic sectors in the input-output table.

Another characteristic of knowledge as a public good is its universality and nonrivalry. However, the input-output relationship as a market good cannot be explicitly expressed in the Japanese input-output table. This is an issue that needs to be addressed in the future.

Despite the limitations of the following analysis, we examine the influence of knowledge asset creation, encompassing public and market goods, on industry productivity and market structure. This refers to changes in each industry's production efficiency or its impact on TFP. The relationship between capital services derived from knowledge stocks as intangible fixed assets in a specific industry and the industry's productive activities can be depicted in a model diagram of the input structure, based on the input-output table framework. We will reconfigure the published input-output table, with special attention to the points below.

(a) Industrial sector activities are categorized into three areas: main production, intra-firm R&D, and intra-firm ICT. Some sectors, because of conceptual and material constraints, cannot be explicitly classified as intra-firm activities. Intra-firm R&D and ICT activities, while not directly profit-generating, support main production through R&D and ITC knowledge accumulation. Prior to the division of each industry's activities into these three areas, intra-firm activities do not directly generate an operating surplus, with the main production activity bearing this activity's capital cost. Capital service costs for main production, intra-firm R&D, and ICT are included in the operating surplus of main production activities. The capital costs of intra-firm activities are consolidated in the operating surplus of the main production activities. The cost of capital is calculated by adding an asset's capital depletion allowance to its operating surplus, for both tangible and intangible assets. The expenses for the three types of capital services need to be allocated among them. Labor input costs should be presented separately for each activity, as the individuals involved in each activity can be specifically identified.

The costs of the three capital service types must be shared.

(b) The input-output table estimates the capital depletion allowance for tangible fixed assets in main production, intra-firm ICT, and intra-firm R&D using tax law depreciation rates. The capital cost of intangible fixed assets is included in the operating surplus of main production. In the input-output table, the capital service costs of intangible fixed assets are included in the operating surplus of main production activity, including the capital depletion allowance. The method for estimating the cost of intangible fixed capital for intra-firm ICT and R&D activities involves allocating the total operating surplus among the activities, using the procedure for estimating capital service costs as described below.

(c) R&D activities of government agencies, universities (both public and private), national institutes, and private university-affiliated research institutes are considered non-profit activities in the current input-output table, assuming zero operating surplus. On the final demand side, capital depletion is recorded as social capital depletion for national and public institutions, and as capital depletion for private non-profit institutions. No distinction is made between tangible and intangible fixed assets. Their output is considered as capital depletion without distinguishing between tangible and intangible assets. In our analytical framework, these R&D activities are seen as efforts to produce and accumulate knowledge assets as intangible fixed assets. They are viewed as actions to produce knowledge services as intangible capital through the input of intermediate goods, labor, and tangible capital services in each period. The tangible capital input is considered exogenously given as a government policy. The output amount is adjusted as an input to produce intangible capital services, in addition to its operating surplus. The output is the government's production of knowledge services as intangible assets, which is recorded and balanced in the final demand for intangible capital investment.

(d) The economic obsolescence rate is calculated for both tangible and intangible assets. The stocks of these assets are estimated using the permanent inventory method. For cost accounting, tangible and intangible fixed assets are estimated separately. The quantity of their capital services is allocated by the attribution calculations to divide the cost of capital, which will be explained in the following section.

(e) The knowledge stock produced by ICT and R&D activities is considered as intangible fixed assets within the firm. The services produced by the asset stock at the beginning of the period are seen as contributing to the production efficiency of the main production activity. This contribution is then related to the capital cost of the main production activity.

To measure TFP and the growth contribution of each factor's output to the industry sector's costs, several steps need to be taken. Identify the cost structure of tangible fixed assets for three activities consolidated in the main production surplus, and the two intangible fixed assets. To calculate the input of capital services, the subsequent section outlines the framework for defining the cost of capital and allocating capital service prices and quantities of capital services.

Table 4.1 Frame of input structure of the *j*-th industry

		Industry *j*-th: productive activity		
		Main activity	Information treatment activity (IT)	Research development activity (RD)
200	Total intermediate input cost	$\Sigma \boldsymbol{p}_{di}\boldsymbol{X}_{dij} + \Sigma \boldsymbol{p}_{mi}\boldsymbol{X}_{mij}$	$\Sigma \boldsymbol{p}_{di}\boldsymbol{X}_{dij} + \Sigma \boldsymbol{p}_{mi}\boldsymbol{X}_{mij}$	$\Sigma \boldsymbol{p}_{di}\boldsymbol{X}_{dij} + \Sigma \boldsymbol{p}_{mi}\boldsymbol{X}_{mij}$
501	Consumption expenditure outside households	*BCHi*		
502	Wages and salaries for employees	*LCj = PEmjhmjEDmj*	*LCitj = PEitjhitjEDitj*	*LCrdj = PErdjhrdjDrdj*
503	Income for employees and family worker	*PSEmjhmjSEDmj*	*PSEitjhitjSEDitj*	*PSErdjhrdjSEDrdj*
504	Total capital surplus for total capital			
504-1	Tangible assets	*PSKmKmj*	*PSKitKitj*	*PSKrdKrdj*
504-2	Intangible assets (IT)	*PSKNitKNitj*		
504-3	Intangible assets (RD)	*PSKNrdKNrdj*		
508	Indirect tax discluding tariff	*IndTaxmj*		
509	(Less) current subsidies	- *SDml*		
600	Gross total value-added	*Vmj*	*Vitj*	*Vrdj*
700	Total of domestic output	*PdZj = PmjXj − PsrtKitj − PSrdKrdj + DEPKNith + DEPKNrdj*	*PditZitj = PitjXitj + PSitKitj − DEPKNith*	*PdrdZrdj = PrdjXrdj + PSrdKrdj − DEPKNrdj*

Note: 1. Composition of factor cost in the *j*-th industry: Total activity in the *j*-th industry is divided into three activities (main, intra-firm activity for information & communication treatment [ICT], and intra-firm activity for research and development treatment (RDT). 2. Definitions: P_{dj}: domestic output price of the *j*-th sector; P_{mj}: imported product price of the *j*-th sector; PSE_{mj}, PSE_{itj}, PSE_{rdj}: wage rate of main, ICT and RDT activity in *j*-th sector; ED_j: number of employee of main, ICT, and RDT activity in *j*-th sector; PSK_{mj}, PSK_{itj}, PSK_{rdj}: capital service price of main, ICT, and RDT activities *j*-th sector; $PSKN_{itj}$, $PSKN_{rdj}$: intangible capital service price in ICT and RDT activity

4.3 Formulation of Capital Service Quantity and the Cost of Capital by Industry

The total cost of capital for each sector is defined as the gross cost of capital, which includes capital depletion cost. Consequently, the capital service price, which is the gross service price, also accounts for the cost of capital depletion. Therefore, the gross capital cost can be expressed as:

$$\textbf{Gross capital cost} = \textbf{Operating surplus} + \textbf{Capital depletion allowance} \\ + \textbf{Property taxes}$$

The gross capital cost encompasses operating surplus, capital depletion allowance, and property tax. Given that property tax is a part of the indirect tax, it needs to be transferred to the operating surplus within the same sector.[1]

Our model assumes that the initial capital services in sector j for year t are proportional to the capital stock volume. We assume that its gross cost of capital is defined as the aggregate of the operating surplus of each industrial sector in the input-output table, the allowance for capital depletion, and the property tax included in indirect taxes. The capital depletion allowance, included in the definition formula for the crude cost of capital, is determined by the obsolescence rate $\delta^{KS}{}_j$. This rate accounts for the economic wear of capital and differs from the capital depletion rate used for tax accounting purposes. Here, the amount of capital depletion corresponds to the economic depletion in the measurement of capital services, which considers economic obsolescence. This is distinct from the value of the capital depletion allowance used for tax accounting purposes, which is mandated by accounting rules for book valuation and differs from the economic obsolescence rate considered here, which accounts for technological change and shifts in industrial behavior. The quantity of capital services and their prices are calculated by also considering this economic capital depletion in the attribution calculation.

Regarding the total crude capital cost, we assume it is adjusted by the operating surplus amount. Therefore, we posit that the operating surplus in the input-output table includes the capital depletion allowance because of economic obsolescence. Consequently, we assume the following capital service amounts.

4.3.1 Capital Service Price

We identify capital stock and capital services, although it is assumed that the latter is proportional to the former. We also posit that the capital would be classified into two types of capital, named into "tangible" and "intangible."

[1] Here, property tax is including in the subsidies by the definition of IO table.

First, we define the capital service price of tangible fixed capital P_j^{KSt} in each industry's main production activity at time *t* as follows. This definition is derived from the assumption that the present value of future after-tax investment returns generated from a unit of investment equals a unit of the current investment amount (i.e., its current investment goods price P_j^{INVKt}).

$$P_j^{KSt} = \left(1-\tau^K\right) r^{Kt}{}_j P_j^{INVKt-1} + \delta_j^{KS} P_j^{INVKt} - \left(P_j^{INVKt} - P_j^{INVKt-1}\right) + \tau^P P_j^{INVKt-1} \quad (4.1)$$

where P_j^{INVKt} is the investment goods' price, rj^{Kt} is the expected pre-tax return rate on investment, and τ^K and τ^P are the investment income tax rate and fixed asset tax rate, respectively. As mentioned earlier, δ_j^{KS} is the economic obsolescence rate of capital. Similarly, we define the capital service prices for intra-firm information activities (*KITE*) and intra-firm R&D activities (*KRDE*).

The capital service price of their tangible fixed capital P_j^{KSITEt} is defined as follows.

$$\begin{aligned} P_j^{KSITE} = {} & \left(1-\tau^{KITE}\right) r_j^{KITEt} P_j^{INVKITEt-1} + \delta_j^{KITE} P_j^{INVKITEt} - \left(P_j^{INVKITEt} - P_j^{INVKIT-1}\right) \\ & + \tau^{PITE} P_j^{INVKITEt-1} \end{aligned} \quad (4.2)$$

where r_j^{KITEt} is the expected rate of return on investment for intra-firm ICT tangible fixed capital investment, and τ^{KITE} and τ^{PITE} are the return on investment tax rate and fixed asset tax rate for intra-firm ICT investment, respectively. Δ_j^{KITE} is the economic obsolescence rate for that capital.

$$\begin{aligned} P_j^{KSRDEt} = {} & \left(1-\tau^{KRDE}\right) r_j^{KRDEt} P_j^{INVKRDEt-1} + \delta_j^{KRDE} P_j^{INVKRDEt} \\ & - \left(P_j^{INVKRDEt} - P_j^{INVKRDEt-1}\right) + \tau^{PRDE} P_j^{INVKRDE}. \end{aligned} \quad (4.3)$$

Similarly, we define the intangible fixed capital services price of intra-firm R&D.

$$\begin{aligned} P_j^{KSNITEt} = {} & \left(1-\tau^{KNITE}\right) r_j^{KNITEt} P_j^{INVSKNITEt-1} + \delta_j^{KNITE} P_j^{INVKNITEt} \\ & - \left(P_j^{INVKNITEt} - P_j^{INVKNITEt-1}\right) + \tau^{SKNITE} P_j^{INVKNITEt-1}, \end{aligned} \quad (4.4)$$

$$\begin{aligned} P_j^{KSNRDEt} = {} & \left(1-\tau^{KNRDE}\right) r_j^{KNRDEt} P_j^{INVKNRDEt-1} + \delta_j^{KNRDE} P_j^{INVKNRDEt} \\ & - \left(P_j^{INVKNRDEt} - P_j^{INVKNRDEt-1}\right) + \tau^{SKNRDE} P_j^{INVKRDNEt-1}, \end{aligned} \quad (4.5)$$

Intra-firm information and R&D intangible fixed asset investments, τ^{KNITE} and τ^{KNRDE} are the investment return tax rates for those investments, τ^{SKNITE} and τ^{SKNRDE} are the fixed asset tax rates for those investments, and δ_j^{KNRDE} and δ_j^{KNITE} are the economic obsolescence rates for the respective intangible fixed capital.

Capital service prices from tangible and intangible fixed capital stocks in the form of public R&D can be similarly defined in the same way as investment goods prices.

$$P_j^{KSGt} = \left(1-\tau^{KG}\right) r_j^{KG} P_j^{INVKGt-1} + \delta_j^{KG} P_j^{INVKGt} - \left(P_j^{INVKGt} - P_j^{INVKGt-1}\right) + \tau^{PKG} P_j^{INVKGt-1}, \tag{4.6}$$

$$P_j^{KNNGt} = \left(1-\tau^{KNG}\right) r_j^{KNG} P_j^{INVKNGt-1} + \delta_j^{KNG} P_j^{INVKNGt} - \left(P_j^{INVKNGt} P_j^{INVKNGt-1}\right) + \tau^{PKNG} P_j^{INVKNGt-1} \tag{4.7}$$

4.3.2 *Foundation of Capital Service: Quantity and the Cost*

As noted, we assume that the quantity of each service is proportional to the quantity of each capital stock of tangible and intangible capital. We also proposed that the total nominal value of the service quantity multiplied by the capital service price equals the total cost of capital for each industrial sector. In the input-output table, the sum of the operating surplus, capital depletion allowance, and property tax for each production sector can be observed for each industrial sector, which is referred to as the gross capital cost. If the cost of capital BS_j^t (net operating surplus + capital depletion allowance + fixed asset tax) for each production sector, represents the gross capital cost of capital services provided by tangible and intangible fixed capital, we can map the observed values to the gross capital cost of capital services.

(a) **Case in which industrial activity is classified as main production activity, intra-firm ICT activity, and intra-firm R&D activity**

$$\begin{aligned} BS_j^t &= K_j^t P_j^{KSt} + KITE_j^t P_j^{KSITEt} + KNITE_j^t P_j^{KSNITEt} + KRDE_j^t P_j^{KSRDEt} + KNRDE_j^t P_j^{KSNRDEt} \\ &= [K_j^t \left\{\left(1-\tau^K\right) r_j^{Kt} P_j^{INVKt-1} + \delta_j^K P_j^{INVKt} - \left(P_j^{INVKt} - P_j^{INVKt-1}\right) + \tau^P P_j^{INVKt-1}\right\} \\ &\quad + KITE_j^t \{\left(1-\tau^{KITE}\right) r_j^{Kt} P_j^{INVKIETt-1} + \delta_j^{KITE}{}_{+1} P_j^{INVKITEt} \\ &\quad - \left(P_j^{INVKITEt} - P_j^{INVKITEt-1}\right) + \tau^{PKITE} P_j^{INVKITEt-1}\} \\ &\quad + KNITE_j^t \left(1-\tau^{KNITE}\right) r_j^{Kt} P_j^{INVKNITEt-1} + \delta_{j+1}{}^{KNITE} P_j^{INVKNITEt} \\ &\quad - \left(P_j^{INVKNITEt} - P_j^{INVKNITEt-1}\right) + \tau^{PKNITE} P_j^{INVKNITEt-1}\} \\ &\quad + KRDE^t{}_j \{\left(1-\tau^{KRDE}\right) r^{Kt}{}_j P^{INVKRDEt-1}{}_j + \delta^{KRDE}{}_j P^{INVKRDEt}{}_j \\ &\quad - \left(P_j^{INVKRDEt} - P_j^{INVKRDEt-1}{}_j\right) + \tau^{PKRDE} P_j^{INVKRDEt-1}\} \\ &\quad + KNRDE_j^t \{\left(1-\tau^{KNRDE}\right) r_j^{Kt} P_j^{INVKNRDEt-1} + \delta_j^{KNRDE} P_j^{INVKNRDEt} \\ &\quad - \left(P_j^{INVKNRDEt} - P_j^{INVKNRDEt-1}\right) + \tau^{PKNRDE} P_j^{INVKNRDEt-1}\}] \cdot P_j^{SK0}, \end{aligned} \tag{4.8}$$

where P_j^{SKO} is the adjustment rate between the actual capital services price per unit at the base time (2005) and the capital services price index in the model (base time index = 1.0). It is an adjustment coefficient that corrects for the difference in the cost of capital at the base time, generated by assuming that real capital services equal real capital stock at the base time, and that both the capital services price index and investment goods price index are 1.0 at the base time. At the base time ($t = 0$), P_j^{KS0} is obtained by assuming that the quantity of capital services equals the quantity of capital stock and it is setting during the whole periods at 1.0. If a firm's activities are categorized into three activities, Main, ICT, and R&D, then in industry j, the tangible fixed capital stock (K_j^t, $KITE_j^t$, $KRDE_j^t$) for these activities within the firm, and the intangible fixed capital stock for ICT ($KNITE_j^t$, $KNRDE_j^t$) and for R&D activities are given. The balance in Eq. (4.8) is assumed to be achieved by integrating the cost of capital of the tangible and intangible fixed capital stock ($KNITE_j^t$, $KNRDE_j^t$) of the government R&D activities and the intra-firm ICT and R&D activities.

(b) **Case for public R&D activity**

As mentioned above, in the published input-output tables, the operating surplus for the public research and development sector is set to zero. However, in our input-output tables, the public research and development sector includes private non-profit research institutions (such as private universities), and the operating surplus from those sectors is added. Additionally, since we consider the research and development activities of government and private non-profit research institutions as production activities of knowledge assets in the form of intangible fixed capital from their research and development during those periods, it should be noted that capital depreciation from the accumulation of intangible fixed assets at the beginning of the period is considered to arise as capital costs.

The research and development activities that serve as policy support by the government during the current term are added as government subsidies. However, the treatment of these subsidies follows the policies of the given period. They are exogenously added either as support for labor costs or capital costs, which results in lowering the assumed factor costs at the beginning of the period as current subsidies. Capital costs consist of both tangible and intangible capital costs, and they are defined as the total of tangible and intangible capital costs as shown below.

$$
\begin{aligned}
BS_j^t &= KSG_j^t P_j^{KSGt} + KSNG_j^t P_j^{KSNGt} \\
&= [KG_j^t \{(1-\tau^K) r_j^{KGt} P_j I^{NVGKt-1} + \delta_j^{KG} P_j^{INVKGt} - (P_j^{INVKGt} - P_j^{INVKGt1}) \\
&\quad -\tau^P P_j^{INVKGt-1}\} + KNG_j^t \{(1-\tau^{KNG}) r_j^{KGt} P_j^{INVKNGt-1} + \delta_j^{KNG} P_j^{INVKNGt} \\
&\quad -(P_j^{INVKNGt} - P_j^{INVKNGt-1}) + \tau_j^{PKNG} P_j^{INVKNGt-l}\}] P_j^{KNG0}
\end{aligned} \tag{4.9}
$$

Here, P_j^{KSG0} is a coefficient used to adjust the capital costs to actual values in the base year. The value calculated for the base year of 2005 is used as a fixed value and applied for each year.

(c) **For the private R&D industrial sector**

For the private R&D industry sector, BS_j^t is defined in the input-output table as the sum of the operating surplus from the value-added portion and the capital depreciation allowance for both tangible and intangible fixed capital:

$$
\begin{aligned}
BS_j^{\,t} &= KSPI_j^{\,t} P_j^{KSPIt} + KSNPI_j^{\,t} P_j^{KSNPIt} \\
&= [K_j^{\,t}\{\left(1-\tau^{K}\right) r^{KPIt}{}_j P_j^{INVKPIt-1} + \delta^{KPI}{}_j P_j^{INVKPIt} - \left(P_j^{INVKPIt} - P_j^{INVKPIt-1}\right) \\
&\quad + \tau^{P} P_j^{INVKPIt-1}\} + KNPI_j^{\,t}\left\{\left(1-\tau^{KNPI}\right) r^{KPIt}{}_j P_j^{INVKNPIt-1} + \delta^{KNPI}{}_j P_j^{INVKNPIt} \right. \\
&\quad \left. - \left(P_j^{INVKNPIt} - P_j^{INVKNPIt-1}\right) + \tau^{KNPI} P_j^{INVKNPI} \quad P_j^{KSPI0}
\end{aligned}
\tag{4.10}
$$

where $P_j^{\ KSPI0}$ is a coefficient by which the estimated capital cost adjusts to the observed capital cost in the base year. In other words, P_j^{KSPI0} is the coefficient that adjusts the cost of capital to the observation prevailing in the base year. This coefficient was calculated in the base year, 2005, and remains fixed in subsequent years.

Estimation of the Adjustment Coefficient of Capital Cost

1. Industrial sector with the main production activity, the ICT activity, and the R&D activity within the enterprise

$$P_j^{KS0} = BS_j^{\,0} / \left\{K_j^{\,0} + KITE_j^{\,0} + KNITE_j^{\,0} + KRDE_j^{\,0} + KNRDE_j^{\,0}\right\} \tag{4.10a}$$

2. Government's public R&D sector

$$P_j^{KSG0} = BS_j^{\,0} / \left\{KG_j^0 + KNG_j^0\right\}. \tag{4.10b}$$

3. Private R&D sector

$$P_j^{KSPI0} = BS_j^{\,0} / \left\{KPI^0{}_j + KN^0{}_j\right\}. \tag{4.10c}$$

4.3.3 *Formulation of the Rate of Return on Capital*

Furthermore, the rate of return on capital is estimated as follows respectively as shown in Eqs. (4.8), (4.9), and (4.10).

4.3.3.1 The Primary Case Activities and Intra-firm of R&D & ICT Activities

$$
\begin{aligned}
DD1 = {} & BS_j^t / P_j^{KS0} - [K_j^t \left\{\delta_j^K - \left(P_j^{INVKt} - P_j^{INVKt-1}\right) + \tau^P P_j^{INVKt-1}\right\} \\
& + KRDE_j^t \left\{\delta_j^{KRTE} P_j^{INVKRDEt} - \left(P_j^{INVKRDEt} - P_j^{INVKRDEt-1}\right) + \tau^{PKRDE} P_j^{IKVKRDEt-1}\right\} \\
& + KITE_j^t \left\{\delta_j^{KITE} P_j^{INVKITEt} - \left(P_j^{INVKITEt} - P_j^{INVKITEt-1}\right) + \tau^{PKITE} P_j^{IKVKITEt-1}\right\} \\
& + KNRDE_j^t \left\{\delta^{KNRDE}{}_j P_j^{INVKNRDEt} - \left(P_j^{INVKNRDEt} - P_j^{INVKNRDEt-1}\right) + \tau^{PKNRDE} P_j^{INVKNRDEt-1}\right\} \\
& + KNITE_j^t \{\delta_j^{KNITE} P_j^{INVKNITEt} - \left(P_j^{INVKNITEt} - P_j^{INVKNITEt-1}\right) + \tau^{PKNRDE} P_j^{INVKNRDEt-1})] \\
DD2 = {} & [K_j^t \left(1-\tau^K\right) P_j^{INVKt-1} + KRDE_j^t \left(1-\tau^{KRDE}\right) P_j^{INVKRDEt-1} \\
& + KITE_j^t \left(1-\tau^{KITE}\right) P_j^{INVKIETt-1} + KNRDE_j^t \{\left(1-\tau^{KNRDE}\right) P_j^{INVKNRDEt-1} \\
& + KNITE_j^t \{\left(1-\tau^{KNITE}\right) P_j^{INVKNITEt-1}] \\
r^{Kt}{}_j = {} & DD1 / DD2
\end{aligned}
\tag{4.11}
$$

$$
\begin{aligned}
\text{Total Capital Cost} = {} & \left\{P^d{}_j X_j - \left(\Sigma_i P^d{}_i X^d{}_{ij} + \Sigma_i P^m{}_i X^m{}_{ij} + BCH_j + LC_j + HOJO_j\right)\right\} \\
& (\text{main production activity}) \\
& + \left\{P^d{}_j X_j - \left(\Sigma_i P^d{}_i X^d{}_{ij} + \Sigma_i P^m{}_i X^m{}_{ij} + BCH_j + LC_j + HOJO_j\right)\right\} \\
& (\text{intra-firm ICT activities}) \\
& + \left\{P^d{}_j X_j - \left(\Sigma_i P^d{}_i X^d{}_{ij} + \Sigma_i P^m{}_i X^m{}_{ij} + BCH_j + LC_j + HOJO_j\right)\right\} \\
& (\text{intra-firm R\&D activities})
\end{aligned}
\tag{4.12}
$$

The input balance of each sector in the input-output table is obtained, where $\Sigma_i P^d{}_i X^d{}_{ij} + \Sigma_i P^m{}_i X^m{}_{ij}$ is intermediate raw material inputs in main production activities, intra-firm ICT activities, and intra-firm R&D activities. $\Sigma_i P^d{}_i X^d{}_{ij}$ represents the intermediate inputs of domestic goods ($P^d{}_i$ is the domestic demand price of *i* good, $X^d{}_{ij}$ is the input of *i* intermediate goods into sector *j*), $\Sigma_i X^m{}_{ij}$ is the intermediate input of imported goods ($P^m{}_i$ is the price of *i*-good imported goods and $X^m{}_{ij}$ is the input of *i* intermediate imported goods into sector *j*), BCH_j is the non-householders consumption expenditure, and LC_j is the total labor cost, which is the sum of employer income, self-employed income, and family employee income. Finally, $HOJO_j$ is the net indirect tax (indirect taxes—subsidies—property taxes).

Estimating $r^{Kt}{}_j$, we can obtain the estimates of gross capital service prices for each capital service.

1) Capital service price for tangible capital in the main productive activity

$$P_j^{KSt} = \left(1-\tau^K\right) r_j^{KPjINVKt-1} + \delta_j P_j^{INVKt} - \left(P_j^{INVKt} - P_j^{INVKt-1}\right) + \tau^P P_j^{INVKt} \quad (4.13a)$$

Capital service price for tangible capital in the intra-firm ICT activities.

$$P_j^{KSITEt} = \left(1-\tau^{KIT}\right) r_j^{KITE} P_j^{INVKITEt-1} + \delta_j P_j^{INVKITEt} - \left(P_j^{INVKITEt} - P_j^{INVKITEt-1}\right) + \tau^{PITE} P_j^{INVKITE} \quad (4.13b)$$

2) Capital service price for tangible capital in the intra-firm R&D activities.

$$P_j^{KSRDEt} = \left(1-\tau^{KRD}\right) r_j^{KRDE} P_j^{INVKRDt-1} + \delta_j P_j^{INVKRDEt} - \left(P_j^{INVKRDEt} - P_j^{INVKRDEt-1}\right) + \tau^{PRDE} P_j^{INVKRDE} \quad (4.13c)$$

3) Capital service price for intangible capital in the intra-firm ICT activities.

$$P_j^{KSNRDEt} = \left(1-\tau^{SKE}\right) r_j^{KNRDE} P_j^{INVSKNRDEt-1} + \delta_j^{KNRDE} P_j^{INVKNRDE} - \left(P_j^{NVKNRDEt} - P_j^{INVKNRDEt-1}\right) + \tau_j^{SKMRDE} P_j^{INVKNRDEt-1} \quad (4.13d)$$

4) Capital service price for intangible capital in the intra-firm R&D activities.

$$P_j^{KSNRDEt} = \left(1-\tau^{SKNRDE}\right) r_j^{KNITE} P_j^{INVSKNRDEt-1} + \delta_j^{KNITE} P_j^{INVKNRDE} - \left(P_j^{NVKNRDEt} - P_j^{INVKNRDEt-1}\right) + \tau_j^{SM(RDE} P_j^{INVKRDTEt-1} \quad (4.13e)$$

where the estimates of rates of capital return r_j^K, r_j^{KITE}, r_j^{KRDE}, r_j^{KNITE}, r_j^{KNRDE} are assumed to be equalized among activities in the industry.

Then the total amount of the capital service cost in the industry is estimated as follows:

$$Cost(K)_j^t = P_j^{KSt} K_j^t + P_j^{KSITEt} KITE_j^t + P_j^{KSRDEt} KRDE_j^t + P_j^{KSNITEt} KNITE_j^t + P_j^{KSNRDEt} KNRDE_j^t, \quad (4.13f)$$

where we can estimate the cost share of the total capital service cost and each capital cost share in the total capital service cost.

Dividing industrial activities into main production, intra-firm *ICT*, and intra-firm *R&D* activities reveals that two hypotheses are valid for cost allocation.

Hypothesis 1 Separation of intra-firm *ICT* and *R&D* activities from main production activities.

Main production capital cost $P_j^{KSt}\, KS_j^t$ = Gross tangible fixed capital cost of main production

Intra-firm *ICT* activity capital cost $P_j^{KSITEt} K_j^{SITEt} + P_j^{KSNITEt} K_j^{SNITEt}$ = Gross tangible capital cost of intra-firm *ICT* activity + Gross intangible capital cost of intra-firm *ICT* activity

Intra-firm R&D activity capital cost $P_j^{KSRDEt} K_j^{SRDEt} + P_j^{KSNRDEt} K_j^{SNRDEt}$ = Gross tangible capital cost of intra-firm R&D activity + Gross intangible capital cost of intra-firm R&D activity

Industrial activities can be categorized into three main types: production, ICT, and R&D. The proportion of capital input costs for each activity can be determined based on this classification, which will subsequently be used to compute the cost shares in the ensuing long-term cost function.

Intra-firm ICT and R&D activities are viewed as intangible assets that generate capital services from both tangible and intangible assets at the start of each period. These services serve as inputs for the current period's ICT or R&D activities. It's important to note that the firm's internal knowledge stock doesn't directly influence the main production activities. Instead, its impact must be internalized on the efficiency of main production activities and on the TFP of the main production. This ensures a comprehensive understanding of the role of knowledge in production efficiency.

Hypothesis 2 The knowledge production from intra-firm ICT and R&D activities directly influences main production.

Main production activity's capital cost = $P_j^{SKt} K_j^t + P_j^{KSNITEt} K_j^{SNITEt} + P_j^{KSNRDEt-1} K_j^{SNRDEt}$ = Gross tangible fixed capital cost of main production + Gross intangible capital cost of intra-firm ICT activity + Gross intangible capital cost of intra-firm R&D activity

Intra-firm ICT activity's capital cost $P^{KSITE}{}_{tj} K_j^{\mathrm{SITEt}}$ = Gross tangible capital cost of intra-firm ICT activity

Intra-firm R&D activity's capital cost $P_j^{KSRDEt} K_j^{\mathrm{SRDE}t}$ = Gross tangible capital cost of intra-firm R&D activity

We posit that the input of capital services into the main production activity stems not only from the gross tangible fixed capital at the start of the period, but also from the accumulated intellectual capital (intangible fixed assets) resulting from intra-firm ICT and R&D activities. This accumulation influences the production activities of the current period. Conversely, the intellectual output of intra-firm ICT and R&D activities is solely influenced by the capital services from the gross tangible fixed capital stock of these two activities. These hypotheses would yield identical results in integrated growth accounting across three activities but would exhibit differences in short-term supply behavior and long-term investment behavior.

This approach remains consistent regardless of whether the activities in the production sector are distinctly separated or divided into two sectors. Moreover, if the production activities include the generation of intellectual knowledge in the scientific and technical sectors of both government public R&D departments and private industrial R&D departments, it is assumed that both the tangible fixed assets provided by policy and the accumulated intangible fixed assets in that sector contribute capital services.

1. **Public R&D Sectors**

$$
\begin{aligned}
DD1 = {} & BS^{t}_{j} / P^{KS0}_{j} - [KG_j^t \ \left\{\delta_j^{KG} P_j^{INVKGt} - \left(P_j^{INVKGt} - P_j^{INVKGt-1}\right) + \tau^P P_j^{INVKGt-1}\right\} \\
& + KNG_j^t \left\{\delta_j^{KNG} P_j^{INVKNGt} - \left(P_j^{INVKNGt} - P_j^{INVKNGt-1}\right) + \tau^{PKNG} P_j^{INVKNGt-1}\right\}] \\
DD2 = {} & [KG_j^t \left(1-\tau^K\right) P_j^{INVGKt-1} + KNG_j^t \ \{\left(1-\tau^{KNG}\right) P_j^{INVKNGt-1}] \\
& r_j^{K*Gt} = DD1 / DD2,
\end{aligned}
$$

where BS^t_j is derived from the value-added side. However, the operating surplus is set to zero, and the capital cost is the sum of the depreciation of both tangible and intangible fixed assets.[2] Using the estimated capital return rate r_j^{K*Gt}, we can estimate the price of tangible and intangible capital services in government public R&D departments.

Price of tangible fixed capital services

$$
P^{KSGKt}_{j} = \left(1-\tau^K\right) r^{K*Gt}_{j} P_j^{INVGKt-1} + \delta_j^{KG} P_j^{INVKGt} - \left(P_j^{INVKGt} - P_j^{INVKGt-1}\right) + \tau^P P_j^{INVKGt-1}
$$

Price of intangible fixed capital services

$$
\begin{aligned}
PKSGN_{tj} = {} & \left(1-\tau^{KNG}\right) r_j^{K*Gt} P_j^{INVKNGt-1} + \delta_j^{KNG} P_j^{INVKNGt} - \left(P_j^{INVKNGt} - P_j^{INVKNGt-1}\right) \\
& + \tau^{KNG} P_j^{INVKNGt-1}
\end{aligned}
$$

2. **Private R&D Sectors**

$$
\begin{aligned}
DD1 = {} & BS^{t}_{j} / P^{KSI0}_{j} - [K_j^t \left\{\delta_j P_j^{INVKIt} - \left(P_j^{INVKIt} - P_j^{INVKIt-1}\right) + \tau^P P_j^{INVKIt-1}\right\} \\
& + KNI_j^t \{\delta^{KNI}_{j} P_j^{INVKNIt} - \left(P_j^{INVKNIt} - P_j^{INVKNIt-1} + \tau^{PKNI} P_j^{INVKNIt-1}\right)] \\
DD2 = {} & \left[KI_j^t \ \left(1-\tau^K\right) P_j^{INVKt-1} + KNI_j^t \left(1-\tau^{KNI}\right) P_j^{INVKNIt-1}\right] \\
r^{KIt}_{j} = {} & DD1 / DD2
\end{aligned}
$$

where BS^t_j is calculated from the value-added side. However, the capital cost is the sum of the operating surplus and the depreciation of tangible and intangible fixed assets. By using the estimated capital return rate, we can estimate the price of tangible and intangible fixed capital services in the private R&D sector.

[2] In Japan's input-output tables, R&D investments are not classified as capital, hence the depreciation of R&D stock as intangible fixed assets is not included in value-added. In our calculations, we account for the depreciation of intangible fixed assets in both government and industry research and development departments, and incorporate it into value-added.

Price of tangible fixed capital services

$$P^{KSIt}{}_{j} = \left(1-\tau^{K}\right) r^{KIt}{}_{j} P_{j}^{INVKIt-1} + \delta_{j} P_{j}^{INVKIt} - \left(P_{j}^{INVKIt} - P_{j}^{INVKIt-1}\right) + \tau^{P} P_{j}^{INVKIt-1}.$$

Price of intangible fixed capital services

$$P^{KSNIt}{}_{j} = \left(1-\tau^{KNI}\right) r^{KIt}{}_{j} P_{j}^{INVKNIt-1} + \delta^{KNI}{}_{j} P_{j}^{INVKNIt} - \left(P_{j}^{INVKNIt} - P_{j}^{INVKNIt-1}\right)$$
$$+\tau^{PKNI} P_{j}^{INVKNIt-1}.$$

We assume that the expected capital return rate for each asset is equal across tangible and intangible fixed capital within the sector. Given that BS_j is observed as the sum of operating surplus and depreciation expenses for each tangible and intangible fixed capital, and the prices of investment goods for each tangible and intangible fixed capital are determined, along with capital tax parameters and depreciation rates, Eqs. (4.8), (4.9), and (4.10) become linear equations for the capital return rate $r^{K}{}_{j}$. This allows for the determination of the current period's capital return rate for each sector. After calculating the capital return rate, the price of each tangible and intangible fixed capital service is determined according to the aforementioned formula.

According to the 2005 input-output tables format, the treatment of operating surplus and capital depreciation for each industrial sector is as follows:

In industrial sectors (excluding public and private research institutions), the operating surplus is aggregated into the main production sector. For intra-firm R&D and intra-firm ICT activity, the operating surplus is treated as zero.

This treatment remains consistent even if industries are not divided into main production, intra-firm ICT, and intra-firm R&D sectors. For capital depreciation allocation (intangible assets), the depreciation cost of intra-firm ICT's intangible fixed assets is grouped under the main production sector's depreciation. For intra-firm R&D, depreciation is not explicitly estimated but is assumed to be included in the main production sector's operating surplus. For capital depreciation allocation (tangible assets), depreciation is separately allocated for main production, intra-firm ICT, and intra-firm R&D. In the estimation of capital depreciation from 1995 to 2011, it is calculated only for tangible assets and intangible assets in intra-firm ICT. depreciation for intra-firm R&D is not specified but is considered part of the main production sector's operating surplus. After determining the capital service price, the capital cost for each activity is calculated using a rental concept that includes depreciation.

The cost balance in the private industry sector is based on the input elements of the input-output tables mentioned above. Its characteristics are as follows:

In the private industry sector, activities are generally classified into three categories. The industrial classification is shown in Table 4.2. Industrial activities are typically divided into three activities: main production, intra-firm information processing activities (ICT activities), and intra-firm research and development activities (R&D

activities). However, in some sectors, activities are not separated, and the three intra-firm activities are integrated into main production activities. When activities are separated into three activities, it is assumed that the main production activity sector bears the capital costs for "tangible capital" of the main activity.

When activities are divided, labor and intermediate material costs are borne by main production, information processing, and R&D activities, respectively. At the same time, it is assumed that tangible capital service costs are borne by each activity, while depreciation and property taxes related to intangible capital services are imposed upon the main production activity. Table 4.1 shows the cost structure of the firm, where it is divided into main production, ICT activities, and R&D activities, in the cases where industrial activities are separated into these three activities.

In the private industry sector, activities are typically classified into three categories: main production activities, intra-firm ICT activities, and intra-firm R&D activities in principle, although, in some sectors, the activity does not segregate into three activities.

When activities are segregated, it is postulated that each activity including the main production activity and both intra-firm activities including ICT and R&D activities sector incurs the capital costs for the tangible capital inputs respectively. These costs for tangible fixed capital (*KM*, *KITE*, and *KRDE*) are assumed to incur in each activity separately. But the capital cost for intangible capital cost in ICT and R&D activities are assumed to incur in the primary sector.

Capital depreciation allocation (intangible assets): The intangible assets derived from intra-firm R&D activities are classified under the operating surplus of the primary production sector. Conversely, the allocation of capital depreciation for intangible assets from intra-firm ICT activities is explicitly stated and classified under the main production sector.

Capital depreciation allocation (tangible assets): Depreciation of tangible assets is categorized separately for main production, intra-firm ICT, and intra-firm R&D activities.

Table 4.1 introduces a framework, based on "Hypothesis 2," for understanding TFP. The capital cost of main production activities encompasses the total capital service cost, which includes the depreciation of tangible fixed capital in main production and the depreciation of intangible fixed capital services from ICT and R&D knowledge assets at the start of the period.

The capital cost of intra-firm ICT activities is composed of the depreciation of tangible fixed capital, forming the tangible capital cost of intra-firm ICT. Likewise, the capital cost of intra-firm R&D activities incorporates the tangible fixed capital cost, which involves the depreciation of tangible capital in intra-firm R&D activities. Consequently, the operating surplus of intra-firm ICT and R&D activities is not presumed to be zero.

4.4 Estimation of Capital Service Price and Capital Rate of Return

In the previous section, we applied the measurement of capital service prices to data from Japan's input-output tables for the years 1995–2011, yielding empirical results. We consolidated the data into 93 industrial sectors based on the industry classification in the input-output tables, as depicted in Table 4.2. Initially, industries were grouped into 33 sectors according to the Japan Standard Industrial Classification (two-digit classification). Of these 33 sectors, 6 industries—agriculture, forestry, fisheries, mining, software industry, information provision services, Internet services, and other services—were identified for intra-firm R&D activities. The main production and ICT sectors were merged, resulting in two sectors. Furthermore, the healthcare and welfare services and education sectors were treated as single industries, without differentiating between main production, R&D, or ICT. For R&D (public and non-profit) and R&D (industry), each was subdivided into five fields (life sciences, information and communication, material sciences, environmental energy, and others), each treated as separate industries. The remaining 23 sectors were split into 3 sectors: main production, intra-firm ICT, and R&D.

The capital service quantities and prices for each industrial sector were estimated using time-series input-output tables, as described in the previous section. The Japanese economy from 1995 to 2011 was marked by a strong yen trend following the 1985 Plaza Accord, a subsequent bubble period, and a transition into a phase of long-term economic stagnation. Assessing the results of R&D and ICT investments during this period is challenging. This section organizes the results of the capital service measurements discussed earlier.

Consider the estimated capital rate of return for each industrial sector. Figure 4.1 presents the estimated capital return rates for each industrial sector. Despite minor differences among sectors, the rates fluctuated annually between approximately 0.01% and 2.5% from 1995 to 2011. There was a gradual improvement from 1995 to 2000 after the collapse of the 1990s real estate bubble. However, the aftermath of the IT bubble from 2000 to 2002 persisted until around 2005. Despite some recovery in certain industries, the capital return rates declined sharply after the 2008 Lehman Shock, but appear to have resumed a recovery trajectory around 2010. Among these trends, the capital return rates of industries related to optics fiber, semiconductors, electronic computers, and their components exhibit distinct patterns. The recovery of capital return rates after 2005 is particularly evident in the ICT and Internet industries. The capital return rates of the public/non-profit R&D sector are generally lower than those of other private sector industries. As previously stated, the public R&D sector, which produces public goods, has zero operating surplus by definition. Therefore, the capital cost is defined by the capital depreciation allocation of tangible/intangible capital in public R&D sectors and the operating surplus, as well as the capital depreciation allocation in non-profit R&D sectors at private universities. The activities of public R&D sectors, defined as the production of non-profit public goods, result in lower capital return rates compared

Table 4.2 Classification of 93 sectors in the input-output tables

	Industry classification (main activity)		Industry classification (ICT activity)		Industry classification (R&D activity)
1	Agriculture/forestry/ fishery			2	Agriculture/forestry/ Fishery
3	Mining including intra-ICT			4	Mining
05	Food product manufacturing	06	Food product manufacturing	7	Food product manufacturing
08	Textile and wood manufacturing	09	Textile and wood manufacturing	10	Textile and wood manufacturing
11	Paper and pulp manufacturing	12	Paper and pulp manufacturing	13	Paper and pulp manufacturing
14	Chemical prod. manufacturing	15	Chemical prod. manufacturing	16	Chemical prod. manufacturing
17	Iron and steel product manufacturing	18	Iron and steel product manufacturing	19	Iron and steel product manufacturing
20	General machinery manufacturing	21	General machinery manufacturing	22	General machinery manufacturing
23	Electric machinery manufacturing	24	Electric machinery products	25	Electric machinery products
26	Fabricated metal products	27	Fabricated metal products	28	Fabricated metal products
29	Semi-conductor equipment products	30	Semi-conductor equipment products	31	Semi-conductor equipment products
32	Communication machinery	33	Communication machinery	34	Communication machinery
35	Electric computer and equipment	36	Electric computer and equipment	37	Electric computer and equipment
38	Semiconductor and integrated circuit	39	Semiconductor and integrated circuit	40	Semiconductor and integrated circuit
41	Other electric parts	42	Other electric parts	43	Other electric parts
44	Heavy electric and transport machinery	45	Heavy electric and transport machinery	46	Heavy electric and transport machinery
47	Robot	48	Robot	49	Robot
50	Other precision machinery	51	Other precision machinery	52	Other precision machinery
53	Oil and coal products	54	Oil and coal products	55	Oil and coal products
56	Other manufacturing	57	Other manufacturing	58	Other manufacturing
59	Energy industry	60	Energy industry	61	Energy industry
62	Construction	63	Construction	64	Construction
65	Transportation	66	Transportation	67	Transportation
68	Communication and broadcasting	69	Communication and broadcasting	70	Communication and broadcasting
71	Trade	72	Trade	73	Trade
74	Software industry including R&D			75	Software industry including R&D

(continued)

Table 4.2 (continued)

	Industry classification (main activity)		Industry classification (ICT activity)		Industry classification (R&D activity)
76	Information and communication service			77	Information and communication service
78	Internet service including ICT			79	Internet service including ICT
80	Medical service				
82	Education				
84	R&D (Life:PB:PNB)				
85	R&D (Information:PNB)				
86	R&D (Material:PNB)				
87	R&D (Environment and Energy: PNB)				
88	R&D (Other R&D: PNB)				
89	R&D (Life: PB)				
90	R&D (Information: PB)				
91	R&D (Material:PB)				
92	R&D (Environment and Energy: PNBusiness)				
93	R&D (Other R&D: PB)				
94	Other service			93	Other service

Note: *PB* private business, *PNB* public non-business

to other profit-oriented sectors. While the capital return rates of highlighted R&D sectors exhibit unique patterns during periods such as 1995–2000 and 2008–2011, the trends in other sectors align with those of other industrial sectors. The impacts of the IT bubble collapse in 2000–2002 and the Lehman Shock in 2007–2008 show a common trend across all industries.

Figure 4.2 depicts the temporal evolution of capital service prices for tangible and intangible fixed assets estimated for each industrial sector. The figure underscores a declining trend in capital service prices for both tangible and intangible fixed assets from 1995 to 2011 across various sectors. Capital service prices are segmented into those for tangible fixed assets in main production activities, intra-firm ICT, and intra-firm R&D activities, as well as those for intangible fixed assets resulting from intra-firm ICT and intra-firm R&D.

In all sectors, tangible and intangible capital service prices have demonstrated nearly parallel trends. Industries showing a downward trend in capital service prices include those related to information equipment, such as semiconductor devices, communication machinery, electronic computers, semiconductor integrated circuits, robots, and other precision machinery industries. Notably, during the IT bubble burst of 2001–2002 and the Lehman Shock of 2007–2009, a common trend of declining capital service prices was observed. This decrease in capital service prices

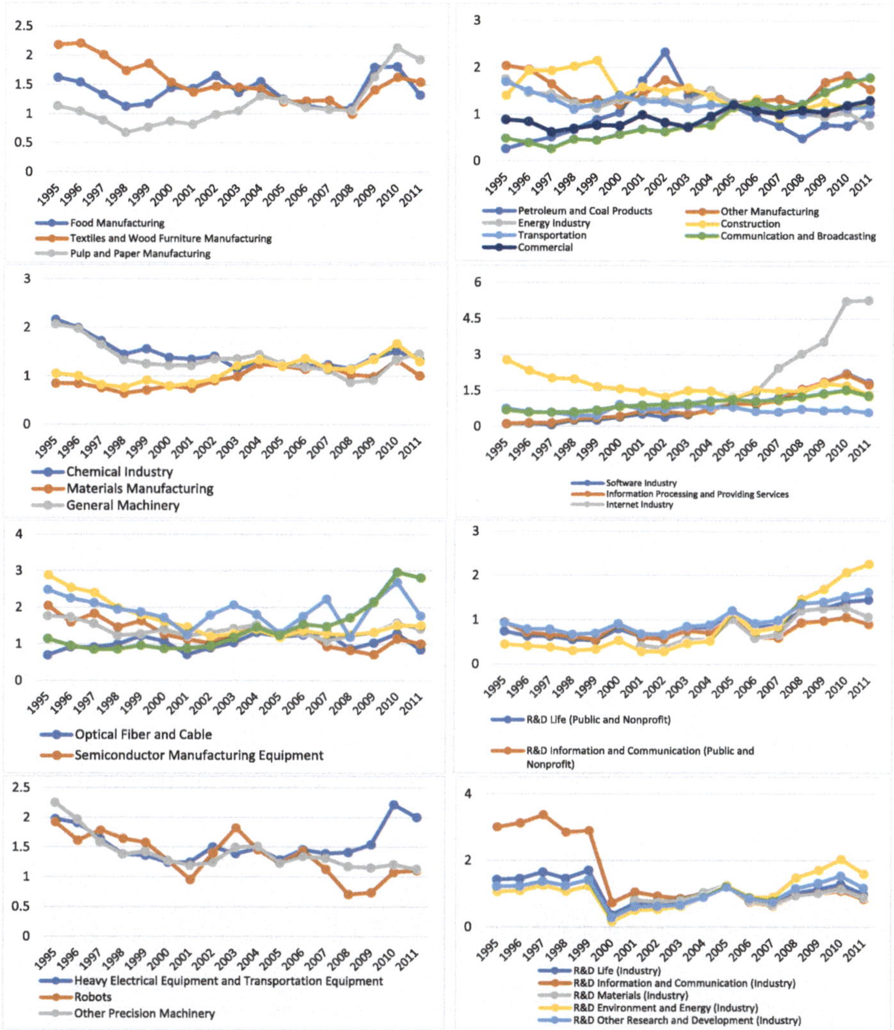

Fig. 4.1 Capital rate of return rates by industry: year 1995–2011

is anticipated to hasten the substitution from labor to capital by increasing the relative factor price of labor to capital (w/r).

Figures 4.3 and 4.4 illustrate the changes in relative factor prices and capital intensity across three periods from 1995 to 2010 for industries where capital service prices have decreased over time. The focus is on tangible fixed assets and intangible fixed assets from intra-firm R&D activities. The changes in factor relative prices (Δ[w/r]) and capital intensity ($\Delta[K/L]$) display a positive correlation, which has intensified since 1995. Industries experiencing a decline in capital service prices

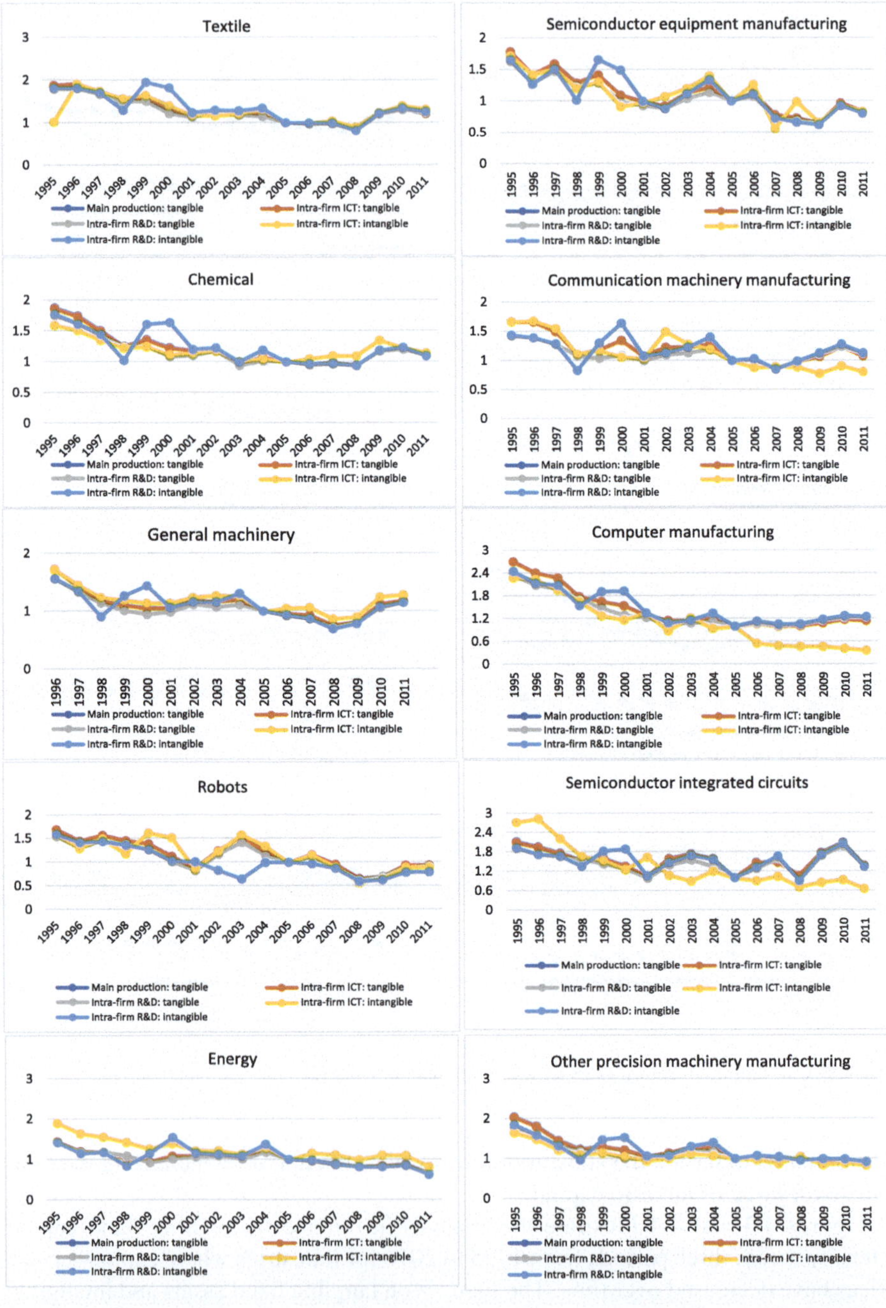

Fig. 4.2 Time-trends of the capital service price: 1995–2011: Tangible and Intangible Capital

tend to increase capital intensity, enhance factor substitution, and improve efficiency in both tangible and intangible fixed assets related to R&D investments.

Next, consider industries where the time-series variation of tangible and intangible fixed capital service prices has shown an upward trend from 1995 to 2011. Figure 4.5 plots the estimated values of capital service prices for these industries. With the exception of the civilian electrical machinery industry, tangible and intangible capital service prices in the included industries have generally increased since 1995. Even in the civilian electrical machinery industry, excluding the intangible fixed capital price related to intra-firm ICT, the upward trend aligns with the aforementioned industries, indicating an overall increase over time.

Figure 4.6 presents the 5-year changes in factor relative prices ($\Delta(w/r)$) and capital intensity ($\Delta(K/L)$) from 1995 for industries with an upward trend in tangible fixed asset service prices since 1995. Compared to Fig. 4.5, the period from 2000 to 2005 does not show a clear positive correlation between changes in factor relative prices and capital intensity.

Figure 4.7 shows the time-series changes in factor relative prices (w/r) and capital intensity (K/L) for two representative industries: the chemical industry and ICT services. These industries were chosen to illustrate the characteristics of industries with declining and rising trends in capital service prices. Both industries display a positive correlation between factor relative prices and capital intensity. Specifically, the increase of the labor service prices compared with the capital service prices tends to correspond with an increase in capital intensity. This is in line with rational business behavior. However, the chemical industry and ICT services show a reversal in the direction of the time-series changes in factor relative prices and capital intensification. In the chemical industry, labor service prices have been rising faster than capital service prices since 1995, leading to increased capital intensity. In contrast, in the ICT services industry, the rise in capital service prices outpaces the increase in labor service prices, resulting in a relative decrease in capital intensity over time. This trend aligns with observations in Figs. 4.2 and 4.5.

The differences in trends of relative factor prices and capital intensity between these industries align with the series changes in capital return rates shown in Fig. 4.1. Industries with declining trends in capital service prices, such as the chemical industry, show capital return rates higher than the industry average, but these rates are decreasing over time. In contrast, industries with rising trends in the capital service prices, like ICT services, often have capital return rates below the industry average, but these rates are increasing over time. As labor service prices, which reflect labor market conditions, do not show significant inter-industry differences, it appears that the levels and trends of capital return rates across industries are driving the differences in changes in capital intensity. This relationship is closely tied to changes in each industry's production efficiency and is also connected to policy guidelines for enhancing overall production efficiency in a country. It is essential to discuss these issues in the context of observed production efficiency facts in the subsequent section.

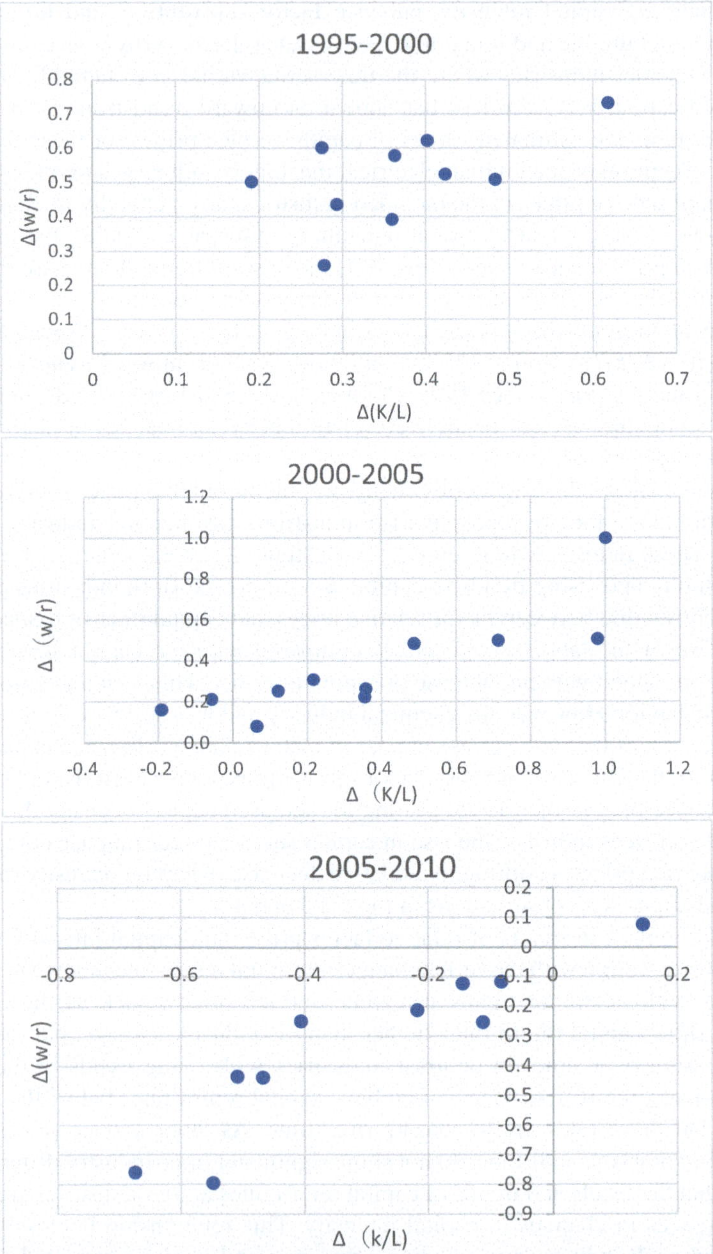

Fig. 4.3 Correlation chart of factor relative prices and capital intensity: decreasing trend in tangible capital service prices

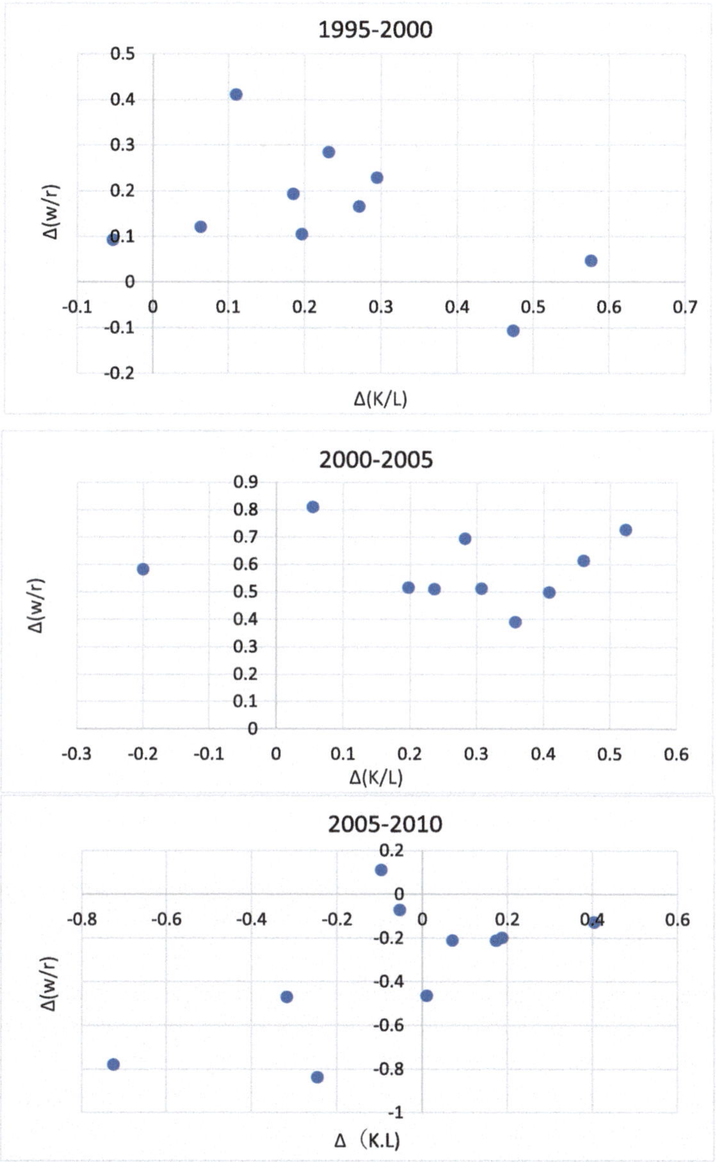

Fig. 4.4 Correlation chart of factor relative prices and capital intensity. Downtrend in the price of intangible capital of R&D

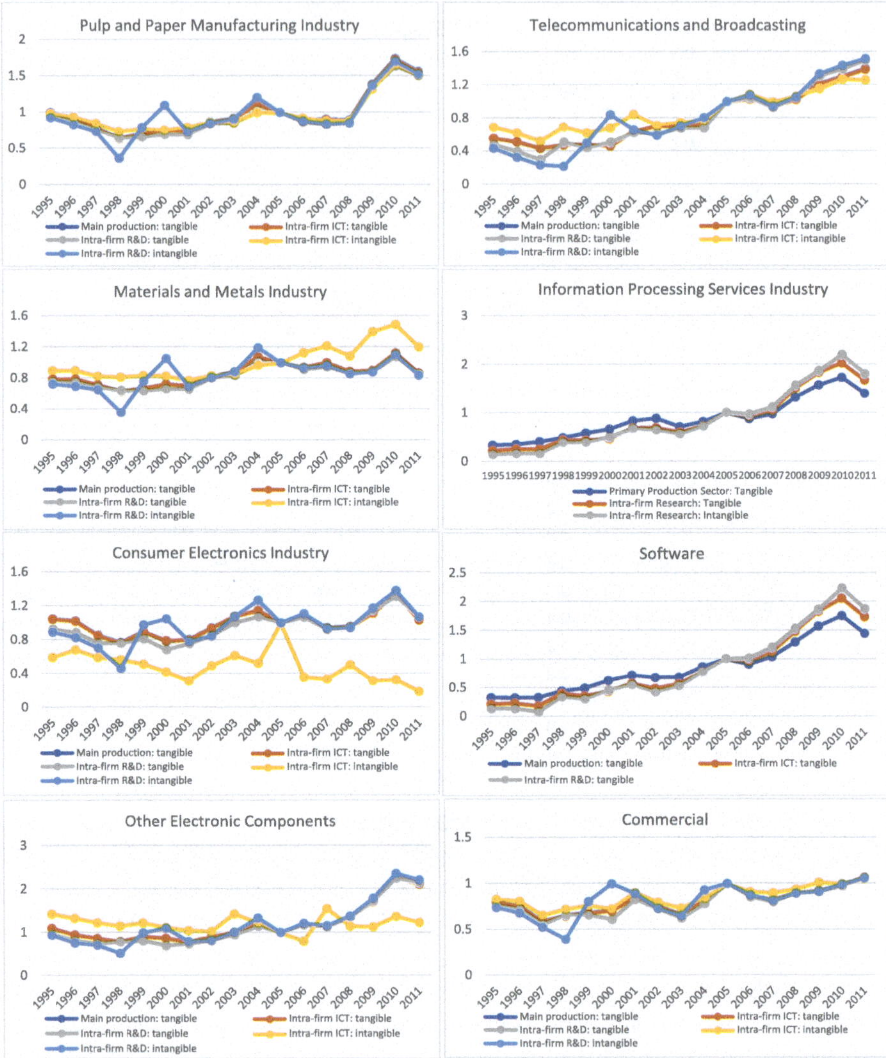

Fig. 4.5 Transition of capital service prices: industries with rising trends in capital service prices

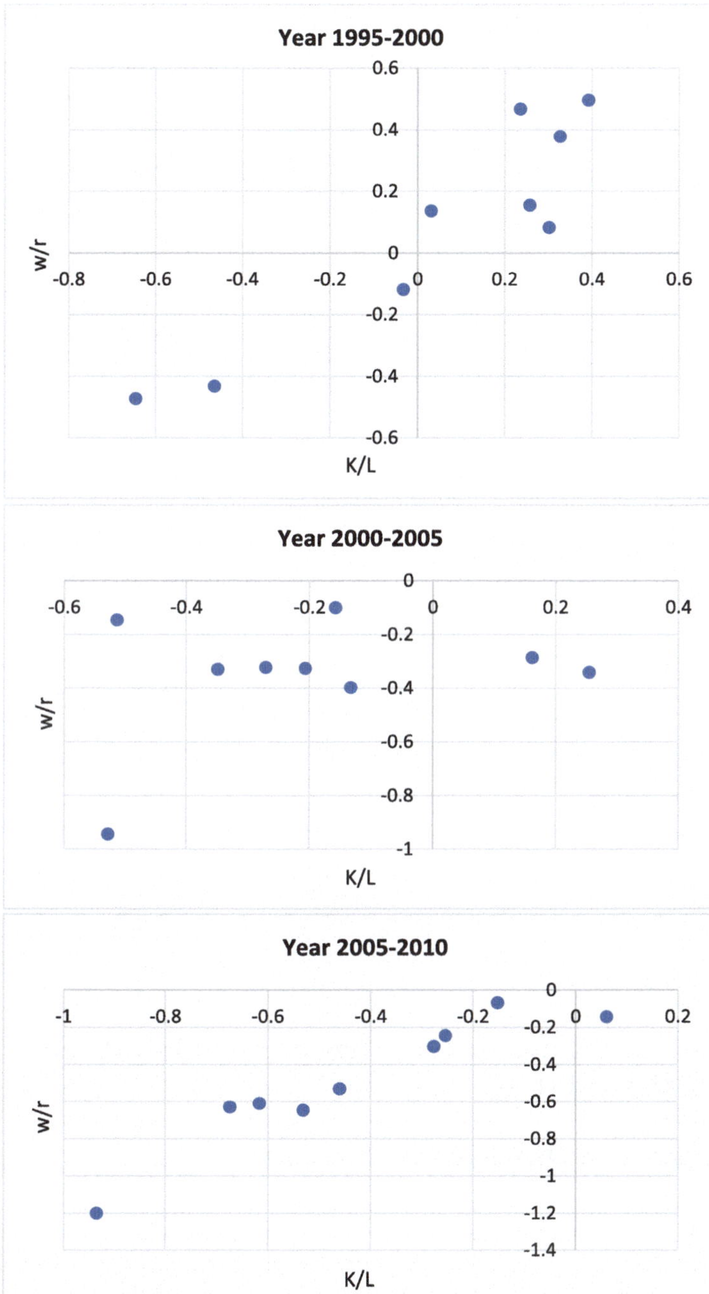

Fig. 4.6 Correlation diagram of factor relative prices and capital intensity: rising trend in tangible capital service prices

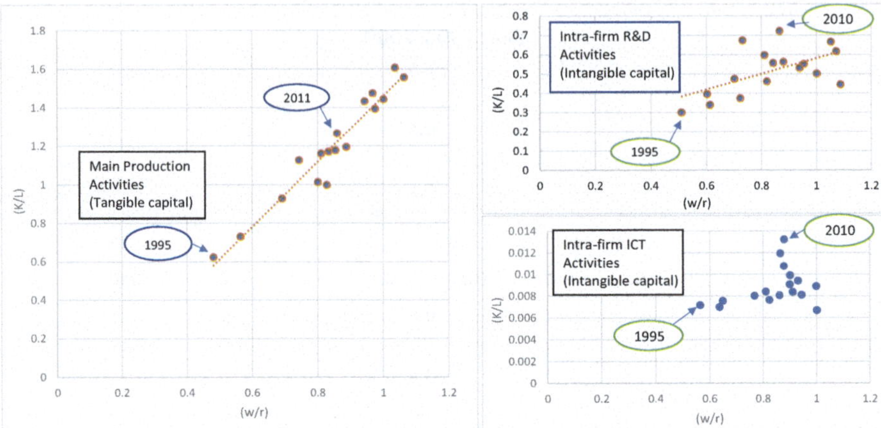

Fig. 4.7 Correlation between factor relative prices and capital intensity for the chemical industry and ICT services

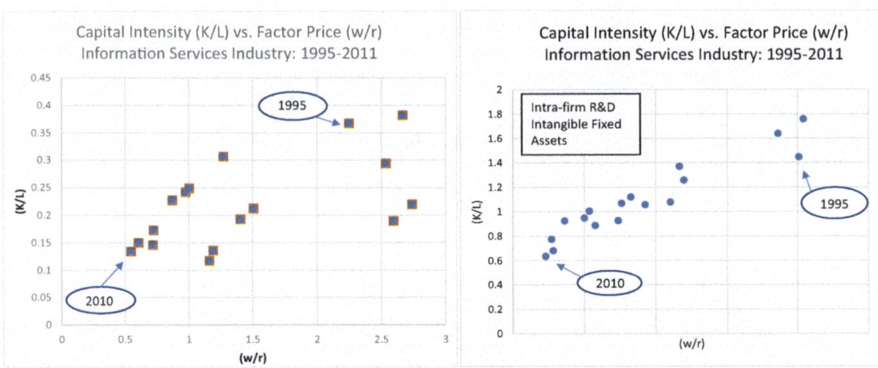

Fig. 4.8 Industries with an Increasing Trend in Capital Service Prices over Time: Information Processing Services Industry

4.5 Static and Dynamic Unit Structure and Investment in R&D and ICT

4.5.1 Static and Dynamic Unit Structure

This section briefly summarizes the concept of our static and dynamic unit structure. In the schematic of the static unit structure, the relationship between the input coefficient matrix A_t at time t, the final demand vector $\boldsymbol{f}_t$, and the output vector $\boldsymbol{x}_t$ is derived as follows.

$$A_t \boldsymbol{x}_t + \boldsymbol{f}_t = \boldsymbol{x}_t.$$

If $\boldsymbol{A}_t$ is non-singular, by providing the identity matrix I, we obtain $\boldsymbol{x}_t = (\boldsymbol{I} - \boldsymbol{A}_t)^{-1}\boldsymbol{f}_t$. Let $\boldsymbol{e}_{(i)}$ be an $n \times 1$ unit vector with only the $\boldsymbol{i}$-th element as 1 and all other elements as 0. When $\boldsymbol{f}^*_t = \boldsymbol{e}_{(i)}$, the input elements for intermediate input, labor input, and capital input required to meet the production of this final demand exactly can be determined as follows:

$$\boldsymbol{X}_t^* = \boldsymbol{A}_t < (\boldsymbol{I} - \boldsymbol{A}_t)^{-1} f_t^*) >$$

$$\boldsymbol{U}^S{}_t \mid_{ft^*=e(i)} = \boldsymbol{L}_t^* = \boldsymbol{B}_t^L < (\boldsymbol{I} - \boldsymbol{A}_t)^{-1} f_t^*) > \tag{4.14}$$

$$\boldsymbol{K}_t^* = \boldsymbol{B}_k^K < (\boldsymbol{I} - \boldsymbol{A}_t)^{-1} f_t^*) >,$$

where <·> denotes a diagonal matrix, and $\boldsymbol{B}_t^L$ and $\boldsymbol{B}_k^K$ represent the labor coefficient matrix ($\boldsymbol{k} \times n$) and capital coefficient matrix ($k \times n$), respectively. $\boldsymbol{l}$ and $\boldsymbol{k}$ denote the number of types of labor and capital service elements, respectively. At this point, Eq. (4.14) is referred to as the static unit structure or simply the unit structure. This unit structure represents a basket of all input elements required to produce one unit of commodity $\boldsymbol{i}$ as final demand.

4.5.2 Dynamic Ripple Structure

The static unit structure concept aims to modularize the production input structure for a specific product at a given time, t, with a focus on the entire production structure. Within this structure, the capital input, representing the capital services input provided by the accumulation of previously invested investment goods, is considered a production factor. All past investment series that form the capital stock at time dynamically constitute the current capital's input factors.

The dynamic unit structure concept seeks to capture the production structure characteristics for a commodity, i, at time t dynamically by tracing the capital accumulation process over time. The technological structure of past investment goods production, as captured by the dynamic unit, mirrors the technology at the time of those investment goods' production. This allows for the evaluation of technology's historical reflection on the production structure of commodity i at time t by looking back in time.

Assuming the following output balance for a commodity at time t,

$$\boldsymbol{A}_t \boldsymbol{x}_t + \boldsymbol{I}_t^P \boldsymbol{i} + \boldsymbol{I}_t^G \boldsymbol{\Gamma} \boldsymbol{i} + \boldsymbol{c}_t = \boldsymbol{x}_t \tag{4.15}$$

where $\boldsymbol{I}_t^P$ and $\boldsymbol{I}_t^G$ denote the fixed capital formation matrices for the private sector ($n \times n$) and the government ($n \times m$ [number of infrastructure sectors]), respectively. We define a transformation matrix $\boldsymbol{\Gamma}(m \times n)$ to convert the government's n (*commodity*) $\times$ m (number of infrastructure sectors) fixed capital formation matrix into an n (commodity) × n (commodity) matrix. This transformation matrix, a rating matrix of 0s and 1s, maps infrastructure sectors to commodity sectors, and the transformed fixed capital matrix becomes the government's fixed capital formation matrix (*in commodity n* commodity). Additionally, I represent the unit vector. The private capital formation matrix I_t^P is defined as follows using the perpetual inventory method, by the capital stock matrix S_t^P and the vector of capital depreciation rates

$$I_t^P = S_{t+1}^P - \left(I - <\delta^P>\right)S_t^P = B_{t+1}^P <x_{t+1}> - \left(I - <\delta^P>\right)B_t^P <x_t> \quad (4.16a)$$

Similarly, the government fixed capital formation matrix is defined as follows.

$$I_t^G\Gamma = S_{t+1}^G\Gamma - \left(I - <\delta^G>\right)S_t^G\Gamma = B_{t+1}^G <x_{t+1}> - \left(I - <\delta^G>\right)B_t^G <x_t> \quad (4.16b)$$

Substituting Eqs. (4.16a) and (4.16b) into Eq. (4.15), using the capital coefficient matrices B_t^P and B_t^G, we get:

$$\left(I - G_t\right)x_t - B_{t;1}^K x_{t+1} = c_t. \quad (4.17)$$

Here, the capital coefficient matrix is defined as

$$B_t^K = B_t^P + B_t^G, \text{ and } G_t = A_t - \left(I - <\delta^P>\right)B_t^P - -\left(I - <\delta^G>\right)B_t^G.$$

When expanded using a constant difference formula, Eq. (4.18) can lead to the following formulation of Leontief's dynamic inverse matrix equation system. Assuming that the past capital coefficient matrices are stably provided, the output balance can be determined by tracing back in time from the base year, ***t***.

$$X = \left(I - D\right)^{-1} c = Mc \quad (4.18)$$

where

$$\mathrm{D} = \begin{bmatrix} G_{t-1} & B_{t-T}^K & & \cdot & \\ \cdot & \cdot & & & \cdot \\ & & & & \\ & x_{t-1} & & \cdot & \cdot \\ & & & \cdot & \\ & & G_{t-1} & B_{t-1}^K & \\ & & & & G_t \end{bmatrix} \quad x = \begin{bmatrix} -r \\ \cdot \\ \\ \cdot \\ \cdot \\ x_{t-1} \\ x \end{bmatrix} \quad c = \begin{bmatrix} c_{t-1} \\ \\ \cdot \\ \cdot \\ -c_{t-1} \\ c_t \end{bmatrix}$$

At the base year t, $G_t = A_t$ holds. Assuming that the elements of final demand at the base year t are exogenously given, Eq. (4.18) calculates the production quantities of each sector required to meet this final demand through both static and dynamic direct and indirect effects. This is referred to as the dynamic unit structure of production for commodity i at the base year t.

When $\boldsymbol{c}^*_t$ is a vector representing $c^*_t = 0$ for $\boldsymbol{\tau} = 0$ and $\boldsymbol{c}^*_t = \boldsymbol{e}_{(j)}$ for $\boldsymbol{\tau} = 1, \cdots \boldsymbol{T}$, Eq. (4.18) becomes as follows:

$$\boldsymbol{x}^* = \boldsymbol{DMc}^* + \boldsymbol{c}^* = \boldsymbol{Mc}^*$$

Thus, the dynamic unit structure can be defined as in Eq. (4.19).

$$\boldsymbol{U}^D \left| \boldsymbol{c}^* \right. = \begin{bmatrix} \boldsymbol{X}^*\_ \\ \boldsymbol{L}^* \\ \boldsymbol{K}^* \end{bmatrix} \begin{bmatrix} \boldsymbol{D} < \boldsymbol{Mc}^* > \\ \boldsymbol{B}^L < \boldsymbol{Mc} \\ \boldsymbol{B}^K < \boldsymbol{Mc}^* > \end{bmatrix} \tag{4.19}$$

Here, $<\boldsymbol{Mc}^*$a is a diagonal matrix of dimensions ($\boldsymbol{n}$ [commodities] $\boldsymbol{x}^{T+1}$ [periods]) × (n [commodities] $\boldsymbol{x}^{T+1}$ [periods]), where the diagonal elements consist of the dynamic output calculated by $\boldsymbol{X}^*$. Therefore, $\boldsymbol{X}^*$ represents the intermediate input matrix of dimensions ($\boldsymbol{n}$ [commodities] × $\boldsymbol{T}$ + 1 [periods]) × (n [commodities] × $\boldsymbol{T}$ + 1 [periods]). Additionally, BL represents the labor input coefficient matrices for each period ($\boldsymbol{l}$ [number of labor types] × $\boldsymbol{T}$ + 1 [periods]). Similarly, $\boldsymbol{B}_K$ represents the capital input coefficient matrices for each period ($\boldsymbol{n}$ [number of capital types] × $\boldsymbol{T}$ + 1 [periods]), where $\boldsymbol{L}^*$ is the labor input matrix of dimensions (l [number of labor types] × $\boldsymbol{T}$ + 1 [periods]), and $\boldsymbol{K}^*$ is the capital input matrix of dimensions (n [number of capital types] × T + 1 [periods]).

4.5.3 R&D Investment Capitalization and Static, Dynamic Unit Structure

The R&D expenditures, undertaken by both private and government sectors, are investments in capital associated with R&D activities. These investments accumulate as intangible fixed assets, contributing to the stock of knowledge. In accordance with the 2008 United Nations (UN) recommendation for the new SNA, such R&D investment should be considered as the formation of a specific type of intangible fixed asset. Pursuant to this UN recommendation, our country has made efforts to reclassify R&D expenditures from intermediate inputs to a component of final demand, specifically intangible capital formation. In 2016, the national economic accounting system was revised in line with the 2008 UN recommendation, necessitating corresponding revisions to the input-output tables. Although preparations for these revisions were incorporated into the 2016 input-output tables of industries, we present here the framework for capitalizing R&D expenditures in the input-output tables prior to these revisions.

In the extended input-output tables framework, each production sector (n) is divided into the main production activity sector n^{NRD} and the R&D sector $\boldsymbol{n}^{RD}$. The R&D sector can be further classified into intra-firm R&D activity $\boldsymbol{n}^{RDF}$ for R&D activities within enterprises, public R&D activity (including government institutes and non-profit R&D activity, $\boldsymbol{n}^{RDG}$), and independent private sector R&D activity $\boldsymbol{n}^{RD}$.

$$n = n^{NRD} + n^{RD} = n^{NRD} + n^{RDF} + n^{RDG} + n^{RDI}, \tag{4.20}$$

$$\boldsymbol{n}^{RD} = \boldsymbol{n}^{RDF} + \boldsymbol{n}^{RDG} + \boldsymbol{n}^{RDI}. \tag{4.21}$$

Here, the number of research and development departments within a company (nRDF) is assumed to be the same as the number of main production activities (nNRD) if each industry sector conducts both main production activities and research and development activities. Additionally, for public and private non-commercial organizations whose primary activity is research, it becomes $n^{NRD} + n^{RDI}$.

The output balance equation corresponding to Eq. (4.2) in Sect. 3.2 is as follows.

$$\begin{aligned} &\boldsymbol{A}_t\boldsymbol{x}_t + \boldsymbol{I}_t^{NRDTA}\boldsymbol{i}_{n(NRD)} + \left(\boldsymbol{I}_t^{RDFTA}\boldsymbol{i}_{n(RDF)} + \boldsymbol{I}_t^{RDGTA}\boldsymbol{i}_{n(RDG)} + \boldsymbol{I}_t^{RDITA}\boldsymbol{i}_{n(RDI)}\right) \\ &+\left(\boldsymbol{I}_t^{RDFINTA}\boldsymbol{i}_{n(RDF)} + \boldsymbol{I}_t^{RDGINTA}\boldsymbol{i}_{n(RDG)} + \boldsymbol{I}_t^{RDIINTA}\boldsymbol{i}_{n(RDI)}\right) + \boldsymbol{I}_t^{G}\boldsymbol{\Gamma}\boldsymbol{i} + \boldsymbol{c}_t = \boldsymbol{x}_t, \end{aligned} \tag{4.22}$$

where $\boldsymbol{I}_t^{NRDTA}(\boldsymbol{n} \times \boldsymbol{n}^{NRD})$, $\boldsymbol{I}_t^{RDFTA}(\boldsymbol{n} \times \boldsymbol{n}^{RDF})$, $\boldsymbol{I}_t^{RDGTA}(\boldsymbol{n} \times \boldsymbol{n}^{RDG})$, and $\boldsymbol{I}_t^{RDITA}$ $(n \times n^{RDI})$ represent the formation of tangible fixed assets in the main production sector, intra-firm R&D sectors, public and non-profit R&D sectors, and private industrial R&D sectors, respectively. Additionally, $\boldsymbol{I}_t^{RDFINTA}$ $(\boldsymbol{n} \times \boldsymbol{n}^{RDF})$, $\boldsymbol{I}_t^{R.DGINTA}$ $(\boldsymbol{n} \times \boldsymbol{n}^{RGD})$, and $\boldsymbol{I}_t^{RDIINTA}(\boldsymbol{n} \times \boldsymbol{n}^{RDI})$ represent the formation of intangible fixed assets resulting from research activities conducted by intra-firm R&D sectors, public and non-profit R&D sectors, and private industrial R&D sectors, respectively. These are represented as diagonal matrices with the output amounts of each R&D activity as diagonal elements. Here, $i_{n(\bullet)}$ denotes unit vectors of sizes $\boldsymbol{n}^{NRD}$, $\boldsymbol{n}^{RDF}$, $\boldsymbol{n}^{RDG}$, and $\boldsymbol{n}^{RDI}$, respectively. Additionally, $\boldsymbol{c}_t$ represents the final demand vector, which sums all final demands excluding the formation of tangible and intangible fixed assets.

As discussed in the previous section, the sum of the columns of each sector's capital formation matrix corresponds to the investment amount in tangible and intangible capital goods of each sector. This is determined by the economic rate of replacement (δ), and is obtained through the perpetual inventory method, corresponding to the matrix of tangible and intangible capital stocks (St). In various R&D sectors, knowledge assets are formed through R&D, accumulating as intangible assets. In this context, the capital coefficient matrices $\boldsymbol{B}_t^{RDFINTA}$, $\boldsymbol{B}_t^{RDGINTA}$, and $\boldsymbol{B}_t^{RDIINTA}$ of intangible fixed assets (knowledge stocks) are derived from intangible fixed asset stocks defined by the perpetual inventory method, similar to tangible fixed assets.

The capital (for tangible and intangible assets) and labor coefficients are defined for main production activities, intra-firm R&D activities, public and private non-profit R&D activities, and dedicated private-sector industrial R&D departments. Specifically, the capital coefficient for tangible fixed assets is defined as B_t^{NRDTA} for main production activities, and as B_t^{RDTA} ($B_t^{RDFTA} + B_t^{RDGTA} + B_t^{RDITA}$) for R&D activities. The aggregate capital coefficient BtK is defined as the sum of these. Further, the capital coefficient for intangible assets (knowledge stock) in the R&D departments is defined separately for intra-firm R&D activities, public R&D activities, private sector industrial R&D activities, and their aggregate as B_t^{RDINTA} ($B_t^{RDFINTA} + B_t^{RDGINTA} + B_t^{RDIINTA}$).

We hereby define static unit structure as follows,

$$U_t^S / f_t^* = e(i) = \begin{bmatrix} X_t^* \\ L_t^{NRD^*} \\ L_t^{RD^*} \\ K_t^{NRDTA^*} \\ K_t^{RDTA^*} \\ K_t^{RDINTA} \end{bmatrix} = \begin{bmatrix} A_t < (I - A_t)^{-1} f_t^*) > \\ B_t^{NRDL} < (I - A_t)^{-1} f_t^*) > \\ B_t^{RDL} < (I - A_t)^{-1} f_t^*) > \\ B_t^{NRDTA} < (I - A)^{-1} \mathbf{f}_t^*) > \\ B_t^{RDTA} < (I - A_t)^{-1} f_t^{**}) > \\ B_t^{RDINTA} < (I - A_t)^{-1} f_t^*) > \end{bmatrix} \tag{4.23}$$

This formulation aligns with Eq. (4.14) from the preceding section. We define the labor and capital coefficients distinctly for main production activities and R&D activities, which include intra-firm, governmental, and private non-profit industrial activities.

We can also introduce the formulation of the dynamic inverse and dynamic structure when labor and capital inputs segregate into non-R&D and R&D inputs.

By substituting Eq. (4.22), we can derive the subsequent balance equations.

$$A_t x_t + I_t^K i + I_t^{RDF} i + I_t^{RDG} i + I_t^{RDI} i + c_t = x_t, \tag{4.24}$$

where I_t^K ($n \times n$) denotes the matrix for tangible fixed capital formation (flow) associated with activities in private enterprises, public and private non-profit research sectors, and public infrastructure development. It also includes capital formation related to activities in private (intra-firm R&D within companies and research institutions as industrial sectors) and public R&D sectors (national and private universities, affiliated research institutions, national research institutes, etc.). Hence, the tangible fixed capital matrix I_t^K encapsulates tangible fixed capital formation pertinent to production activities and R&D activities across all private and public sectors.

Similarly, $I_t^{RDF}(n \times n^{RDF})$ signifies the flow matrix of intangible fixed capital formation by the R&D departments within private enterprises. It outlines the formation of intangible fixed capital as knowledge capital for the n^{RDF} sectors that have generated intangible fixed capital through intra-firm research activities among the n sectors. All other elements are zero. Likewise, $I_t^{RDG}(n \times n^{RDG})$ and $I_t^{RDI}(n \times n^{RDI})$ represent the matrices of intangible fixed capital formation in public R&D sectors and private industrial R&D sectors, respectively. Thus, it is inferred that intangible fixed capital forms through intra-firm research activities within private enterprises, R&D activities in public and private non-profit organizations, and industrial R&D activities in the private sector.

These formations of tangible and intangible fixed capital, $\boldsymbol{I}_t^K$, $\boldsymbol{I}_t^{RDF}$, $\boldsymbol{I}_t^{RDG}$, and $\boldsymbol{I}_t^{RDI}$, are defined to correlate with the economic depreciation rates of each asset. The constant economic depreciation rate row vectors $\boldsymbol{\delta}^K$, $\boldsymbol{\delta}^{RDF}$, $\boldsymbol{\delta}^{RDG}$, and $\boldsymbol{\delta}^{RDI}$, determined by the perpetual inventory method, relate to the stocks of tangible and intangible fixed capital $\boldsymbol{S}_t^K$, $\boldsymbol{S}_t^{RDF}$, $\boldsymbol{S}_t^{RDG}$, and $\boldsymbol{S}_t^{RDI}$, respectively.

$$\boldsymbol{I}_t^K = \boldsymbol{S}_{t+1}{}^K - \left(\boldsymbol{I} - <\delta^K>\right)\boldsymbol{S}_t^K = \boldsymbol{B}_{t+1}{}^K <\boldsymbol{x}_{t+1}> - \left(\boldsymbol{I} - <\delta^K>\right)\boldsymbol{B}_t^K <\boldsymbol{x}_t>, \quad (4.25)$$

$$\boldsymbol{I}_t^{RDF} = \boldsymbol{S}_{t+1}{}^{RDF} - \left(\boldsymbol{I} - <\delta^{RDF}>\right)\boldsymbol{S}_t^{RDF} = \boldsymbol{B}_{t+1}{}^{RDF} <\boldsymbol{x}_{t+1}> - \left(\boldsymbol{I} - <\delta^{RDF}>\right)\boldsymbol{B}_t^{RDF} <\boldsymbol{x}_t>,$$

(4.26)

$$\boldsymbol{I}_t^{RDG} = \boldsymbol{S}_{t+1}{}^{RDG} - \left(\boldsymbol{I} - <\delta^{RDG}>\right)\boldsymbol{S}_t^{RDG} = \boldsymbol{B}_{t+1}{}^{RDG} <\boldsymbol{x}_{t+1}> - \left(\boldsymbol{I} - <\delta^{RDG}>\right)\boldsymbol{B}_t{}^{RDG} <\boldsymbol{x}_t>,$$

(4.27)

$$\boldsymbol{I}_t^{RDI} = \boldsymbol{S}_{t+1}{}^{RDI} - \left(\boldsymbol{I} - <\delta^{RDI}>\right)\boldsymbol{S}_t{}^{RDI} = \boldsymbol{B}_{t+1}{}^{RDI} <\boldsymbol{x}_{t+1}> - \left(\boldsymbol{I} - <\delta^{RDI}>\right)\boldsymbol{B}_t{}^{RDI} <\boldsymbol{x}_t>.$$

(4.28)

By substituting Eqs. (4.25), (4.26), (4.27), and (4.28) into Eq. (4.24) and using the capital coefficient matrices $\boldsymbol{Bt}^K$, $\boldsymbol{Bt}^{RDF}$, $\boldsymbol{B}_t^{RDG}$, and $\boldsymbol{B}_t^{RDI}$, it can be reformulated as follows:

$$\left(\boldsymbol{I} - \boldsymbol{G}_t\right)\boldsymbol{x}_t - \left(\boldsymbol{B}_{t+1}{}^K + \boldsymbol{B}_{t+1}{}^{RDF} + \boldsymbol{B}_{t+1}{}^{RDG} + \boldsymbol{B}_{t+1}{}^{RDI}\right)\boldsymbol{x}_{t+1} = \boldsymbol{c}_t. \quad (4.29)$$

Here we define $\boldsymbol{G}_t$

$$\begin{aligned}\boldsymbol{G}_t = \boldsymbol{A}_t &- \left(\boldsymbol{I} - <\delta^K>\right)\boldsymbol{B}_t^K - \left(\boldsymbol{I} - <\delta^{RDF}>\right)\boldsymbol{B}_t^{RDF} - \left(\boldsymbol{I} - <\delta^{RDG}>\right)\boldsymbol{B}_t^{RDG} \\ &- \left(\boldsymbol{I} - \delta^{RDI}>\right)\boldsymbol{B}_t^{RDI}\end{aligned}$$

From this point, as defined in the previous section, we can establish the dynamic unit structure using Leontief's dynamic inverse matrix. This offers an analytical tool that enables us to trace the history of production structures by retracing past labor, tangible, and intangible capital formation and input structures. Specifically, by illustrating the process of knowledge accumulation through R&D as intangible fixed assets, we can trace the historical process of knowledge accumulation and the resultant changes in production structures.

4.6 Framework for Total Factor Productivity and Growth Accounting Using Input-Output Tables Including R&D Activities

4.6.1 Static Unit Structure and Total Factor Productivity

The rate of technological progress in specific production activities, or the growth rate of TFP, is defined by the difference between the growth rate of gross production resulting from those activities and the weighted average of input quantities of various production factors. This is determined by the growth rate of the total input quantity. We previously referred to the indirect and direct production spillover structure accompanying a production activity of the j-th sector as the "static unit structure." Based on this structure, we define the total factor productivity reflecting the structure of a production activity in the j-th sector as "static unit total factor productivity (static unit TFP)."

First, the traditional TFP growth rate of a j-th sector's production activity at time t is defined based on the following input cost balance:

$$p_{j,t}X_{j,t} = \Sigma_{(i=1,n)}p_{ij}X_{ij,t} + \Sigma_{(l=1,l)}p^{L}{}_{lj,t}L_{lj,t} + \Sigma_{(k=1,k)}p^{K}{}_{kj,t}K_{kj,t}, \tag{4.30}$$

Here, X_j represents the gross output of sector j, and X_{ij}, L_{lj}, K_{kj}, respectively, denote the intermediate input, labor input of type l, and capital input of type k for sector j. The price of the output of sector j is denoted by, $p_{j,t}$, and the prices of each input factor are denoted by p_j, $p^L{}_{lj,}$ $p^K{}_{kj}$. In this context, capital input involves aggregating capital stocks using a tangible and intangible fixed capital matrix (number of investment goods × number of production activities) and allocating tangible and intangible investment goods to production activities. Economic depreciation rates are considered, and perpetual inventory methods are applied to calculate tangible and intangible capital stocks for production activities. Thus, the types of capital inputs are separated into tangible fixed capital stock and intangible fixed capital stock for each production activity sector.

At this point, the traditional TFP growth rate is defined by the following equation.

$$\begin{aligned}(\dot{T}_j / T_j) = (\dot{X}_j / X_j)_t &- \Sigma_i \left(p_{i,t}X_{ij,t} / p_{j,t}X_{j,t}\right)\left(\dot{X}_{ij,t} / X_{ij,t}\right) \\ &- \Sigma_l \left(p_{lj,t}L_{lj,t} / p_{j,t}X_{j,t}\right)\left(\dot{L}_{lj,t} / L_{lj,t}\right) \\ &- \Sigma_k \left(p_{kj,t}K_{kj,t} / p_{j,t}X_{j,t}\right)\left(\dot{K}_{kj,t} / K_{kj,t}\right),\end{aligned} \tag{4.31}$$

If we consider the costs associated with R&D activities, the cost definition equation can be modified from Eq. (4.30) to the following:

$$p_{j,t}X_{j,t} = \Sigma_{(i=1,n)}p_{ij}X_{ij,t} + \Sigma_{(l=1,lNRD)}p^{LNRD}{}_{lj,t}L^{NRD}{}_{lj,t} + \Sigma_{(l=1,lRD)}p^{LRD}{}_{lj,t}L^{RD}{}_{lj,t}$$
$$+ \Sigma_{(k=1,KNRD)}p^{KNRD}{}_{kj,t}K^{NRD}{}_{kj,t} + \Sigma_{(k=1,kRD)}p^{KRD}{}_{kj,t}K^{RD}{}_{kj,t}. \quad (4.32)$$

If the activities in sector *j* are divided into the main production activities (***j*** = ***m***), R&D (***j*** = ***r***), and ICT (***j*** = ***d***) activities, as in Table 4.1, each of them would have the following cost structures:

$$p_{m,t}X_{m,t} = \Sigma_{(i=1,n)}p_{im}X_{im,t} + p^{L}{}_{m,t}L^{NRD}{}_{\mathrm{lm},t} + p^{K}{}_{m,t}K_{mt} + p^{K}{}_{rt}K_{rt} + p^{K}{}_{dt}K_{dt} \quad (4.32a)$$

$$p_{r,t}X_{rt} = \Sigma_{(i=1,n)}p_{ir}X_{irt} + p^{L}{}_{r,t}L^{RD}{}_{rt} + p^{K}{}_{rt}K_{rt}, \quad (4.32b)$$

$$p_{d,t}X_{dt} = \Sigma_{(i=1,n)}p_{id}X_{idt} + p^{L}{}_{dt}L^{IT}{}_{dt} + p^{K}{}_{dt}K_{dt}. \quad (4.32c)$$

Equations (4.32a), (4.32b), and (4.32c) correspond to the cost input balances for the main production activities, intra-firm R&D, and intra-firm ICT activities in sector *j*, respectively. In Eq. (4.32), for the main production activities in this sector, the costs include intermediate inputs $\boldsymbol{\Sigma_{(i=1,n)}p_{im}X_{im,t}}$, labor services inputs $\boldsymbol{p^{L}{}_{m,t}L^{NRD}{}_{m,t}}$, tangible capital services inputs $\boldsymbol{p^{K}{}_{m,t}K_{mt}}$, intangible capital services inputs for R&D $\boldsymbol{p^{K}{}_{rt}K_{rt}}$, and intangible capital services inputs for ICT $\boldsymbol{p^{K}{}_{dt}K_{dt}}$. Similarly, for intra-firm R&D and ICT activities, the costs include labor services, $\boldsymbol{p^{L}{}_{r,t}L^{RD}{}_{rt}}$ and $\boldsymbol{p^{L}{}_{dt}L^{IT}{}_{dt}}$, and tangible capital services, $\boldsymbol{p^{K}{}_{rt}K_{rt}}$ and $\boldsymbol{p^{K}{}_{dt}K_{dt}}$, with intangible capital services inputs assumed to be zero.

In Eq. (4.32), $\boldsymbol{L^{NRD}{}_{lj}}$, $\boldsymbol{L^{RD}{}_{lj}}$, $\boldsymbol{K^{NRD}{}_{kj}}$, and $\boldsymbol{K^{RD}{}_{kj}}$ represent the costs in sector *j* for labor and tangible/intangible capital services inputs, distinguishing between main production activities and intra-firm R&D or ICT activities. For labor and capital inputs, main production activities and other intra-firm activities are distinguished, and the quantities of labor and capital services inputs are denoted as $\boldsymbol{L^{NRD}{}_{lj}}$, $\boldsymbol{L^{RD}{}_{lj}}$, $\boldsymbol{K^{NRD}{}_{kj}}$, and $\boldsymbol{K^{RD}{}_{kj}}$, respectively. The service prices for each input element are represented as $\boldsymbol{p^{LNRD}{}_{lj}}$, $\boldsymbol{p^{LRD}{}_{lj}}$, $\boldsymbol{p^{KNRD}{}_{lj}}$, *and* $\boldsymbol{p^{KRD}{}_{lj}}$, respectively.

At that point, the growth rate of TFP, considering both the main production activities and the intra-firm activities of R&D and ICT, can be defined as follows. In main production activities, where the fourth term on the right side of Eq. (4.33) becomes zero, and in the intra-firm R&D and ICT activities, where the third and sixth terms on the right side of Eq. (4.33) become zero, each corresponds to the growth rate of TFP for the respective activities, as indicated by Eqs. (4.32a), (4.32b), and (4.32c). Here, intra-firm R&D and ICT activities are combined and represented as intangible fixed capital, denoted as $\boldsymbol{K^{RD}{}_{kj,t}}(\boldsymbol{k} = 1,\boldsymbol{KRD})$.

$$\begin{aligned}
\left(\dot{T}_j / T_j\right) &= \left(\dot{X}_j / X_j\right)_t - \Sigma_i \left(p_{i,t} X_{ij,t} / p_{j,t} X_{j,t}\right)\left(\dot{X}_{ij,t} / X_{ij,t}\right) \\
&- \Sigma_{(l=1,lNRD)} \left(p^{LNRD}{}_{lj,t} L^{NRD}{}_{lj,t} / p_{j,t} X_{j,t}\right)\left(\dot{L}^{NRD}_{lj,t} / L^{NRD}{}_{lj,t}\right) \\
&- \Sigma_{(t=1,lRD)} \left(p^{LRD}{}_{lj,t} L^{RD}{}_{lj,t} / p_{j,t} X_{j,t}\right)\left(\dot{L}^{RD}_{lj,t} / L^{RD}{}_{lj,t}\right) \\
&- \Sigma_{(k=1,KNRD)} \left(p^{KNRD}{}_{kj,t} K^{NRD}{}_{kj,t} / p_{j,t} X_{j,t}\right)\left(\dot{K}^{NRD}_{kj,t} / K^{NR}{}_{kj,t}\right)
\end{aligned} \tag{4.33}$$

Next, based on the SNA, it is possible to derive the aggregate-level TFP growth rate corresponding to this formulation for the entire country. By aggregating the sectoral balances from formula (4.32) at the national level, we can obtain the balance equation for output and input at the aggregate level:

$$\begin{aligned}
\Sigma_{(j=1,n)} p_{j,t} X_{j,t} &= \Sigma_{(j=1,n)} \Sigma_{(i=1,n)} p_{ij} X_{ij,t} + \Sigma_{(j=1,n)} \Sigma_{(l=1,lNRD)} p^{LNRD}{}_{lj,t} L^{NRD}{}_{lj,t} \\
&+ \Sigma_{(j=1,n)} \Sigma_{(l=1,lRD)} p^{LRD}{}_{lj,t} L^{RD}{}_{lj,} \\
&+ \Sigma_{(j=1,n)} \Sigma_{(k=1,KNRD)} p^{KNRD}{}_{kj,t} K^{NRD}{}_{kj,t} + \Sigma_{(j=1,n)} \Sigma_{(k=1,KRD)} p^{KRD}{}_{kj,t} K^{RD}{}_{kj,t}.
\end{aligned} \tag{4.34}$$

Incorporating the next defining equation into Eq. (4.34) yields the following.

$$\Sigma_{(j=1,n)} p_{j,t} X_{j,t} - \Sigma_{(j=1,n)} \Sigma_{(i=1,n)} p_{ij} X_{ij,t} = \Sigma_{(i=1,n)} p_{i,t} f_{i,t}$$

Ultimately, the cost balance at the aggregate level of the entire country is equal to the total input factor cost corresponding to the total final demand of the country's aggregate.

$$\begin{aligned}
\Sigma_{(i=1,n)} p_{i,t} f_{i,t} &= \Sigma_{(i=1,n)} \Sigma_{(l=1,lNRD)} p^{LNRD}{}_{lj,t} L^{NRD}{}_{lj,t} + \Sigma_{(j=1,n)} \Sigma_{(l=1,lRD)} p^{LRD}{}_{lj,t} L^{RD}{}_{lj,t} \\
&+ \Sigma_{(i=1,n)} \Sigma_{(k=1,KNRD)} p^{KNRD}{}_{kj,t} K^{NRD}{}_{kj,t} + \Sigma_{(j=1,n)} \Sigma_{(k=1,KRD)} p^{KRD}{}_{kj,t} K^{RD}{}_{kj,t}.
\end{aligned} \tag{4.35}$$

Both sides of Eq. (4.35) are differentiated with respect to time to obtain the TFP growth rate at the aggregate level.

$$\begin{aligned}
\left(\dot{T} / T\right)_t &= \Sigma_{(i=1,n)} \left(p_{i,t} f_{i,t} / P_t' F_t\right)\left(f_i / f_i\right)_t \\
&- \Sigma_{(j=1,n)} \Sigma_{(l=1,LNRD)} \left(p^{LNRD}{}_{lj,t} L^{NRD}{}_{lj,t} / P_t' F_t\right)\left(\dot{L}^{NRD}{}_{lj} / L^{NRD}{}_{lj}\right)_t \\
&- \Sigma_{(j=1,n)} \Sigma_{(l=1,LRD)} \left(p^{LRD}{}_{lj,t} L^{RD}{}_{lj,t} / P_t' F_t\right)\left(\dot{L}^{RD}_{lj} / L^{RD}{}_{lj}\right)_t \\
&- \Sigma_{(j=1,n)} \Sigma_{(k=1,KNRD)} \left(p^{KNRD}{}_{kj,t} K^{NRD}{}_{kj,t} / P_t' F_t\right)\left(\dot{K}^{NRD}_{kj} / K^{NRD}{}_{kj}\right) \\
&- \Sigma_{(j=1,n)} \Sigma_{(k=1,KRD)} \left(p^{KRD}{}_{kj,t} K^{RD}{}_{kj,t} / P_t' F_t\right)\left(\dot{K}^{RD}_{kj} / K^{RD}{}_{kj}\right)
\end{aligned} \tag{4.36}$$

Thus,

$$\left(\dot{T}/T\right)_t = \Sigma_{(j=1,n)}\left(p_{j,t}X_{jt}/P_t'F_t\right)\left(\dot{T}_j/T_j\right)_{t.} \tag{4.37}$$

Here, $P'_t\,F_t$ represents the nominal aggregate amount of final demand, which also corresponds to the total value added generated in a country. The aggregated TFP growth rate $\left(\dot{T}/T\right)_t$ for a country's activities is a weighted average of the TFP growth rates of each sector, weighted by the so-called Domar weights ($p_{j,t}$ $X_{jt}/P'_t\,F_t$).

In Eq. (4.36), labor and capital inputs can be considered as all direct and indirect factor inputs necessary to meet the final demand. Corresponding to the "static unit structure" defined in Eq. (4.23) of the previous section, introducing this concept into Eq. (4.36) allows for determining the growth rate of "static unit TFP" as a productivity indicator when producing a specific final demand i-good per unit, as defined by the static unit structure.

$$\begin{aligned}\left(\dot{T}_i/T_i\right)^U_t = &-\Sigma_{(j=1,n)}\Sigma_{(l=1,lNRD)}\left(p^{LNRD}_{lj,t}L^{NRD^*}_{lj,t}/p_{it}\right)\left(\dot{L}^{NRD^*}_{lj}/L^{NRD^*}_{lj}\right)_t \\ &-\Sigma_{(j=1,n)}\Sigma_{(l=1,lRD)}\left(p^{LRD}_{lj,t}L^{RD^*}_{lj,t}/p_{it}\right)\left(\dot{L}^{RD^*}_{lj}/L^{RD^*}_{lj}\right)_t \\ &-\Sigma_{(j=1,n)}\Sigma_{(k=1,KNRD)}\left(p^{KNRD}_{kj,t}K^{NRD^*}_{kj,t}/p_{it}\right)\left(\dot{K}^{NRD^*}_{kj}/K^{NRD^*}_{kj}\right) \\ &-\Sigma_{(j=1,n)}\Sigma_{(k=1,KRD)}\left(p^{KRD}_{kj,t}K^{RD^*}_{kj,t}/p_{it}\right)\left(\dot{K}^{RD^*}_{kj}/K^{RD^*}_{kj}\right)_{t.}\end{aligned} \tag{4.38}$$

Here, $L^{NRD*} = B_t^{NRDL} <(I - A_t)^{-1}e_t^*)>$, $L^{RD*} = B_t^{RDL} <(I - A_t)^{-1}e_t^*)>$, $K^{NRD*} = B_t^{NRDK} <(I - A_t)^{-1}e_t^*)>$, $K^{RD*} = B_t^{RDK} <(I - A_t)^{-1}e_t^*)>$ are defined. Notably, they are related to the productivity indices at the national aggregate level defined in Eq. (4.36).

$$\left(\dot{T}/T\right)_t = \Sigma_{(i=1,n)}\left(p_i f_i/P_t'F\right)\left(\dot{T}_i/T_i\right)^U. \tag{4.39}$$

Notably, the indicator for TFP growth rate at the national aggregate level is equivalent to the growth rate of the productivity index "Static Unit TFP" when producing each final demand ***i***-good with a weight given by the nominal share of each good's final demand. The total weight sum is 1.0.[3]

The indicator defined in Eq. (4.38) can be extended to any final demand f^*_t as desired.

[3]The static unit TFP growth rate defined in this context aligns with the "effective rate of TFP" as introduced by Hulten (1978).

$$
\begin{aligned}
\left(\dot{T}/T\right)_t^{U(f*t)} &= \Sigma_{(i=1,n)} p_{it} f_i^* / P_t' F^*{}_t)\left(\dot{f}^* / f^*\right) \\
&-\Sigma_{(j=1,n)}\Sigma_{(l=1,lNRD)} p^{LNRD}{}_{lj,t} L^{NRD^*}{}_{lj,t} / P_t F^*{}_t)\left(\dot{L}^{NRD^*}{}_{lj} / L^{NRD^*}{}_{lj}\right)_t \\
&-\Sigma_{(j=1,n)}\Sigma_{(l=1,lRD)} p^{LRD}{}_{lj,t} L^{RD^*}{}_{lj,t} / P_t F^*{}_t)\left(\dot{L}^{RD^*}{}_{lj} / L^{RD^*}{}_{lj}\right)_t \\
&-\Sigma_{(j=1,n)}\Sigma_{(k=1,KNRD)} p^{KNRD}{}_{kj,t} K^{NRD^*}{}_{kj,t} / P_t F^*{}_t)(\dot{K}^{NRD^*}{}_{kj} / K^{NRD^*}{}_{kj} \\
&-{}_{(j=1,n)}\Sigma_{(l=1,KRD)} p^{KRD}{}_{kj,t} K^{RD^*}{}_{kj,t} / P_t F^*{}_t)\left(\dot{K}^{RD^*}{}_{kj} / K^{RD^*}{}_{kj}\right)_{t.}
\end{aligned}
\tag{4.40}
$$

This is the aggregated TFP growth rate corresponding to the compositional changes in the growth rates of any final demand goods. It is distinguished from the individual goods' static unit TFP and referred to as compound unit TFP. When $\boldsymbol{f^*t} = \boldsymbol{e(i)}$ in Eq. (4.40), it reduces to Eq. (4.38). In this case, $\left(\dot{T}_i / T_i\right)U^t = \left(\dot{T}/T\right)^t U(e(i)$, and especially when the final demand vector represents the aggregated vector of actual final demand items, the growth rate of this compound unit TFP becomes the productivity growth rate at the aggregated level (macro gross domestic product) as defined in Eq. (4.36).

4.6.2 Dynamic Unit Total Factor Productivity

The production activities each year are influenced by the historical accumulation of tangible and intangible fixed capital, which encapsulates the technology available at the time of investment. The "dynamic unit TFP" concept, introduced by Kuroda & Nomura (2004), assesses the efficiency of present production activities by considering the dynamic nature of past technology, as embodied in the historical accumulation of tangible and intangible capital stocks. We aim to clarify the historical progression of past investments in tangible and intangible capital and the technological attributes they embody. Tangible fixed capital refers to the accumulated investment goods such as buildings, structures, machinery, and transport equipment used in main production, intra-firm research, and intra-firm ICT activities across industry sectors. On the other hand, intangible fixed capital represents the accumulated intangible capital investment goods for intra-firm R&D and ICT activities in each industry sector. This includes knowledge capital investment goods resulting from R&D activities conducted by public institutions, private non-profit organizations, and private industrial research institutions, as well as the accumulation of intangible information-related investment goods resulting from the main production activities of information and communication-related industries.

We define dynamic unit TFP by associating the growth of TFP with the dynamic unit structure. To accomplish this, we adopt several theoretical assumptions.

Assumption 1: We posit that the services provided by the capital stock are proportional to the stock quantity for both tangible and intangible fixed capital. For simplicity, we assume a proportional relationship between the stock quantity and service quantity of capital across various levels of capital stock aggregation. Thus, we propose the following relationship:

$$\left(\dot{S}/S\right)_t = \left(\dot{K}/K\right)_t = \Sigma_j\Sigma_k\left(p^K{}_{kj,t}K_{kj,t}/\Sigma_j\Sigma_k p^K{}_{kj,t}K_{kj,t}\right)\left(\dot{K}_{kj}/K_{kj}\right)_t \tag{4.41a}$$

Here, $\left(\dot{S}/S\right)_t$ and $\left(\dot{K}/K\right)_t$ denote the growth rates of aggregate-level capital service quantity and capital stock quantity, respectively. Moreover, when capital is divided into tangible and intangible components, the aggregate quantities are defined as follows:

$$\begin{aligned}\left(\dot{S}^{NRD}/S^{NRD}\right)_t &= \left(\dot{K}^{NRD}/K^{NRD}\right)\\ &= \Sigma_j\Sigma_k\left(p^{KNRD}{}_{kj,t}K^{NRD}{}_{kj,t}/\Sigma_j\Sigma_k p^{KNRD}{}_{kj,t}K^{NRD}{}_{kj,t}\right)\left(\dot{K}^{KNRD}{}_{kj}/K^{KNRD}{}_{kj}\right)_t,\end{aligned} \tag{4.41b}$$

$$\left(\dot{S}^{RD}/S^{RD}\right)_t = \left(\dot{K}^{RD}/K^{RD}\right)_t = \Sigma_j\Sigma_k\left(p^{KRD}{}_{kj,t}K^{RD}{}_{kj,t}/\Sigma_j\Sigma_k p^{KRD}{}_{kj,t}K^{RD}{}_{kj,t}\right)\left(\dot{K}^{KRD}{}_{kj}/K^{KR}{}_{kj}\right) \tag{4.41c}$$

The aggregate quantities of labor input and final demand are defined as follows:

$$\left(\dot{L}/L\right)_t = \Sigma_j\Sigma_l\left(p^L{}_{lj}L_{lj,t}/\Sigma_j\Sigma_l p^L{}_{lj}L_{lj,t}\right)\left(\dot{L}_{lj}/L_{lj}\right)_t, \tag{4.41d}$$

$$\left(\dot{f}/f\right)_t = \Sigma_i\left(p_{ij}f_{ij,t}/P_t'F_t\right)\left(\dot{f}_i/f_i\right)_t. \tag{4.41e}$$

In line with the division of capital into tangible and intangible capital, labor input is also divided into non-R&D activity and R&D activity. The aggregate quantities for each are defined as follows:

$$\left(\dot{L}^{NRD}/L^{NRD}\right)_t = \Sigma_j\Sigma_l\left(p^{LNRD}{}_{lj}L^{NRD}{}_{lj,t}/\Sigma_j\Sigma_l p^{LNRD}{}_{lj}L^{NRD}{}_{lj,t}\right)\left(\dot{L}^{NRD}{}_{lj}/L^{NRD}{}_{lj}\right)_t \tag{4.41f}$$

$$\left(L^{RD}/L^{RD}\right)_t = \Sigma_j\Sigma_l\left(p^{LRD}{}_{lj}K^{RD}{}_{lj,t}/\Sigma_j\Sigma_l p^{LRD}{}_{lj}L^{RD}{}_{lj,t}\right)\left(\dot{L}^{RD}_{lj}/L^{RD}{}_{lj}\right)_t \tag{4.41g}$$

Assumption 2: The dynamic unit structure is described under the assumption that the historical process can be approximated by a discrete-time transition of a continuous type. Additionally, it is assumed that the following relationship exists between the stock of tangible and intangible fixed capital and gross investment:

$$S^{NRD}{}_{t-\tau} = \left(1-\delta^{NRD}\right)S^{NRD}{}_{t-\tau-1} + I^{NRD}{}_{t-\tau-1}\left(\tau,1,2,\ldots\right), \tag{4.42a}$$

$$S^{RD}{}_{t-\tau} = \left(1-\delta^{RD}\right)S^{RD}{}_{t-\tau-1} + I^{RD}{}_{t-\tau-1}\left(\tau = 0,1,2,\ldots\right). \tag{4.42b}$$

Here, $\boldsymbol{\delta}^{NRD}$ and $\boldsymbol{\delta}^{RD}$ denote the depreciation rates of tangible and intangible fixed capital, respectively, and are assumed to be constant parameters. Differentiating with respect to time $\boldsymbol{t}$,

$$\begin{aligned}\left(\dot{S}^{NRD}/S^{NRD}\right)_{t-\tau} &= \left(1-\delta^{NRD}\right)\left(S^{NRD}{}_{t-\tau-1}/S^{NRD}{}_{t-\tau}\right)\left(\dot{S}^{NRD}/S^{NRD}\right)_{t-\tau-1} \\ &+ \left(I^{NRD}{}_{t-\tau-1}/S^{NRD}{}_{t-\tau}\right)\left(\dot{I}^{NRD}/I^{NRD}\right)_{t-\tau-1} \quad \left(\tau = 0,1,2,\ldots\right),\end{aligned} \tag{4.43a}$$

$$\begin{aligned}\left(\dot{S}^{RD}/S^{RD}\right)_{t-\tau} &= \left(1-\delta^{RD}\right)\left(S^{RD}{}_{t-\tau-1}/S^{RD}{}_{t-\tau}\right)\left(\dot{S}^{RD}/S^{RD}\right)_{t-\tau-1} \\ &+ \left(I^{RD}{}_{t-\tau-1}/S^{RD}{}_{t-\tau}\right)\left(\dot{I}^{RD}/I^{RD}\right)_{t-\tau-1} \quad \left(\tau = 0,1,2,\ldots\right).\end{aligned} \tag{4.43b}$$

Consider the dynamic production structure at time t when supplying one unit of a specific product (goods or services). This production activity involves not only the direct intermediate inputs and labor inputs invested at time t, but also the capital inputs influenced by the accumulation of past investments in tangible and intangible capital. To clarify this relationship, let's revisit the definition of the compound unit TFP. The compound unit TFP at time t-τ is defined by Eq. (4.40). Based on premises (1) and (2), the growth rates of labor, capital, final demand, and the compound unit TFP as aggregate quantities at time t-τ can be expressed as follows:

$$\begin{aligned}\left(\dot{T}/T\right)_{t-\tau}^{U(f^*_{t-\tau})} &= \left(\dot{f}^*/f^*\right)_{t-\tau} - \sigma^*{}_{LNRD,t-\tau}\left(\dot{L}^{NRD^*}/L^{NRD^*}\right)_{t-\tau} - \sigma^*{}_{LRD,t-\tau}\left(\dot{L}^{RD^*}/L^{RD^*}\right)_{t-\tau} \\ &-\sigma^*{}_{KNRD,t-\tau}\left(\dot{K}^{NRD^*}/K^{NRD^*}\right)_{t-\tau} - \sigma^*{}_{KRD,t-\tau}\left(\dot{K}^{RD^*}/K^{RD^*}\right)_{t-\tau}, \quad \left(\tau = 0,1,2,\ldots\right)\end{aligned} \tag{4.44}$$

here,

$$\sigma^*{}_{LNRD,t-\tau} = \Sigma_{(j=1,n)}\Sigma_{(l=1,lNRD)}p^{LNRD}{}_{lj,t-\tau}L^{NRD^*}{}_{lj,t-\tau}/P'_{it-\tau}f^*{}_{t-\tau}$$

$$\sigma^*{}_{LRD,t-\tau} = \Sigma_{(j=1,n)}\Sigma_{(l=1,lRD)}p^{LRD}{}_{lj,t-\tau}L^{RD^*}{}_{lj,t-\tau}/P'_{it-\tau}f^*{}_{t-\tau}$$

$$\sigma^*{}_{KNRD,t-\tau} = \Sigma_{(j=1,n)}\Sigma_{(k=1,KNRD)}p^{KNRD}{}_{kj,t-\tau}K^{NRD^*}{}_{kj,t-\tau}/P'_{it-\tau}f^*{}_{t-\tau}$$

$$\sigma^*{}_{KRD,t-\tau} = \Sigma_{(j=1,n)}\Sigma_{(k=1,KRD)}p^{KRD}{}_{kj,t-\tau}K^{RD^*}{}_{kj,t-\tau}/P'_{it-\tau}f^*{}_{t-\tau},$$

Equation (4.44) states that when $\tau = 0$, it equals the current composite unit TFP defined in Eq. (4.40). Now, considering a dynamic production process to meet the production of one unit of a specific product's final demand f^*_t at time t ($\tau = 0$), it is a necessary condition that the gross investment required to generate the capital service input quantity needed for this production is realized as the previous year's final demand. Hence, for tangible and intangible fixed capital, the realization of the required gross investment amount in the previous year necessitates a series of accumulations of gross investment going back in time to produce it. This can be expressed as a series of past required final demand as follows:

$$\left(I^{*NRD}/I\right)_{t-\tau}\left(\dot{I}^{*NRD}/I^{*NRD}\right)_{t-\tau}+\left(I^{*RD}/I\right)_{t-\tau}\left(\dot{I}^{*RD}/I^{*RD}\right)_{t-\tau}=\left(\dot{f}^{*}/f^{*}\right)_{t-\tau} \quad (\tau=0,1,2\ldots). \tag{4.45}$$

Equation (4.45) indicates that the growth rate of the series of final demand at time t-τ ($\tau = 1, 2,\ldots$) necessary to fulfill the production of one unit of a specific product as final demand at time t ($\tau = 0$) is equal to the aggregate growth rate of tangible and intangible gross investment at that time point. Substituting Eqs. (4.43a) and (4.43b) into Eq. (4.44) for $\tau = 1$, the growth rate of the capital service input quantity at time t can be organized as follows. Expanding using the third and fourth terms on the right side of Eq. (4.44), we get:

$$\begin{aligned}
&\left(\dot{K}^{*}/K^{*}\right)_{t-\tau+1}=\left(\dot{K}^{*NRD}/K^{*NRD}\right)_{t-\tau+1}+\left(\dot{K}^{*RD}/K^{*RD}\right)_{t-\tau+1}\\
&=s^{*NRD}{}_{t-\tau}[\left(\dot{T}/T\right)_{t-\tau}^{U(f^{*}{}_{t-\tau})}+\sigma^{*}{}_{LNRDt-\tau}\left(\dot{L}^{NRD^{*}}/L^{NRD^{*}}\right)_{t-\tau}+\sigma^{*}{}_{LRD,t-\tau}\left(\dot{L}^{RD^{*}}/L^{RD^{*}}\right)_{t-\tau}\\
&\quad+\left\{\left(1-\delta^{NRD}\right)S^{NRD}{}_{t-\tau}/I^{NRD^{*}}{}_{t-\tau}+\sigma^{*}{}_{KNRDt-\tau}\right\}\left(\dot{S}^{NRD^{*}}/S^{NRD^{*}}\right)_{t-\tau}]\\
&\quad+s^{*RD}{}_{t-\tau}[\left(\dot{T}/T\right)_{t-\tau}^{U(f*RD\quad)}{}_{t-\tau}+\sigma^{*}{}_{LRDt-\tau}\left(\dot{L}^{RD^{*}}/L^{RD^{*}}\right)_{t-\tau}\\
&\quad+\left\{\left(1-\delta^{RD}\right)S^{RD}{}_{t-\tau}/I^{RD^{*}}{}_{t-\tau}+\sigma^{*}{}_{KRDt-\tau}\right\}\left(\dot{S}^{RD^{*}}/S^{RD^{*}}\right)_{t-\tau}]\\
&\qquad(\tau=1).
\end{aligned} \tag{4.46}$$

Here,

$$\left(\dot{K}^{NRD^*}/K^{NRD^*}\right)_{t-\tau+1}$$
$$= s^{*NRD}{}_{t-\tau}[\left(\dot{T}/T\right)_{t-\tau}{}^{U(f*NRD_{t-\tau})} + \sigma^*{}_{LNRDt-\tau}\left(\dot{L}^{NRD}/L^{NRD^*}\right)_{t-\tau}$$
$$+\left\{\left(1-\delta^{NRD}\right)S^{*NRD}{}_{t-\tau}/I^{NRD^*}{}_{t-\tau} + \sigma^*{}_{KNRDt-\tau}\right\}\left(\dot{S}^{NRD^*}/S^{NRD^*}\right)_{t-\tau}$$
$$\left(\dot{K}^{RD^*}/K^{RD^*}\right)_{t-\tau+1}$$
$$= s^{*RD}{}_{t-\tau}[\left(\dot{T}/T\right)_{t-\tau}{}^{U(f*RD_{t-\tau})} + \sigma^*{}_{LRDt-\tau}\left(\dot{L}^{RD^*}/L^{RD^*}\right)_{t-\tau}$$
$$+\left\{\left(1-\delta^{RDINTA}\right)S^{*RD}{}_{t-\tau}/I^{RD^*}{}_{t-\tau} + \sigma^*{}_{KRDt-\tau}\right\}\left(\dot{S}^{RD^*}/S^{RD^*}\right)_{t-\tau}$$
$$(\tau = 1)$$
$$s^{*NRD}{}_{t-\tau} = I^{*NRD}{}_{t-\tau}/S^{*NRD}{}_{t-\tau-1}, s^{*RD}{}_{t-\tau} = I^{*RD}{}_{t-\tau}/S^{*RD}{}_{t-\tau-1}.$$

This equation demonstrates that the growth rate of the capital service input, required to produce one unit of a specific product as final demand at time t, is broken down into a dynamic sequence of past capital accumulation. The second equation in Eq. (4.46) decomposes the growth rate of the capital service input quantity at time into three historical elements on its right-hand side. The first element is the growth rate of the composite unit TFP at time t-τ, which generates the investment series for $\tau = 1, 2, \ldots$. The second and third elements represent the contribution of the labor service input quantity needed for investment at time t-τ. The fourth and fifth elements signify the contribution of the accumulation of tangible and intangible capital service quantities. By sequentially expanding Eqs. (4.43a), (4.43b), (4.44), and (4.45) into the third and fourth terms on the right side of Eq. (4.46) for $\tau = 2$, we can derive the subsequent equation.

$$s^{*NRD}{}_{t-\tau+1}\left\{\left(1-\delta^{NRD}\right)\left(S^{NRD}{}_{t-\tau+1}/I^{NRD^*}{}_{t-\tau+1}\right)+\sigma^*{}_{KNRDt-\tau+1}\left(\dot{S}^{*NRD}/S^{*NRD}\right)_{t-\tau+1}\right\}$$
$$= \Phi^{*NRD}{}_{t-\tau}\left\{\left(\dot{T}/T\right)_{t-\tau}{}^{U\left(f^*(NRD)\right)} + \Phi^{*NRD}{}_{t-\tau}\sigma^*{}_{LNRDt-\tau}\left(\dot{L}^{NRD^*}/L^{NRD^*}\right)_{t-\tau}\right\}$$
$$+\Phi^{*NRD}{}_{t-\tau}\{\left(1-\delta^{NRD}\right)(S^{NRD}{}_{t-\tau}/I^{*NRD}{}_{t-\tau} + \sigma^*{}_{KNRDt-\tau}\left(\dot{S}^{NRD^*}/S^{NRD^*}\right)_{t-\tau}\}$$
$$(\tau = 2). \tag{4.47a}$$

Here,

$$\Phi^{*NRD}{}_{t-\tau} = s^{*NRD}{}_{t-\tau}s^{*NRD}{}_{t-\tau+1}\{\left(1-\delta^{NRD}\right)(S^{NRD}{}_{t-\tau+1})/I^{*NRD}{}_{t-\tau+1} + \sigma^*{}_{KNRDt-\tau+1}\}$$

and

$$s^{*RD}_{t-\tau+1}\left\{\left(1-\delta^{RD}\right)\left(S^{RD}_{t-\tau+1}/I^{RD^*}_{t-\tau+1}\right)+\sigma^*_{KRDt-\tau+1}\left(\dot{S}^{*RD}/S^{*RD}\right)_{t-\tau+1}\right\}$$
$$=\Phi^{*RD}_{t-\tau}\left\{\left(\dot{T}/T\right)_{t-\tau}^{U(f*RD)}+\Phi^{*RD}_{t-\tau}\sigma^*_{LRDt-\tau}\left(\dot{L}^{RD^*}/L^{RD^*}\right)_{t-\tau}\right\}$$
$$+\Phi^{*RD}_{t-\tau}\{\left(1-\delta^{RD}\right)(S^{RD}_{t-\tau}/I^{RD^*}_{t-\tau}+\sigma^*_{KRDt-\tau}\left(\dot{S}^{RD^*}/S^{RD^{**}}\right)_{t-\tau}\}$$
$$(\tau=2). \tag{4.47b}$$

Here,

$$\Phi^{*RD}_{t-\tau}=s^{*RD}_{t-\tau}s^{*RD}_{t-\tau+1}\{\left(1-\delta^{RD}\right)(S^{RD}_{t-\tau+1)}/I^{RD}_{t-\tau+1}+\sigma^*_{KRDt-\tau+1}\}$$

Equation (4.45) extends the contribution of capital services in period $t-2$, akin to the expansion in Eq. (4.46) for the contribution in period $t-1$. By repeating these steps, it becomes feasible to trace the direct and indirect ripple effects of past dynamic production activities, reaching back in time, which are required as necessary conditions to expand any arbitrary final demand by 1 unit in period t. In this dynamic process, the accumulated capital stock from the past, which reflects the technological conditions at the time of investment, allows for their sum to be considered as the growth rate of TFP.

$$\left(\dot{T}/T\right)_t^{D(f*_t)}=\left(\dot{T}/T\right)_t^{U\left(f*(NRD)_t\right)}+\left(\dot{T}/T\right)_t^{U(f*(RD_t)}$$
$$+\sigma^*_{KNRDt-\tau}\Sigma_{(\tau=1,\ldots\infty)}\Phi^{*NRD}_{t-\tau}\left(\dot{T}/T\right)_{t-\tau}^{U(fNRD*)}{}_{t-\tau}$$
$$+\sigma^*_{KRDt-\tau}\Sigma_{(\tau=1,\ldots\infty)}\Phi^{*RD}_{t-\tau}\left(\dot{T}/T\right)_{t-\tau}^{U(fRD)}{}_{t-\tau}. \tag{4.48}$$

Here,

$\Phi^{*NRD}_{t-\tau}$

if $\tau=1,\quad \Phi^{*NRD}_{t-1}=s^{*NRD}_{t-\tau}$

if $\tau=2,\ldots T,$

$$\Phi^{*NRD}_{t-\tau}=s^{*NRD}_{t-\tau}\Phi^{*NRD}_{t-\tau+1}\left\{\left(1-\delta^{NRD}\right)S^{NRD}_{t-\tau+1}/I^{*NRD}_{t-\tau+1}+\sigma*_{KNRDt-\tau+1}\right\}$$

$\Phi^{*RD}_{t-\tau}$

if $\tau=1,\Phi^{*RD}_{t-1}=s^{*RD}_{t-\tau}$ and *if* $\tau=2,\ldots T,$

$$\Phi^{*RD}_{t-\tau}=s^{*RD}_{t-\tau}\Phi^{*RD}_{t-\tau+1}\left\{\left(1-\delta^{RD}\right)S^{RD}_{t-\tau+1}/I^{*RD}_{t-\tau+1}+\sigma*_{KRDt-\tau+1}\right\}$$

The growth rate of the dynamic unit TFP is delineated by Eq. (4.48). This equation facilitates the decomposition and quantification of the dynamic contributions of tangible and intangible fixed assets to productivity. The analysis thus far has concentrated on two categories of fixed assets: tangible fixed assets (Non-Research and

Development or NRD) and intangible fixed assets derived from R&D activities. However, the analysis can be extended to include ICT activities as a component of intangible fixed assets.

The growth rate of the dynamic unit TFP defined by Eq. (4.48) as $\left(\dot{T}/T\right)_t^{D(f^*_t)}$, represents the weighted average of the growth rates of the composite unit TFP associated with the formation of tangible and intangible capital at past points $t-\tau$ ($\tau = 1, 2, \ldots$) and the static unit TFP at time t. The composite unit TFP at each point in time encapsulates the direct and indirect inter-sectoral spillover of inputs of tangible and intangible fixed asset services and intermediate goods, as illustrated in Eq. (4.48). Thus, the growth rate of the dynamic unit TFP mirrors the state of all direct and indirect technological spillovers when manufacturing a specific commodity unit at a particular point in time.

Until now, the expansion of dynamic unit TFP has been centered on the dynamic production structure at time t, where the provision of capital services at that point is a consequence of the accumulation of past investments in tangible and intangible fixed assets. This investment is determined by the technological structure at the time of investment, thereby capturing how current capital accumulation is a compilation of these past technological structures at each point in time. Given that the production of a specific commodity involves the collaboration of the main production, intra-firm research activities, and intra-firm ICT activities in a specific sector, we can consider the composite unit structure that sets the composition of final demand for these three activities in a specific production sector as 1. By providing a vector that sums up the three activities to 1 in the final demand vector $\boldsymbol{f}^*$ of the above expansion, we can commence from the composite unit TFP and perform the above expansion to obtain the dynamic unit TFP of these three activities in this production sector, which are interconnected for this commodity production.

This approach enables us to derive an indicator to assess the synergy effect of the three activities within the sector, namely the mutual dependence relationship of the three activities within the sector, by comparing it with the dynamic units of the sector. Moreover, by evaluating the composite unit structure based on the vector of final demand composition, we can derive the macro dynamic unit TFP. Consequently, we can dynamically decompose and evaluate the growth components of GDP, and determine the contribution of labor, tangible capital, and intangible capital (R&D, ICT) at the aggregate and industry sector levels.

4.7 Measurement of TFP Through Capitalization of R&D and ICT Investment

This section, utilizing the TFP measurement framework, aims to delineate the characteristics of Japan's economic structure during the post-bubble era of low growth (1995–2011). Despite demographic challenges such as an aging population, the economy experienced significant transformations because of the information revolution, marked by advancements in Artificial Intelligence and social media. This revolution, often referred to as the Fourth Industrial Revolution, is reshaping the dynamic between science and society by transforming information dissemination. Therefore, assessing the impact of knowledge accumulation through information and R&D investments on societal knowledge services is crucial for understanding this dynamic.

ICT services and R&D services, both intangible fixed assets, have distinct impacts on the economy and society. ICT technology, evolving over time, enhances the efficiency of economic structures and acts as a market good within market economies. The "information" processed and disseminated by this technology, while some are valued and traded as market goods, often enhances user convenience and utility without direct market transactions. ICT technology contributes to the expansion of the "volume of societal information," indirectly affecting the efficiency of various industries. Similarly, knowledge services generated by R&D activities provide intangible information to society, with industry-conducted R&D enhancing the production efficiency of their primary activities by providing intangible services.

In contrast, knowledge services generated by government R&D investments in public research institutions are considered public goods, differing from the nature of ICT services as market goods. Understanding the impact of these activities on the economy and society, and exploring their structure, can provide how scientific and technological advancements shape society and meet societal needs. It also aids in policymaking by providing insights into fostering scientifically and technologically beneficial advancements for society. The impact of intangible assets such as information and scientific knowledge on the economy and society is expected to differ from the tangible fixed asset investments that historically drove economic growth in Japan's postwar economic growth patterns.

From this perspective, we categorize the formation of knowledge stock into two activities: R&D investment and ICT investment, which generate two types of

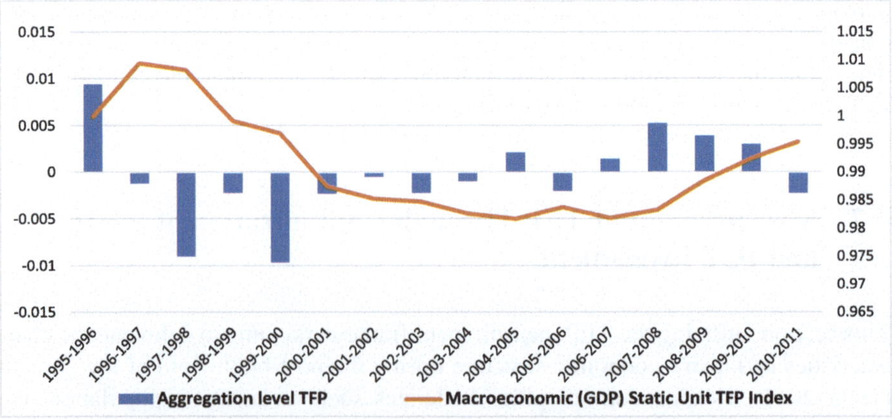

Fig. 4.9 Estimation of the static TFP growth: Mactor level (1995–2011)

intellectual assets. We are restructuring the input-output tables accordingly. R&D investment is divided into three categories based on previous discussions: intra-firm R&D activities within industries, activities of public and non-profit universities and research institutions, and activities of private R&D institutions operating independently as commercial entities. We further categorize the activities of public and non-profit universities and research institutions and private R&D institutions into five research fields each. For ICT activities, we highlight industries engaged in intra-firm ICT activities within each sector and industries providing ICT services as private enterprises, including software, Internet, and communication and broadcasting sectors. The industrial sectors are classified according to the 93-sector classification presented in Table 4.2.

We organize the observed facts from the input-output tables of connected industries through the capitalization of R&D investment and ICT investment in the 93-sector classification for the period from 1995 to 2011. Firstly, we examine the evolution of macroeconomic productivity during this period. Using the calculation formula for macro static TFP defined in Eq. (4.39) of Sect. 4.6, we obtained the growth rate of macro static TFP and the productivity index with a base year of 2005 set to 1.0 for the macroeconomy from 1995 to 2011, as represented in Fig. 4.9. The growth rate of macro TFP is aggregated from the growth rates of static unit TFP of each industry weighted by the Domar weight.

During the period from 1995 to 2011, the macro TFP growth rate averaged a low −0.05% annually. Despite some recovery from the burst of the bubble economy in the 1990s, productivity growth remained nearly stagnant following the burst of the so-called IT bubble in the early 2000s and the impact of the 2007–2008 financial crisis, often referred to as the Lehman Shock.

Figure 4.10(I)~(III) plots the growth rates of industry-static Unit TFP by industry for every 5 years since 1995, alongside the Domar weights of each industry at those points, to examine each industry's TFP contribution to macro TFP by Main, intra-firm R&D, and intra-firm ICT activities. The lines represent the growth rates of industry-specific TFP at each activity in time, while the bar graphs indicate the magnitude of Dormer weight. Industries are classified into main production activities, intra-firm R&D activities, and intra-firm ICT activities.

Significant contributions to macro growth were observed in industries such as tele-communications, broadcasting, software, and other service industries. However, the growth rate slowed down from 2000 to 2005. Despite this, industries such as semiconductor devices, computers, and electronic components continued to improve industrial productivity. In 2010, recovery from the financial crisis progressed, indicating an increase in industrial productivity, particularly notable in the information industry such as the Internet.

The middle section of the graph illustrates the impact of intangible fixed asset formation as intra-firm R&D activities on productivity. The trend in productivity does not show significant changes compared to the main production activity. The bottom section depicts the effect of intra-firm ICT activities on productivity improvement. The enhancement in productivity because of intra-firm R&D activities does not exhibit a notable trend across various industry sectors. The impact of

R&D on productivity increase in different scientific and technological fields also does not appear to be substantial. However, it can be observed that industries specialized in R&D activities have become more active from 2005 to 2010.

To clarify the significance of these macro trends in productivity over time, it is necessary to examine the characteristics of main production activities, intra-firm ICT activities, and intra-firm R&D activities alongside industry-specific TFP. By analyzing the trends of Static Unit TFP, which is calculated according to the formula

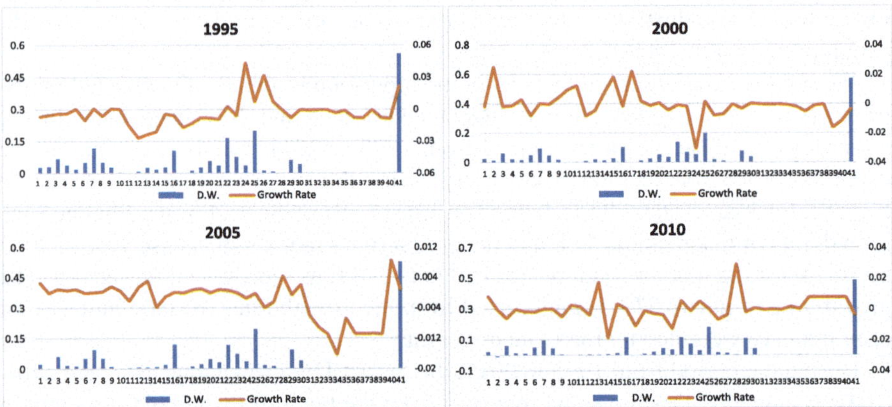

Note:

1. Agriculture/Forestry/Fishery
2. Mining including Intra-ICT
3. Food Product Manufacturing
4. Textile & Wood Manufacturing
5. Paper & Pulp Manufacturing
6. Chemical Prod. Manufacturing
7. Iron & Steel Product Manufacturing
8. General Machinery Manufacturing
9. Electric Machinery Products
10. Fabricated Metal Products
11. Semi-conductor Equipment Products
12. Communication Machinery
13. Electric Computer &Equipment
14. Semiconductor & Integrated Circuit
15. Other Electric Parts
16. Heavy Electric &Transport Machinery
17. Robot
18. Other Precision Machinery
19. Oil & Coal Products
20. Other Manufacturing
21. Energy Industry
22. Construction
23. Transportation
24. Communication & Broadcasting
25. Trade
26. Software Industry including R&D
27. Information & Communication Service
28. Internet Service including ICT
29. Medical Service
30. Education
31. R&D (Life: Public & Non-Business)
32. R&D (Information: Public & Non-Business)
33. R&D (Material: Public & Non-business)
34. R&D (Environment & Energy: Public & Non-Business)
35. R&D (Other R&D: Public & Non-Business)
36. R&D (Life: Private & Business)
37. R&D (Information: Private & Non-Business)
38. R&D (Material: Private & Business)
39. R&D (Environment & Energy: Private & Non-Business)
40. R&D (Other R&D: Private & Business)
41. Other Service

Fig. 4.10 (I) Unit TFP by industrial sectors of main production activities and its contribution to macro TFP growth

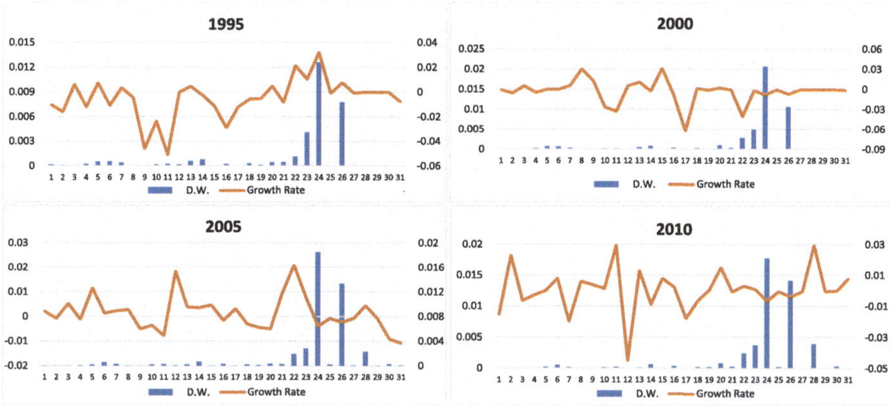

Note:
1. Food Manufacturing
2. Textile & Wood Manufacturing
3. Paper & Pulp Manufacturing
4. Chemical Prod. Manufacturing
5. Iron & Steel Product Manufacturing
6. General Machinery Manufacturing
7. Electric Machinery Products
8. Fabricated Metal Products
9. Semi-conductor Equipment Products
10. Communication Machinery
11. Electric Computer & Equipment
12. Semiconductor & Integrated Circuit
13. Other Electric Parts
14. Heavy & Transportation Mach.
15. Robot
16. Other Precision Mach.
17. Oil & Coal Product
18. Other Manufacturing
19. Energy Industry
20. Construction
21. Transportation
22. Communication & Broadcasting
23. Trade
24. Software
25. Information Services
26. Internet
27. Medical Service
28. Education
29. R&D (Life: Public)
30. R&D (Information: Public)
31. R&D (Material: Public)]
32. R&D (Environment: Public)
33. R&D (Others: Public)
34. R&D (Life: Industry)
35. R&D (Information: Industry)
36. R&D (Material: Industry)
37. R&D (Environment: Industry)
38. R&D (Others: Industry)
39. Other Services

Fig. 4.10 (II) Unit TFP by industrial sectors of information processing activities and its contribution to macro TFP growth

presented in Sect. 4.6 (4.38), we can detect the characteristics of these activities. Static Unit TFP serves as an indicator that considers changes in the productivity of all intermediate inputs required to produce one unit of a specific industry's product, in relation to the TFP of that industry. This analysis will provide a more nuanced understanding of the productivity trends during this period.

The TFP trends for each industry were calculated for three periods: 1995–2000, 2000–2005, and 2005–2010. The TFP growth rate for each industry signifies productivity changes within that industry over the given periods. Conversely, Static Unit TFP represents productivity shifts across industries during these periods,

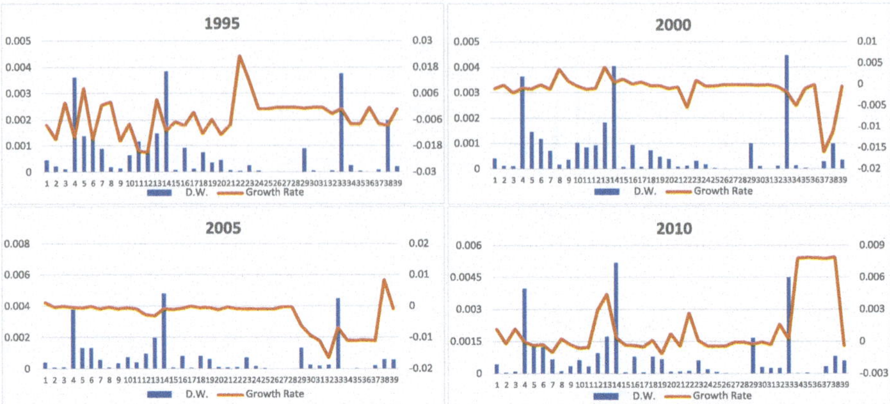

Note:
1. Food Manufacturing
2. Textile & Wood Manufacturing
3. Paper & Pulp Manufacturing
4. Chemical Prod. Manufacturing
5. Iron & Steel Product Manufacturing
6. General Machinery Manufacturing
7. Electric Machinery Products
8. Fabricated Metal Products
9. Semi-conductor Equipment Products
10. Communication Machinery
11. Electric Computer & Equipment
12. Semiconductor & Integrated Circuit
13. Other Electric Parts
14. Heavy & Transportation Mach.
15. Robot
16. Other Precision Mach.
17. Oil & Coal Product
18. Other Manufacturing
19. Energy Industry
20. Construction
21. Transportation
22. Communication & Broadcasting
23. Trade
24. Software
25. Information Services
26. Internet
27. Medical Service
28. Education
29. R&D (Life: Public)
30. R&D (Information: Public)
31. R&D (Material: Public)]
32. R&D (Environment: Public)
33. R&D (Others: Public)
34. R&D (Life: Industry)
35. R&D (Information: Industry)
36. R&D (Material: Industry)
37. R&D (Environment: Industry)
38. R&D (Others: Industry)
39. Other Services

Fig. 4.10 (III) Unit TFP by industrial sectors of R&D activities and its contribution to macro TFP growth

encompassing both direct and indirect effects on goods production. Industrial productivity changes involve not only productivity enhancement within a specific industry by itself, but also productivity variations across all related industries. Typically, the growth rate of Static Unit TFP is anticipated to exceed that of industry-specific TFP.

Table 4.3 showed the industry-specific TFP (Unit TFP) and its Static Unit TFP growth rates for the main production activities of each industry sector in 5-year intervals from 1995 to 2010. Industries with positive Unit TFP and where Static Unit TFP surpasses Unit TFP are denoted with an asterisk (*). From 1995 to 2000, only three industry sectors—mining, materials manufacturing, and petroleum and coal products—exhibited this trend. Although positive Unit TFP values were

Table 4.3 Unit TFP and static unit TFP growth rate per year by industry (%): main activity

Main industry	1995–2000			2000–2005			2005–2010		
	U.TFP vs. static U. TFP (%)			U.TFP vs. static U.TF (%)			U.TFP vs. static U.TFP (%)		
Agriculture/forestry/ fishery including intra-ICT	−0.45	−0.53		0.03	−0.07		0.18	0.23	※
Mining including intra-ICT	0.66	0.78	※	0.20	0.05		−0.67	−0.65	
Food product Manufacturing	−0.15	−0.19		0.08	−0.13		0.04	0.14	※
Textile and wood manufacturing	−0.17	−0.32		0.01	−0.06		0.16	0.26	※
Paper and pulp manufacturing	0.01	−0.05		−0.02	−0.06		0.31	0.50	※
Chemical prod. manufacturing	−0.30	−0.36		−0.13	−0.20		0.04	0.09	※
Iron and steel product manufacturing	0.01	0.08	※	0.02	0.00		−0.06	−0.16	
General machinery manufacturing	−0.44	−0.57		0.00	−0.03		0.11	0.22	※
Electric machinery products	−0.20	−0.53		−0.15	−0.38		0.06	0.42	※
Fabricated metal products	0.33	0.31		0.22	0.14		−0.05	−0.06	
Semi-conductor equipment products	−0.41	−0.54		0.16	0.10		0.09	−0.15	
Communication machinery	−0.74	−1.20		−0.28	−0.34		0.30	0.78	※
Electric computer and equipment	−0.70	−2.01		−0.37	−0.75		0.25	0.83	※
Semiconductor and integrated circuit	−0.58	−1.29		0.12	0.25	※	1.13	1.33	※
Other electric parts	−0.36	−0.89		−0.11	−0.23		1.15	1.67	※
Heavy electric and transport machinery	−0.34	−0.59		0.13	0.16	※	0.23	0.46	※
Robot	−0.37	−0.62		0.15	0.21	※	0.11	0.06	
Other precision machinery	−0.50	−0.70		−0.15	−0.21		0.00	0.41	※
Oil and coal products	0.34	0.64	※	0.30	0.28		0.28	−0.14	
Other manufacturing	−0.42	−0.51		0.01	−0.06		0.13	0.23	※
Energy industry	−0.36	−0.22		0.19	0.21	※	0.00	−0.06	
Construction	−0.07	0.00		−0.04	−0.10		0.36	0.36	※
Transportation	−0.04	−0.13		−0.01	−0.07		0.04	0.08	※
Communication and broadcasting	−0.10	0.15		−1.25	−1.78		0.02	0.10	※

(continued)

Table 4.3 (continued)

Main industry	1995–2000 U.TFP vs. static U. TFP (%)			2000–2005 U.TFP vs. static U.TF (%)			2005–2010 U.TFP vs. static U.TFP (%)		
Trade	0.22	0.20		0.08	−0.02		−0.09	−0.05	
Software industry including R&D	−1.31	−1.87		−0.23	−0.42		0.53	0.59	※
Information and communication service	−0.93	−1.18		0.30	0.20		0.38	0.42	※
Internet service including ICT	0.00	−0.22		0.00	0.56		2.20	0.69	
Medical service	−0.44	−0.38		0.01	−0.15		0.00	0.05	※
Education	0.11	0.10		0.01	−0.01		0.00	0.02	※
Other service	−0.08	−0.15		−0.12	−0.26		0.11	0.15	※

Note: Unit TFP $\gg$ 0 and static unit TFP > unit TFP: ※

observed in sectors such as pulp and paper, optical fibers, commerce, and education, efficiency, inclusive of spillover effects, diminished when considering Static Unit TFP. From 2000 to 2005, production efficiency enhancements were observed in sectors such as semiconductor and integrated circuits, heavy electrical machinery and transportation equipment, robotics, energy, and Internet industries. However, from 2005 to 2010, production efficiency improvements were observed in sectors excluding mining, materials manufacturing, optical fibers, semiconductor manufacturing equipment, robotics, petroleum and coal products, energy, commerce, and Internet-related industries. Among these industries, mining, materials manufacturing, and optical fibers exhibited negative Unit TFP values, indicating a vicious cycle of further decreasing production efficiency because of interdependence with other industries. Nevertheless, during the period from 2005 to 2010, an increase in production efficiency was observed in many industries, leading to a synergistic effect and the recovery of production efficiency.

Table 4.4 evaluates the impact of efficiency in the production of intellectual assets on the growth rates of TFP for corporate and independent R&D activities. These independent activities include those of national/public universities, research institutions, and the R&D industry. The independent R&D sector is subdivided into national/public universities, research institutions, and industrial research institutions, each further categorized into five scientific fields. National/public universities and research institutions generate knowledge stock as a public good. In the published input-output tables, their operating surplus is defined as zero. However, in our constructed tables, allowances for capital consumption of tangible and intangible fixed assets, such as research facilities, are included in the operating surplus. In our industry classification, the activities of non-profit private universities and their affiliated research institutions are merged with those of national/public universities and research institutions. Consequently, their operating surplus and capital consumption allowances are incorporated into the operating surplus of national/public activities, leading to the inclusion of Unit TFP. While the definition of operating surplus

Table 4.4 Unit TFP and static unit TFP growth rate per year by industry (%): R&D activity

Unit TFP vs. static unit TFP	1995–2000		2000–2005		2005–2010		
R&D activity	U.TFP vs. static U.TF (%)		U.TFP vs. static U.TFP (%)		U.TFP vs. static U.TFP (%)		
Agriculture/forestry/fishery	−0.44	0.00	0.06	0.00	0.18	0.00	
Mining including intra-ICT	0.28	0.25	0.44	0.37	−0.46	−0.41	
Food product manufacturing	−0.22	−0.33	−0.02	−0.11	0.24	0.34	※
Textile and wood manufacturing	−0.35	−0.45	0.13	0.04	0.12	0.18	※
Paper and pulp manufacturing	−0.08	−0.17	−0.10	−0.20	0.21	0.30	※
Chemical prod. manufacturing	−0.28	−0.38	0.03	−0.05	0.07	0.14	※
Iron and steel product manufacturing	−0.20	−0.32	−0.15	−0.26	−0.10	−0.04	
General machinery manufacturing	−0.41	−0.52	0.08	−0.01	−0.12	−0.06	
Electric machinery products	−0.22	−0.30	0.03	−0.06	0.07	0.13	※
Fabricated metal products	−0.09	−0.17	0.06	−0.02	−0.11	−0.04	
Semi-conductor equipment products	−0.23	−0.33	0.08	−0.01	−0.33	−0.27	
Communication machinery	−0.14	−0.23	0.09	0.00	0.06	0.12	※
Electric computer and equipment	−0.41	−0.52	0.06	−0.04	0.12	0.18	※
Semi-conductor and integrated circuit	0.00	−0.08	0.28	0.19	0.46	0.50	※
Other electric parts	−0.32	−0.45	0.00	−0.09	0.61	0.64	※
Heavy electric and transport machinery	−0.28	−0.37	0.09	0.01	0.18	0.24	※
Robot	−0.06	−0.14	0.13	0.04	−0.17	−0.10	
Other precision machinery	−0.20	−0.28	0.08	−0.01	0.03	0.10	※
Oil and coal products	−0.99	−1.27	0.30	0.20	0.18	0.23	※
Other manufacturing	−0.31	−0.41	0.10	0.01	0.16	0.23	※
Energy industry	−0.13	−0.21	0.07	−0.01	−0.13	−0.06	
Construction	−0.17	−0.28	0.03	−0.06	−0.03	0.02	
Transportation	−0.19	−0.26	0.05	−0.04	−0.06	0.01	
Communication and broadcasting	−0.25	−0.36	−0.31	−0.43	0.22	0.29	※

(continued)

Table 4.4 (continued)

Unit TFP vs. static unit TFP	1995–2000		2000–2005		2005–2010		
R&D activity	U.TFP vs. static U.TF (%)		U.TFP vs. static U.TFP (%)		U.TFP vs. static U.TFP (%)		
Trade	0.14	0.13	0.04	−0.03	−0.26	−0.26	
Software industry	−1.21	−1.82	−0.26	−0.41	0.84	1.03	※
Information and communication service	−1.14	−1.97	−0.35	−0.55	0.94	1.24	※
Internet service including ICT	0.00	0.00	0.00	0.00	1.13	0.44	
R&D (life: public/non-profit)	0.06	0.04	0.14	0.10	−0.23	−0.16	
R&D (information: public/nonprofit)	0.01	−0.01	0.28	0.22	−0.38	−0.30	
R&D (material: public/nonprofit)	0.00	0.00	0.26	0.22	−0.39	−0.31	
R&D (environment: public/nonprofit)	−0.07	−0.04	0.11	0.04	−0.55	−0.45	
R&D (others: public/nonprofit)	0.10	0.07	0.13	0.08	−0.24	−0.17	
R&D (life: industry)	−0.90	−0.35	12.93	0.99	−3.62	−0.51	
R&D (information: industry)	−3.59	−0.42	4.00	1.05	−0.08	−0.51	
R&D (material: industry)	0.00	0.00	2.88	1.21	0.24	−0.60	
R&D (environment: industry)	−3.58	−0.09	2.39	0.80	−2.77	−0.56	
R&D (others: industry)	−0.05	−0.32	5.07	1.31	−1.64	−0.45	
Other service	−0.02	−0.08	−0.15	−0.26	0.34	0.45	※

Note: Unit TFP $\gg$ 0 and static unit TFP > unit TFP

differs from that of other industry sectors, prohibiting a comparison of the absolute level of TFP with other industries, it is feasible to observe the time-series variation of TFP in this sector. During 1995–2000, the Unit TFP of national/public and non-profit research institutions increased, reflecting the influence of government policies such as the expansion of competitive funding. However, from 2005 to 2010, negative values were observed, suggesting that the productivity enhancement effect of R&D investment was not evident.

The performance of independent R&D departments, viewed as industries, exhibits variations in Unit TFP in relation to their primary production activities. Between 1995 and 2000, production efficiency was negative, shifted to positive from 2000 to 2005, but reverted to negative from 2005 to 2010. Although Static Unit TFP values suggest a synergistic effect on the enhancement of R&D productivity, this effect is not markedly evident.

Assessing the outcomes of intellectual stock accumulation in national public and non-profit research institutions, which possess a significant public dimension, might necessitate a distinct approach compared to evaluating the functionality of ICT services, which are internalized within the market economy mechanism as a market commodity.

The appraisal of the influence of knowledge stock, considered as intangible fixed assets, must be deferred until the measurement of dynamic TFP and evaluated in that context. The impacts of intra-firm R&D have been consistently escalating since 1995. Notably, from 2005 to 2010, they significantly contributed to the surge in production efficiency in information-related industries such as electronic equipment, semiconductor devices, integrated circuits, other electronic components, the software industry, ICT services, and the Internet, demonstrating a significant synergistic effect.

Table 4.5 displays the measurement outcomes of the spillover effects on the production efficiency of the market commodity, specifically ICT activities.

The spillover effects from the independent sectors of the information industry (e.g., the telecommunications and broadcasting industry, software industry, and ICT service provision industry) to other industry sectors are mainly because of the provision of information services as intermediate inputs, functioning as market commodities. The impacts of these market commodities can be identified through their spillover into the intermediate inputs of various industry sectors from these information-related industries.

Figures 4.11, 4.12, 4.13, and 4.14 illustrate the time-series progression of the intermediate input coefficients from the telecommunications and broadcasting industry, software industry, and ICT service provision industry to various industry sectors, respectively.

Figure 4.11 illustrates the time-series transition of intermediate inputs from the telecommunications and broadcasting sectors to other industries. It is evident that inputs from these sectors are rapidly expanding in all industry sectors, except for the electronic-related manufacturing sector (including civilian electronics/electronic equipment, optical fibers, semiconductor manufacturing equipment, communication equipment, computers, semiconductors/integrated circuits, other electronic components, heavy electrical machinery/transport machinery, robots, and other precision machinery). This trend is particularly pronounced in R&D departments operating as independent industry sectors, highlighting the role of the telecommunications and broadcasting sectors as platform industries. In the electronic-related manufacturing sector, fluctuations in the structure of inputs from the telecommunications and broadcasting sectors suggest these activities within these industries, reflecting the internalization of internalization ICT services within each sector.

Figure 4.12 presents a different trend in the intermediate inputs from the software industry to other industries compared to the telecommunications and broadcasting sectors. Many sectors are experiencing a decrease in input coefficients, which appears to be related to the internalization through intra-firm ICT in software companies and the proliferation of packaged software. However, sectors such as public and non-profit R&D departments and industrial R&D departments exhibit a

Table 4.5 Unit TFP and static unit TFP growth rate per year (%): information (ICT) activity

Unit TFP vs. Static Unit TFP	1995–2000			2000–2005			2005–2010		
Information and treatment activity	U.TFP vs. static U. TFP (%)			U.TFP vs. static U. TFP (%)			U.TFP vs. static U. TFP (%)		
Agriculture/forestry/ fishery	−0.45	−0.53		0.03%	−0.07		0.18	0.23	※
Mining including intra-ICT	0.66	0.78	※	0.20	0.05		−0.67	−0.65	
Food product manufacturing	−0.22	−0.32		−0.25	−0.48		0.50	0.90	※
Textile and wood manufacturing	−0.91	−1.15		−0.68	−1.06		0.36	0.53	※
Paper and pulp manufacturing	0.13	−0.01		−0.10	−0.26		0.46	0.61	※
Chemical prod. manufacturing	−0.72	−0.88		−0.19	−0.38		0.10	0.21	※
Iron and steel product manufacturing	0.29	0.34	※	−0.13	−0.33		0.02	0.11	※
General machinery manufacturing	−0.61	−0.79		−0.31	−0.60		−0.14	−0.10	
Electric machinery products	−0.13	−0.34		−0.17	−0.41		0.95	1.30	※
Fabricated metal products	−0.37	−0.57		0.06	−0.22		0.24	0.36	
Semi-conductor equipment products	−0.81	−1.14		−0.06	−0.32		−0.12	−0.66	
Communication machinery	−0.93	−1.31		−0.74	−0.99		0.53	0.79	※
Electric computer and equipment	−2.45	−5.15		−1.39	−2.21		1.62	2.34	※
Semi-conductor and integrated circuit	−0.52	−1.23		−1.10	−1.20		1.33	1.95	※
Other electric parts	−0.02	−0.22		−0.66	−0.94		2.32	2.90	※
Heavy electric and transport machinery	−0.61	−0.88		−0.09	−0.25		0.93	1.22	※
Robot	−0.43	−0.82		0.07	0.06		0.34	−0.11	
Other precision machinery	−1.19	−1.51		−0.33	−0.54		0.21	0.35	※
Oil and coal products	−2.09	−3.28		−1.62	2–2.46		−0.32	−0.42	
Other manufacturing	−0.54	−0.72		−0.21	−0.38		0.33	0.52	※
Energy industry	−0.28	−0.46		−0.19	−0.39		−0.24	−0.22	
Construction	−0.30	−0.46		−0.19	−0.36		0.09	0.16	※
Transportation	−0.32	−0.46		0.02	−0.12		−0.29	−0.31	
Communication and broadcasting	0.61	0.85	※	−1.16	−1.73		0.05	0.13	※

(continued)

Table 4.5 (continued)

Unit TFP vs. Static Unit TFP	1995–2000			2000–2005			2005–2010		
Information and treatment activity	U.TFP vs. static U. TFP (%)			U.TFP vs. static U. TFP (%)			U.TFP vs. static U. TFP (%)		
Trade	0.49	0.62	※	−0.11	−0.34		−0.04	0.02	
Software industry	−1.31	−1.87		−0.23	−0.42		0.53	0.59	※
Software industry including R&D	−1.21	−1.82		−0.26	−0.41		0.84	1.03	※
Internet service (main activity)	0.00	−0.22		0.00	0.56	※	2.20	0.69	
Internet service including ICT	0.00	0.00		0.00	0.00		1.13	0.44	
Information and treatment	−0.93	−1.18		0.30	0.20		0.38	0.42	※
Information and Treatment (R&DActivity	−1.14	−1.97		−0.35	−0.55		0.94	1.24	※
Medical service	−0.44	−0.38		0.01	−0.15		0.00	0.05	※
Education	0.11	0.10		0.01	−0.01		0.00	0.02	※

Note:Unit TFP ≫ 0 and static unit TFP > unit TFP

steady expansion trend in intermediate input coefficients. This suggests that digitization and electronicization are advancing in various fields of science and technology, leading to the externalization of software and system development to drive R&D.

The expanding trend of intermediate inputs from the ICT service industry to other industries is universally observed across all industries, signifying the increasing platformization of information provision services.

The trends in intermediate goods input from various information industries to other industries clearly underscore the role of the information sector as a platform. They highlight the importance of increased productivity in the information sector and the advancement of the supporting electronics industry, both of which play vital roles in enhancing efficiency across all industries. Collaboration with the information industry is crucial for activating the activities of public research institutions and private research facilities. Furthermore, it indicates the essential policy need for extending the effects of such collaboration through the market.

Figure 4.13 illustrates the societal impact of the government's scientific and technological knowledge development, as a public good, contrasting it with the social impact of ICT as a market good. The government's R&D investment, acting as a public good, is distributed as research funding to public research institutions, including national and public universities, research institutions, and non-profit organizations. These activities generate a knowledge stock as a public good, disseminate knowledge developed by private sector entities such as corporate R&D and for-profit research institutions, and stimulate efficiency improvements in the private sector. However, this knowledge dissemination does not occur through market

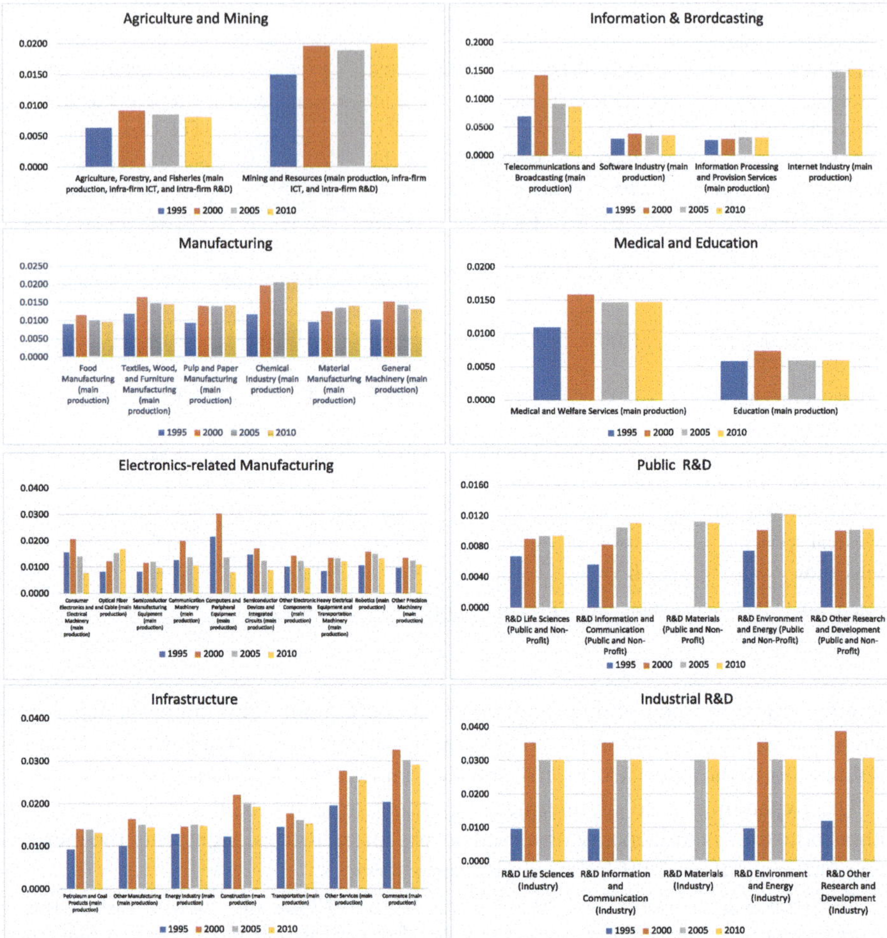

Fig. 4.11 Time-series variation of intermediate input coefficients from the telecommunication Service industry to other industries

goods transactions but rather diffuses among private entities essentially without cost, not explicitly manifesting as transactions of intermediate goods.

Knowledge services, as public goods, exhibit non-rivalrous and non-excludable characteristics, implying they are accessible to all without exclusion, and their consumption by one does not reduce their availability to others. It is crucial to consider how the government's R&D investment, as a form of public good, dynamically accumulates knowledge stock and influences society through R&D deepening.

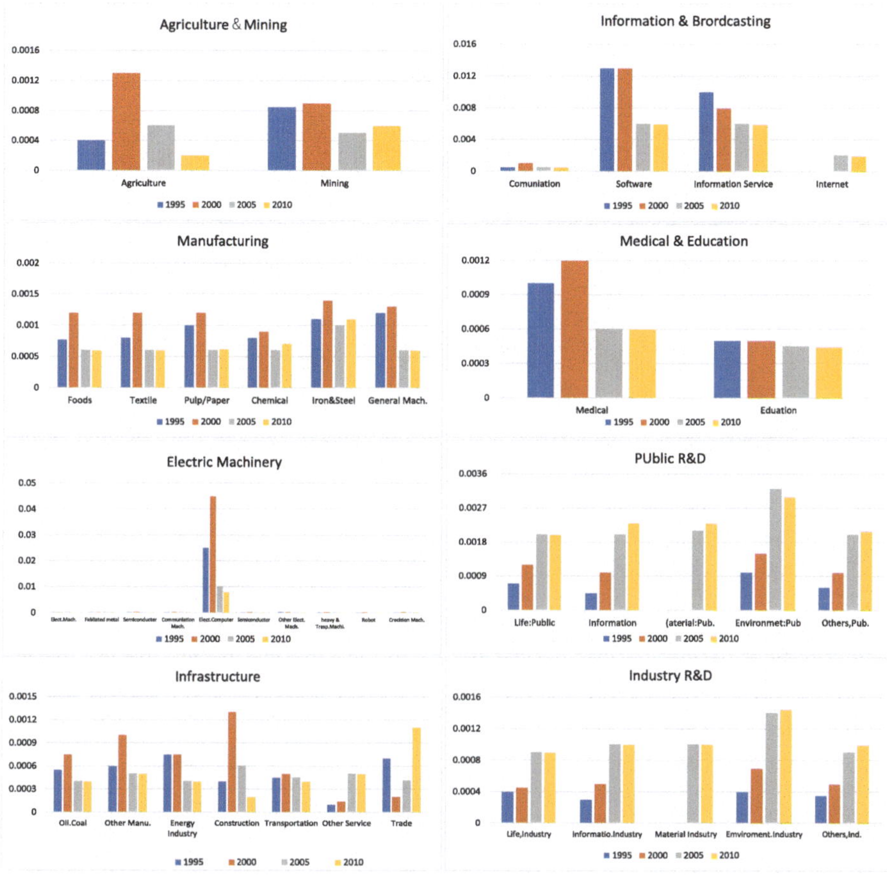

Fig. 4.12 Time-series variation of intermediate input coefficients from the Information and Comunication Service industry to other industries

The deepening of R&D leads to knowledge expansion, which has significant effects, including the accumulation of intellectual services that influence the accumulation of human capital through education.

The concept of dynamic unit TFP, proposed in the previous section, underscores the importance of measuring dynamic TFP, taking into account the technological conditions at the time of fixed asset accumulation when evaluating the productivity of tangible and intangible fixed assets. This method allows for the observation of the effects of both tangible fixed capital accumulation and intangible fixed capital accumulation, such as R&D investment and ICT investment, on production efficiency through the indicator of dynamic unit TFP.

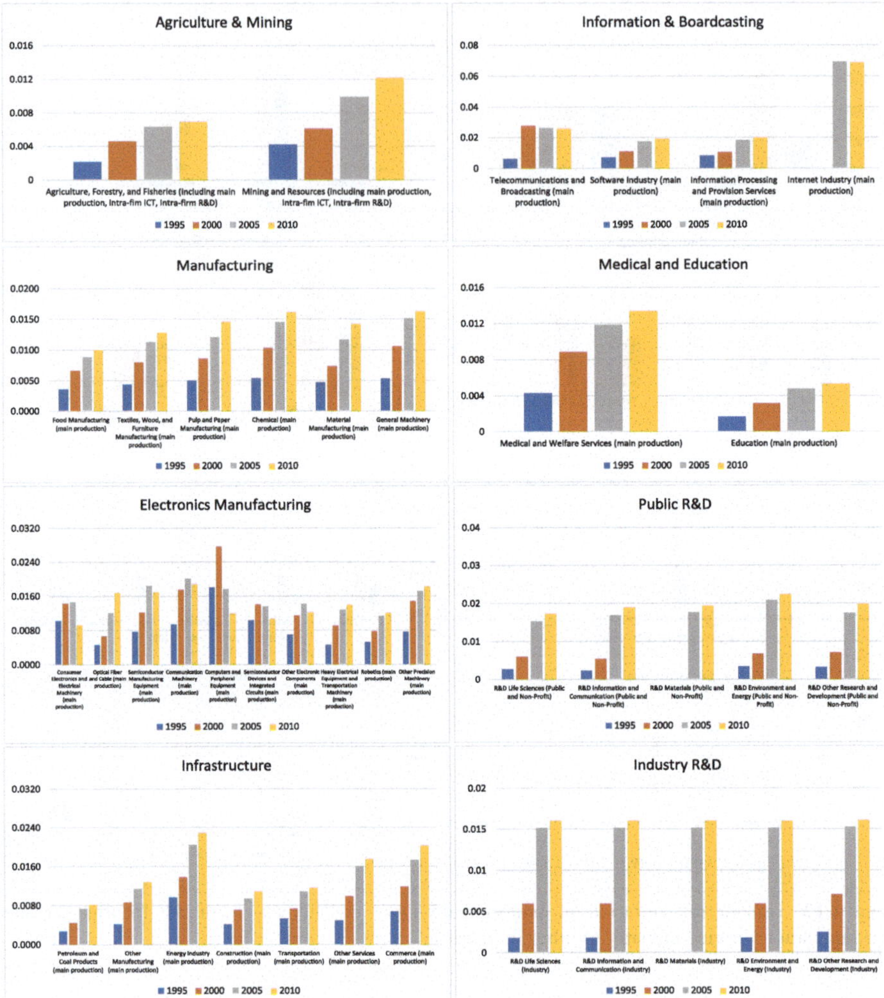

Fig. 4.13 Change in input coefficients of Software Industry to other industries

Figure 4.14 depicts the dynamic unit TFP at the macro level, measured from 1995 to 2011, considering the accumulation of capital stock from tangible and intangible fixed assets. Intangible fixed assets include the accumulation of capital from intra-industry ICT activities, R&D activities, private sector ICT industries, government R&D departments, and independent private sector R&D departments across various industry sectors. At the macro level, the dynamic TFP effects from

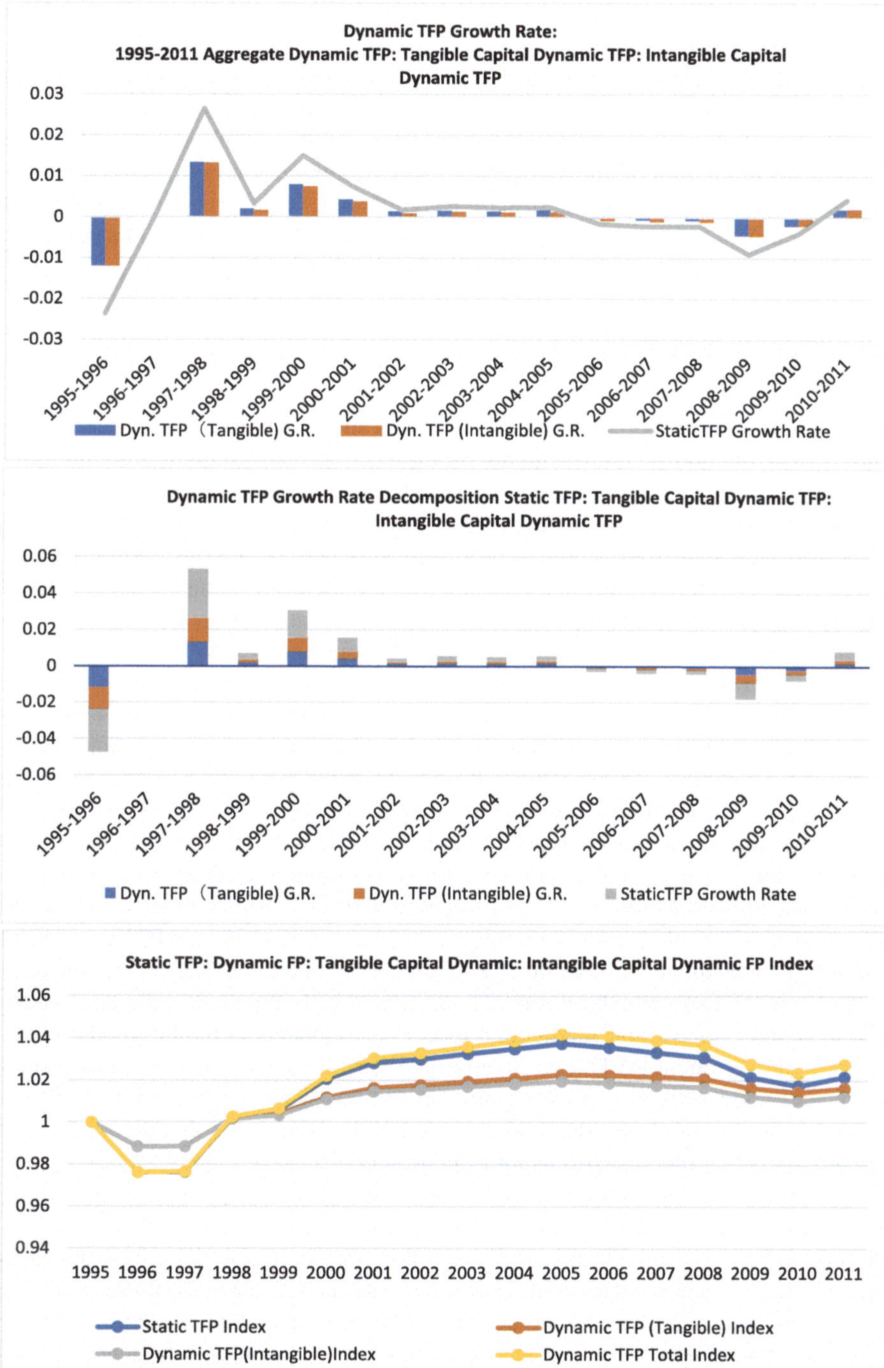

Fig. 4.14 Macro dynamic total factor productivity: effects of tangible and intangible fixed assets

the accumulation of both tangible and intangible capital stocks show growth rates comparable to the static TFP growth rates annually, indicating a significant impact. It is clear that the effect of intangible fixed assets as knowledge stock has a comparable annual rate of impact on TFP as the accumulation of tangible fixed assets.

The impact of government R&D investment as a public good is measured via the dynamic accumulation of knowledge stock, providing indicators to assess its effectiveness. Further analysis on the structural significance of the productivity effects of R&D investment by sectoral deepening is provided in the subsequent chapters.

4.8 Concluding Remark

In this chapter, we propose a framework within input-output tables to capture the impact of R&D and ICT investments on the economy and society, treating these investments as capital. We have restructured Japan's input-output tables from 1995 to 2011 in a time-series format. Using these results, we measure the impact of these investments on industrial productivity during this period by considering them as accumulations of intangible fixed assets. We assess their effects on the economy and society through changes in the TFP of industries. We demonstrate the accumulation of intangible assets such as ICT and R&D and propose methods for measuring the resulting knowledge stock and services. We further suggest indicators to measure how these investments contribute to the expansion of knowledge services and efficiency improvements across various industrial sectors and macroeconomic conditions. We attempt to empirically demonstrate these outcomes in the Japanese economy during the recession period from 1995 to 2011.

Investments, when viewed as accumulations of intangible fixed assets, lead to the accumulation of intellectual assets. The provision of intellectual services from these assets measures their impact on the economic and social spheres. However, a significant disparity in their economic ripple effects is evident between ICT investment and R&D investment, both evaluated as intangible fixed services. This disparity arises from the distinction between ICT services, which impact the economic structure as market goods, and intellectual assets accumulated as public goods, represented by government R&D investments and their intellectual services. The former are market goods, affecting the economy like other typical services, while the latter, as non-exclusive public goods, represent intangible fixed assets that do not directly engage with price mechanisms as knowledge services.

The production of intellectual services through R&D activities includes those with market value, such as patents, and those without market value, such as government R&D investments. These contribute to the formation of social infrastructure as part of society's knowledge stock. The latter intellectual services, through educational activities, contribute to the accumulation of human capital, and the service value of intellectually advanced human capital can contribute to the market economy. Furthermore, the advancement of knowledge services in one field may have intangible effects on the efficiency of R&D in other sectors as a result of knowledge advancement. These effects, as externalities of the market, require a separate analytical framework from the input-output analysis presented here. The development of indicators to measure the qualitative changes in knowledge services provided by knowledge stock is essential.

References

Aulin-Ahmavara, P. (1990). Efficient rates of sectoral productivity change. *Economic Systems Research, 11*(4), 349–363.

Hulten, C. R. (1978). Growth accounting with intermediate inputs. *Review of Economic Studies, 45*, 511–518.

Kuroda, M., & Nomura, K. (2004). Technological change and accumulated capital: A dynamic decomposition of Japan's growth. In E. Dietzenbacher & M. L. Lahr (Eds.), *Wassily Leontied and input-output economics*. Cambridge University Press.

Peterson, W. (1979). A total factor productivity growth in the U.K.: Disaggregated analysis. In W. Patterson & M. Scott (Eds.), *The measurement of capital theory and practice*. London, Macmillan.

Chapter 5
Long-Term Technology Choice Behavior and Cost Function Formulation: Measurement of the Trans-Log Cost Function

Masahiro Kuroda

Abstract This chapter presents an in-depth analysis of long-term technology choice behavior in industrial sectors through the formulation and empirical measurement of the trans-log cost function. It examines the role of government science and technology policies in shaping producers' technological decisions, thus influencing both tangible and intangible capital investments. The chapter formalizes a cost minimization framework, exploring how firms optimize production costs in response to evolving market conditions and policy interventions. The analysis focuses on the effects of government-driven Research and Development investments, allocation of subsidies, and technological progress on industrial cost structures. By estimating trans-log price functions across multiple sectors, the chapter evaluates the relationship between technological advancements, input prices, and production efficiency. The findings offer insights into the dynamics of long-term industrial growth, demonstrating the critical influence of policy on innovation and market transformation in various industries.

Keywords Trans-log cost function · Long-term technology choice · Intra-firm R&D investment · Government-driven Research & Development Activity (R&D) and Information & Communication Treating Activity processing activity (ICT) investment · Evaluation of the dynamics of long-term industrial growth

M. Kuroda (✉)
Professor Emeritus, Keio University, Tokyo, Japan

M. Kuroda (ed.), *Intangible Capital Investment and Economic Growth*, Translational Systems Sciences 45,
https://doi.org/10.1007/978-981-96-7675-0_5

5.1 Long-Term Technical Choice Activity of Producers

The establishment of government-created science and technology policy scenarios, along with the activities to implement these plans, leads the private sector to the appropriate technology choice and contributes to the evolution of the industrial technological structure. The government's science and technology policy significantly affect the long-term technological decisions of producers in related industries. This influence extends to main production activities, intra-firm research and development (R&D) and information & Communication Treating activities (ICT), and investments in tangible and intangible capital and labor inputs. The technological structure of intra-firm the entire industry alters the supply and demand structures of goods and services across all related industrial sectors. This alteration affects the quantity and price structure of the goods and services market, ultimately impacting the long-term demand for goods and services. This chapter seeks to empirically examine the impact of the government's science and technology policy on the market and determine whether this policy drives changes in the market structure (See Haskel & Weslake (2018) and Kuroda (1993)).

In light of the government's science and technology policy, various forms of support, such as subsidies, are provided to the public R&D sector in line with policy scenarios, thereby promoting scientific and technological R&D. Current policy provides for research subsidies, a portion of which ($\alpha\%$) is allocated to tangible capital, such as research facilities and experimental equipment, at the start of the period, forming tangible fixed assets for the subsequent period. Conversely, the output of the R&D sector in the current period, representing the knowledge produced by current R&D, forms the stock of intangible fixed assets for the next period when added to the intellectual assets at the start of the period. The remaining portion $(1-\alpha)$ % is introduced as recurrent subsidies influencing short-term supply behavior across various industrial sectors. The intra-firm allocation of such recurrent subsidies depends on the policy objectives of the government's research subsidies. The total amount of subsidies, whether directed toward expanding investment in tangible capital, such as R&D equipment, or allocated to recurrent subsidies for other current R&D costs, will result in different policy effects, such as R&D support. When $\alpha\%$ is 1.0, it implies that all current period subsidies are allocated to tangible fixed assets in the public R&D sector. However, when $\alpha\%$ is 0.0, it indicates that the subsidies affect short-term supply behavior in the current period, altering the market's demand–supply structure as support for the short-term supply behavior of subsidized industrial sectors in the current period.

Government policies support tangible fixed capital equipment in the public R&D sector. This exogenous support leads to an expansion of tangible fixed capital in the supported public R&D sector. When added to the equipment present at the start of the current period, this support increases the tangible fixed capital at the beginning of the subsequent period, thereby influencing future R&D activities. The production of knowledge services determines the future stock of intangible fixed assets in the public research sector through current period R&D activities. However, the expansion of tangible fixed capital owing to government support in the current period affects output by expanding knowledge services as intangible capital in the next period.

Excluding the government-supported public R&D sector, it is assumed that the investment behavior in various industrial sectors follows long-term cost functions formulated using trans-log cost functions (Jorgenson, 1987, 1995; Jorgenson et al., 1987). Under the assumption of rationality and minimizing long-term production costs while providing forecasts for planned production levels, it is assumed that the optimal amounts of tangible and intangible fixed capital, labor inputs, and intermediate inputs for minimizing costs in the next period are determined. Based on static economic assumptions, it is assumed that the current prices of investment goods persist into the next period. Furthermore, labor demand, derived from long-term cost minimization behavior, and the wage rate for the next period in the labor market correspond to the assumed labor supply for the next period.

This chapter aims to formalize the structure of long-term technological choices and clarify the decision-making structure of capital (both tangible and intangible) and labor derived from cost minimization (Haskel & Westleke, 2018).

5.1.1 Formulation of Long-Term Technological Choices

5.1.1.1 Main Production Activities, Intra-firm ICT Activities, and Intra-firm R&D Activities

(a-1). Technological choices in the main production activities of industrial sectors

In our framework, producers must strategically balance investments across these different activities to optimize long-term productivity and cost efficiency. Integrating ICT and R&D in a firm improves production adaptability and supports ongoing knowledge development and use. Such integration is es-sential for maintaining technological competitiveness and responding effectively to evolving market and policy environments. The industry is assumed to have three interconnected activities structured as a tree, in which the main activity is supported by two intra-firm activities representing the Information and Communication Treatment activity (ICT) and the research and development activity (R&D). It is suggested that actions are undertaken to minimize long-term costs vis-à-vis the industry's technological choices. The long-term cost function is formulated using a trans-log cost function (Christensen et al., 1973; Jorgenson, 1987, 1995; Jorgenson et al, 1987).

$$\begin{aligned}\ln C_j^* &= \alpha_j^{\,0} + \Sigma_k \alpha_j^{\,k} \ln P_j^{k*} + \alpha_j^{\,x} \ln X_j^* + \alpha_j^{\,t} g\left(KNG^t{}_{\theta}, KNRDE^t{}_j\right) \\ &\quad + \left(1/2\right) \Sigma_k \Sigma_l \ln \beta_j^{\,kl} \ln P_j^{k*} \ln P_j^{l*} + \Sigma_k \beta_j^{\,kx} \ln P_j^{k*} \ln X^*{}_j \\ &\quad + \Sigma_k \beta_j^{\,kt} \ln P_j^{k*} g\left(KNG^t{}_{\theta}, KNRDE^t{}_j\right) \\ &\qquad \left(k = SKE, SKNRDE, SKNITE, LE, ME\right)\end{aligned} \tag{5.1}$$

P_j^{k*} denotes the prices of input elements in main production activities, including tangible capital services (*SKE*), intangible capital services related to intra-firm *R&D* (*SKNRDE*), intangible capital services related to intra-firm *ICT* (*SKNITE*), labor services (*LE*), and intermediate input (*ME*). X^*_j represents the planned production quantity of the main product in the next period, and $g\,(KNG_\theta^t, KNRDE_j^t)$ represents the technological transition resulting from the accumulation of knowledge in the government's θ science field and the intra-firm R&D. The prices of tangible and intangible fixed capital services are assumed to remain constant into the next period. Moreover, planned production quantities are considered exogenous.

Under these conditions, Shephard's lemma provides the share functions based on the long-term cost minimization principle.

$$\begin{aligned} v_j^{SKE} = \partial \ln C_j^* / \partial \ln P_j^{SKE*} = \alpha_j^{SKE} + \Sigma_i \beta_j^{SKEi} \ln P_j^i + \beta_j^{SKEx} \ln \\ + \beta_j^{SKET} g\left(KNG^t{}_\theta, KNRDE^t{}_j\right), \end{aligned} \tag{5.2}$$

$$\begin{aligned} v_j^{SKRDE} = \partial \ln C_j^* / \partial \ln P_j^{SKRDE*} = \alpha_j^{SKRDE} + \Sigma_i \beta_j^{SKRDEi} \ln P_j^i + \beta_j^{SKRDEx} \ln X_j^* \\ + \beta_j^{SKRDET} g\left(KNG^t{}_\theta, KNRDE^t{}_j\right), \end{aligned} \tag{5.3}$$

$$\begin{aligned} v_j^{SKITE} = \partial \ln C_j^* / \partial \ln P_j^{SKITE*} = \alpha_j^{SKITE} + \Sigma_i \beta_j^{SKITEi} \ln P_j^i + \beta_j^{SKITEx} \ln X_j^* \\ + \beta_j^{SKITET} g\left(KNG^t{}_\theta, KNRDE_j{}^t\right), \end{aligned} \tag{5.4}$$

$$\begin{aligned} v_j^{LE} = \partial \ln C_j^* / \partial \ln P_j^{LE*} = \alpha_j^{LE} + \Sigma_i \beta_j^{LEi} \ln P_j^i + \beta_j^{LEx} \ln X_j^* \\ + \beta_j^{LET} g\left(KNG^t{}_\theta, KNRDE^t{}_j\right), \end{aligned} \tag{5.5}$$

$$\begin{aligned} v_j^{ME} = \partial \ln C_j^* / \partial \ln P_j^{ME*} = \alpha_j^{ME} + \Sigma_i \beta_j^{MEi} \ln P_j^i + \beta_j^{MEx} \ln X_j^* \\ + \beta_j^{MET} g\left(KNG^t{}_\theta, KNRDE^t{}_j\right), \end{aligned} \tag{5.6}$$

$$\begin{aligned} v_j^{X} = \partial \ln C_j^* / \partial \ln X_j^* = \alpha_j^{x} + \Sigma_i \beta_j^{xi} \ln P_j^i + \beta_j^{xx} \ln X_j^* \\ + \beta_j^{xT} g\left(KNG^t{}_\theta, KNRDE^t{}_j\right) \end{aligned} \tag{5.7}$$

Here, v_j^{SKE}, v_j^{SKRDE}, v_j^{SKITE}, v_j^{LE}, and v_j^{ME} represent the shares of input costs given the changes in each input element, such as tangible capital services, intangible R&D capital services, intangible ICT capital services, labor input services, and intermediate material input. Additionally, v_j^X represents the rate of change in costs, given changes in the assumed demand scale, that is, the rate of technological progress due to changes in the output scale.

As the sum of the input shares for each input element is 1.0, the parameter constraints are given by $\Sigma_i\beta_j^{ix} = 0.0$ and $\Sigma_i\beta_j^{iT} = 0.0$. These equations represent the constraints on the parameters.

Equation (5.1) determines the total long-term cost C_j^*. Asuming the constant return to sale, unit cost of outout is equaled to the output price. From Eqs. (5.2) to (5.7), the input shares of each input element are calculated, allowing for the determination of the optimal input quantities for the next period (t + 1) that minimize long-term costs. Assuming that the capital stock quantity is proportional to the capital service quantity, we can estimate the tangible fixed capital stock (*KE*), intangible R&D asset stock within the firm (*KNRDE*), and intangible *ICT* asset stock within the firm (*KNITE*) in the main production sector. Additionally, the labor input quantity (*LE*) and intermediate input quantity (*ME*) in the main production sector can be determined.

Regarding tangible and intangible fixed assets,

$$KE_j^{t+1*} = \left(1-\delta_j^{KE}\right)KE_j^t + INVKE_j^t \,/\, PINVKE_j^t \tag{5.8}$$

Here, δ_j^{KE} represents the economic depreciation rate, $INVKE_j^t$ is the nominal investment amount, and $PINVKE_j^t$ is the investment goods price for the current period. Given the optimal capital stock for the next period, the beginning capital stock for the current period, and the investment goods price, we can determine the investment amount for the current period. This represents the flow amount of tangible capital formation in the main production activities of sector j for the current period. It is transformed into a vector based on the share of the capital formation matrix of the final demand for each commodity, becoming an element of the final demand for the current period.

Similarly, for intangible assets associated with intra-firm R&D and ICT, we can also determine the investment amount for the current period.

$$KNRDE_j^{t+1*} = \left(1-\delta^{KNRDE}\right)KNRDE_j^t + INVKNRDE_j^t \,/\, PINVKNRDE_j^t. \tag{5.9}$$

$$KNITE_j^{t+1*} = \left(1-\delta_j^{KNITE}\right)KNITE_j^t + INVKNITE_j^t \,/\, PINVKNITE_j^t. \tag{5.10}$$

From Eq. (5.8), we can determine the current investment amount for intra-firm R&D and ICT activities, which are like the calculation for tangible fixed assets, and which is based on the optimal intangible fixed asset stock for the next period, the intangible fixed asset stock at the beginning of the current period, and the price of investment goods. The investment amount for intangible fixed capital becomes the final demand element for the current period, with prices determined by short-term market supply and demand equilibrium.

The supply curves for the industry's leading production activities and intra-firm R&D and ICT activities are derived from the assumption that price equals average cost (P = AC). Equilibrium prices and quantities are derived from the condition whereby the supply and demand for tangible and intangible fixed assets, determined by the principle of long-term cost minimization, coincide in the current market. The investment quantity contributes to accumulating intangible fixed assets for the next period, thereby tracing the economy's dynamic path.

(a-2). In the industrial sector, technological choices regarding intra-firm R&D and ICT activities

(a-2-1). In the case of intra-firm R&D activities

As previously stated, the optimal tangible capital stock for the next period and the optimal intangible fixed capital stock for intra-firm R&D and ICT activities are selected based on long-term cost minimization in the industry's main production sector. Simultaneously, the optimal labor and intermediate inputs for the next period's main production activities are chosen. The current period's investment amount for intangible fixed assets is reflected in the supply and demand of intangible fixed capital goods, which determine the equilibrium quantity and price. However, to ascertain the supply level of intangible fixed assets for subsequent periods, it is necessary to determine the tangible fixed capital for intra-firm R&D and ICT activities for the next period.

The long-term cost function for intra-firm R&D activities is formulated as follows:

$$
\begin{aligned}
\ln C_j^{RD*} = {} & \alpha_j^{RD0} + \Sigma_k \alpha_j^{kRD} \ln P_j^{k*} + \alpha_j^{xRD} \ln X_j^{*} + \alpha_j^{tRD} g\left(KNG^t{}_\theta, KNRDE_j^t\right) \\
& + (1/2)\, \Sigma_k \Sigma_l \ln \beta_j^{kl\,RD} nP_j^{k*} \ln P_j^{l*} + \Sigma_k \beta_j^{kx\,RD} \ln P_j^{k*} \ln X^*{}_j \\
& + \Sigma_k \beta_j^{kt\,RD} \ln P_j^{k*} g\left(KNG^t{}_\theta, KNRDE_j^t\right) \\
& \qquad (k = SKRDE, LRDE, MRDE).
\end{aligned}
\tag{5.11}
$$

Here, P_j^k* represents the prices of tangible fixed capital services, labor input, and intermediate input in intra-firm R&D, if the current prices persist into the next period based on a static economic assumption. X_j signifies the planned production quantity for intra-firm R&D activities in the next fiscal year. However, considering the technological selection behavior for the next period, together with the main production activities, the optimal intangible fixed capital stock quantity $SKNRDE_j^{t+1}$ required for R&D activities in the next period is determined by Eq. (5.9), yielding the determination of the current investment amount (real) $INVKNRDE_j^t$. This investment amount corresponds to the formation of intangible fixed assets in intra-firm R&D for the current period and equals the output quantity X_j of intra-firm research activities in this industry sector, as determined at the equilibrium of supply and demand for the current period. Therefore, the planned production quantity X_j* for intra-firm research activities in the next period will be assumedly linked to the output quantity X_j^t of intra-firm research activities in this industry sector for the current period.

From Eq. (5.11), the share functions are derived from Shephard's lemma, based on the long-term cost minimization condition for intra-firm research activities.

$$v_j^{SKRDE} = \partial \ln C_j^{RD*} / \partial \ln P_j^{SKRDE*} = \alpha_j^{SKRDE} + \Sigma_i \beta_j^{SKRDEi} \ln P_j^i + \beta_j^{SKRDEx} \ln X_j + \beta_j^{SKRDET} g\left(KNG^t{}_\theta, KNRDE^t{}_j\right), \tag{5.12}$$

$$v_j^{LRDE} = \partial \ln C_j^{RD*} / \partial \ln P_j^{LRDE*} = \alpha_j^{LRDE} + \Sigma_i \beta_j^{LRDEi} \ln P_j^i + \beta_j^{LRDEx} \ln X_j^* + \beta_j^{LRDET} g\left(KNG^t{}_\theta, KNRDE^t{}_j\right) \tag{5.13}$$

$$v_j^{MRDE} = \partial \ln C_j^{RD*} / \partial \ln P_j^{MRDE*} = \alpha_j^{MRDE} + \Sigma_i \beta_j^{MRDEi} \ln P_j^i + \beta_j^{MRDEx} \ln X_j^* + \beta_j^{MRDET} g\left(KNG^t{}_\theta, KNRDE^t{}_j\right), \tag{5.14}$$

$$v_j^X = \partial \ln C_j^{RD*} / \partial \ln X_j^* = \alpha_j^x + \Sigma_i \beta_j^{xi} \ln P_j^i + \beta_j^{xx} \ln X_j^* + \beta_j^{xT} g\left(KNG^t{}_\theta, KNRDE^t{}_j\right). \tag{5.15}$$

Here, as the total of input shares for each input element is 1.0, the parameter constraints are given by $\Sigma i \beta j^{ix} = 0.0$ and $\Sigma i \beta_j^{iT} = 0.0$.

These equations serve as constraints on the parameters. Equations (5.11) and (5.12) to (5.15) determine the total long-term cost and input shares for each input element. Assuming that the stock quantities of each input element are proportional to the service quantities, the optimal input quantities for the next period (t + 1) are determined. These quantities pertain to the tangible fixed capital stock in the intra-firm *R&D* sector (*KRDE*), labor input quantity in the intra-firm *R&D* sector (*LRDE*), and intermediate input quantity in the intra-firm R&D sector (*MRDE*). Regarding tangible fixed assets,

$$KRDE_j^{t+1*} = \left(1 - \delta_j^{KRDE}\right) KRDE_j^t + INVKRDE_j^t / PINVKRDE_j^t. \tag{5.16}$$

Here, δ_j^{KRDE} represents the economic depreciation rate, $INVKRDE_j^t$ is the nominal investment amount, and $PINVKRDE_j^t$ is the investment goods price for the current period. Given the optimal capital stock for the next period and the beginning capital stock of the current period, along with the investment goods price, we can calculate the investment amount for the current period. This signifies the flow amount of tangible fixed capital formation in the intra-firm R&D activities of sector j for the current period. It is transformed into a vector based on the share of the capital formation matrix of the final demand for each commodity, thereby becoming an element of the final demand for the current period.

(a-2-2) In the case of intra-firm ICT activities

For intra-firm ICT activities, the process mirrors that of intra-firm R&D activities. The investment amount for the current period is derived from the level of optimal intangible fixed assets for the next period, as determined by the intra-firm ICT activities in the main production activities of the industry. The equilibrium price and quantity for the current period are determined through short-term supply and demand equilibrium. Accordingly, the optimal tangible fixed asset stock ($SKITE_j^{t+1}$), labor input quantity ($LITE_j^{t+1}$), and intermediate input quantity ($MITE_j^{t+1}$) associated with the next period's intra-firm ICT activities are determined. These are conditioned on the assumed planned production quantity (X_j^{IT*}) for intra-firm ICT activities.

$$\begin{aligned}\ln C_j^{IT*} &= \alpha_j^{IT0} + \Sigma_k \alpha_j^{kIT} \ln P_j^{k*} + \alpha_j^{xIT} \ln X_j^{IT*} + \alpha_j^{tIT} g\left(KNG^t_{\theta}, KNRDE_j^t\right) \\ &+ (1/2)\, \Sigma_k \Sigma_l \ln \beta_j^{klIT}{}_j \ln P_j^{k*} \ln P_j^{l*} + \Sigma_k \beta_j^{kxIT} \ln P_j^{k*} \ln X^{IT*}{}_j \\ &+ \Sigma_k \beta_j^{ktIT} \ln P_j^{k*} g\left(KNG^t_{\theta}, KNRDE_j^t\right) \\ &\left(k = SKITE, LITE, MITE\right).\end{aligned} \tag{5.17}$$

However, k is formulated as an aggregate function of current prices for *SKITE* (tangible fixed capital services for intra-firm ICT activities), *LITE* (labor input for intra-firm ICT activities), and *MITE* (aggregated domestic and imported intermediate inputs for intra-firm ICT activities). P_j^{k*} represents the prices for capital services from tangible fixed capital related to intra-firm ICT (P_j^{SKITE}), labor input (P^{LITE}), and intermediate input goods (P_j^{MITE}). X_j^* signifies the planned production quantity for intra-firm ICT activities in the subsequent period, determined by the optimal intangible fixed asset stock ($KNITE_j^{t+1}$) and the current investment amount associated with the intra-firm ICT activities in the next period. This is derived from the selection of technology in the main production activities for the upcoming period. Moreover, KGN^t_{θ} represents the government's intellectual knowledge stock related to information science and technology at the start of period t. It is postulated that the accumulation of past government R&D investments enhances the technological level of ICT activities as a public good, thereby influencing productivity in intra-firm ICT activities. Similarly, $KNRDE_j^t$ is thought to represent the accumulation of knowledge stock in the respective sector at the start of the current period, which affects the current technological level.

Regarding the technological selection for intra-firm ICT activities in the upcoming period, it is assumed that the current market equilibrium determines the prices of capital services and intermediate inputs, which will persist into the next period under static price assumptions. However, the labor service price is influenced by the labor market conditions in the upcoming period, considering the anticipated labor supply and demand. From the perspective of firms' long-term cost minimization behavior in relation to intra-firm ICT activities, the optimal quantity of tangible fixed capital services for the next period is determined. Assuming that the stock quantity of services is proportional to the stock quantity, the investment amount needed to achieve the tangible capital stock quantity for the next period is reflected in the current final demand, thus contributing to the equilibrium of supply and demand in the capital goods market for investment. Moreover, the input of labor services is established concurrently with determining the labor service price for the next period, considering the anticipated labor supply volume and its relationship with the labor market in the upcoming period. Furthermore, concerning the optimal intermediate goods, the real input quantities are determined based on the optimal nominal intermediate goods input share and static price assumptions for the next period's intermediate goods.

From the perspective of Shephard's lemma of cost minimization, the input shares of capital, labor, and intermediate inputs for intra-firm ICT activities are determined.

$$v_j^{SKITE} = \partial \ln C_j^{IT*} / \partial \ln P_j^{SKITE*} = \alpha_j^{SKITE} + \Sigma_i \beta_j^{SKITEi} \ln P_j^i + \beta_j^{SKITEx} \ln X_j^* + \beta_j^{SKITET} g\left(KNG^t_{\theta}, KNITE^t_{\ j}\right), \tag{5.18}$$

$$v_j^{LITE} = \partial \ln C_j^{IT*} / \partial \ln P_j^{LITE*} = \alpha_j^{LITE} + \Sigma_i \beta_j^{LITEi} \ln P_j^i + \beta_j^{LITEx} \ln X_j^* + \beta_j^{LITET} g\left(KNG^t_{\theta}, KNRDE^t_{\ j}\right), \tag{5.19}$$

$$v_j^{MITE} = \partial \ln C_j^{IT*} / \partial \ln P_j^{MITE*} = \alpha_j^{MITE} + \Sigma_i \beta_j^{MITEi} \ln P_j^i + \beta_j^{MITEx} \ln X_j^* + \beta_j^{MITET} g\left(KNG^t_{\theta}, KNRDE^t_{\ j}\right), \tag{5.20}$$

$$v_j^{X} = \partial \ln C_j^{IT*} / \partial \ln X_j^* = \alpha_j^{x} + \Sigma_i \beta_j^{xi} \ln P_j^i + \beta_j^{xx} \ln X_j^* + \beta_j^{xT} g\left(KNG^t_{\theta}, KNRDE^t_{\ j}\right)\left(i = KRDE, LRDE, MRD\right). \tag{5.21}$$

Here, as the total of the input shares for each input element is 1.0, the following parameter constraints are applied:

$$\Sigma_i \beta_j^{ix} = 0.0 \quad \text{and} \quad \Sigma_i \beta_j^{iT} = 0.0$$

The total cost C_j^* is derived from Eq. (5.17). For each input cost share, the input shares of tangible capital, labor, and intermediate goods are obtained from the share functions. These shares represent the cost-minimizing scales of capital, labor, and intermediate raw material inputs for intra-firm ICT activities in the next period. The nominal shares of each input at the reference time are provided.

In certain industry sectors, intra-firm R&D or ICT activities may not be explicitly mentioned. In such instances, if a division exists between the main production activity and R&D or ICT activities, only the main production activity may be referenced, and other intra-firm activities may not be explicitly enumerated. In these situations, undivided intra-firm activities are assumed to be incorporated into the main production activity.

(b). In the Private R&D Institute (*j* = 87–91)

Long-term cost function is derived as:

$$\begin{aligned}\ln C_j^{PI*} &= \alpha_j^0 + \Sigma_k \alpha_j^k \ln P^k{}_j^* + \alpha_j^X{}_j \ln X_j^* + \alpha_j^T g\left(KNG^t{}_\theta, KNPI_j^t\right)\\ &+ (1/2)\,\Sigma_k \Sigma_l \ln \beta_j^{kl} \ln P_j^{k*} \ln P_j^{l*} + \Sigma_k \beta_j^{kX} \ln P_j^{k*} \ln X_j^*\\ &+ \Sigma_k \beta_j^{kT} \ln P_j^{k*} g\left(KNG^t{}_\theta, KNPI_j^t\right)\\ &\left(k = KPI, KNPI, LNPI, MNPI\right).\end{aligned} \tag{5.22}$$

Here, *k* (=*KPI*, *KNPI*, *LNPI*, *MNPI*) corresponds to the tangible fixed asset service input, intangible fixed asset service, labor service, and intermediate input, respectively, and $P^k{}_j^*$ represents the service price of each input element. As mentioned above, regarding prices, it is assumed that the prices of tangible and intangible fixed capital services will continue into the next period based on a static assumption. Conversely, the labor service price is expected to be set by the equilibrium between supply and demand in the forthcoming labor market. KNG_t^θ represents the quantity of knowledge stock provided as a public good, accumulated from the start of the period in the government's θ field R&D stock, and $KNPI_j^t$ represents the quantity of knowledge stock originating from the intangible capital stock because of this industry sector's R&D activities at the start of the period. This serves as a precedent variable in this sector's technology advancement function. In line with the main production activities in the industrial sectors, it is assumed that the input quantities of tangible and intangible capital stock, labor, and intermediate goods are determined based on the share functions. These functions are derived from the long-term cost function's cost minimization conditions, given the anticipated output level.

$$\begin{aligned}v_j^{KPI} &= \partial \ln C_j^{PI*} / \partial \ln P_j^{SKPI*}\\ &= a_j^{KPI*} + \sum_i \beta_j^{ki} \ln P_j^{i*} + \beta_j^{KPIX} \ln X_j^* + \beta_j^{KPIT} g\left(KNGt\theta, KNPI\,j\ \right)\end{aligned} \tag{5.23}$$

$$\begin{aligned}v_j^{KNPI} &= \partial \ln C_j^{PI*} / \partial \ln P_j^{SKNPI*}\\ &= a_j^{KNPI*} + \sum_i \beta_j^{ki} \ln P_j^{i*} + \beta_j^{KNPIX} \ln X_j^* + \beta_j^{KNPIT} g\left(KNGt\theta, KNPIj\ \right)\end{aligned} \tag{5.24}$$

$$\begin{aligned}v_j^{LPI} &= \partial \ln C_j^{PI*} / \partial \ln P_j^{LPI*}\\ &= a_j^{LPI} + \sum_i \beta_j^{Li} \ln P_j^{i*} + \beta_j^{LPIX} \ln X_j^* + \beta_j^{LPIT} g\left(KNGt\theta\right)\end{aligned} \tag{5.25}$$

$$\begin{aligned}v_j^{MPI} &= \partial \ln C_j^{PI*} / \partial \ln P_j^{MPI*}\\ &= a_j^{MPI} + \sum_i \beta_j^{Mi} \ln P_j^{i*} + \beta_j^{MPIX} \ln X_j^* + \beta_j^{MPIT} g\left(KNGt\theta\right)\\ &\left(i = KPI, KNPI, LPI, MPI\right).\end{aligned} \tag{5.26}$$

If the cost function is assumed to be of the Cobb–Douglas type, then the parameters β_j^{kl} ($k = KPI, KNPI, LPI, MPI, l = KPI, KNPI, LPI, MPI$) for the trans-log function are all zero. Given the input shares a_j^{k*} for each input element at the reference time, the next period's input shares for each production factor satisfy the constraint that their sum equals 1.0. Therefore, the parameters and βj^{kT} ($k = KPI, KNPI, LPI, MPI$) are subject to the following constraints:

$$\sum_k \beta_j^{kX} = 0 \text{ and} \sum_k \beta_j^{kT} = 0$$

These constraints ensure that the sum of the distribution rates of the input elements equals 1.0, which is a necessary and sufficient condition. Under these constraints, the input shares $v_j^{k*\ (k = KPI, KMPI, LNPI. MNPI)}$ for each production factor that satisfies the cost minimization conditions for the next period ($t + 1$) can be derived from the share functions. Thus, the optimal input quantities for the next period are determined for *KPI* (tangible fixed assets), *KNPI* (intangible fixed assets), *LPI* (labor input), and *MPI* (intermediate input). Additionally, the total long-term cost $\boldsymbol{C}_j^{PI*}$ can be obtained from Eq. (5.22).

$$KPI_j^{t+1} = v_j^{KPI*} \cdot \left(\mathbf{C}_j^{PI*} / P^{SKPI}{}_j^{*} \right) \tag{5.27}$$

$$KNPI_j^{t+1} = v_j^{KNPI*} \cdot \left(\mathbf{C}_j^{PI*} / P_j^{SKNPI*} \right) \tag{5.28}$$

$$LPI_j^{t+1} = v_j^{LPI*} \cdot \left(C_j^{PI*} / P_j^{LPI*} \right) \tag{5.29}$$

$$MPI_{ij}^{t+1} = v_j^{MPI^*} \cdot \left(P_{j\,i}^{d\,0} / \Sigma P_i^{d0} X_{ij}^0 \right) \cdot \left(P_j^{d^*} X_j^* / P^d{}_i^{*} \right) \tag{5.30}$$

where P_j^{d*} (the assumed domestic price of good j for the next period), P_j^{SKPI*} (the assumed domestic price of tangible fixed capital services for sector j for the next period), P_j^{SKNPI*} (the assumed domestic price of intangible fixed capital services for sector j for the next period), and P_j^{MPI*} (the assumed domestic price of intermediate materials for sector j for the next period) are assumed to remain at the current market prices for the next period, based on the static assumption that current market prices persist. However, P_j^{LPI+1} represents the optimal input quantity of labor services (in person-hours) for the next period. Given the standard labor hours h^*, the optimal number of workers is determined, which is then linked to the labor market block, assuming consistency with the expected labor force population and total demand for workers from each sector.

For intermediate materials, the total intermediate goods input v_j^{MPI*} is allocated among sector j's nominal input shares for goods at the reference time to determine the real input coefficients of sector j's goods at time $t + 1$. Furthermore, regarding capital input, assuming that the next period's optimal tangible and intangible capital stock quantities are proportional to the capital service quantity,

$$KPI_j^{t+l} - \left(1 - \delta_j^{KPI}\right) KPI_j^t = INVKPI_j^t / P_j^{INVKPIt}, \tag{5.31}$$

$$KNPI_j^{t+1} - \left(1 - \delta_j^{KNPI}\right) KNPI_j^t = INVKNPI_j^t \,/\, P_j^{INVKNPI}. \tag{5.32}$$

Thus, the optimal investment quantities, $INVKPI_j^t$ and $INVKNPI_j^t$, required to achieve the next period's optimal capital stock levels can be estimated. These investment amounts are reflected in the elements of the current final demand. δ_j^{KPI} and δ_j^{KNPI} represent the economic depreciation rates of tangible and intangible capital.

(c). Public R&D Institute (*j* = 82–86)

Reflecting the government's science and technology policy, the nominal subsidy for the public R&D sector is determined by exogenously provided policy scenarios. At the start of each period, the government determines the amount of research support based on its science and technology policy. This subsidy is then added to the tangible capital stock of the sector to form the tangible fixed assets for the current period. The output of the R&D sector in the current period comprises the knowledge output from the current period's R&D, which is added to the intangible fixed assets at the start of the period, forming the intangible fixed asset stock for the next period. The public research department has the discretion to either add the subsidy to the tangible capital at the start of the period or allocate the subsidy for cost reduction in the current period's R&D. If the ratio is 1.0, it implies that all current period subsidies are added to the tangible fixed assets at the start of the period, altering the value of tangible fixed assets for the next period. Conversely, a ratio of 0.0 indicates that all current period subsidies are added to intangible fixed assets, thus subsidizing the current period's output. The tangible fixed asset stock of the government's R&D sector for the next period is determined by exogenous policy factors, while the intangible fixed asset stock is determined by the production of knowledge services through R&D activities. The price of investment goods is determined by the current investment goods, weighted by the nominal share of capital goods in the public fixed asset matrix.

The fixed capital of public research institutions, including national, public, and non-profit private institutions, is assumed to be provided exogenously as policy variables based on government policies and allocated for each R&D purpose or sector θ. This is adjusted for thinflation using the current period's investment goods prices, P_j^{INVKGt} and $P_j^{INVKNGt}$, to determine the initial tangible fixed capital stock for the next period. Simultaneously, given the investment goods shares of the current fixed capital matrix for the relevant sector, this contributes to the expansion of private final demand in the fixed capital formation vector of the current period for that sector.

Capital stock for the next *t* period is defined as follows,

$$KG_j^{t+1} = \left(1 - \delta_j^{KG}\right) KG_j^t + INVKG_j^t \,/\, P_j^{INVKGt}, \tag{5.33}$$

$$KNG_j^{t+1} = \left(1 - \delta_j^{KNG}\right) KNG_j^t + INVKNG_j^t \,/\, P_j^{INVNKGt} \tag{5.34}$$

where δ_j^{KG} and δ_j^{KNG} are the economic capital depreciation rates used to calculate capital service prices.

We substitute this into the capital cost

$$\begin{aligned}
BS_j^{t+1} &= \left\{SKG_j^{t+1} P_j^{SKG} + SKNG_j^{t+1} P_j^{SKNG}\right\} \cdot P_j^{SKG0} \\
&= \Big[KG_j^{t+1} \left\{ \left(1-\tau^{K}\right) r^{KG}{}_j P_j^{INVGKt-1} + \delta_j^{KG} P_j^{INVKGt} - \left(P_j^{INVKGt} - P_j^{INVKGt-1}\right) \right. \\
&\quad \left. + \tau^{P} P_j^{INVKGt-1} \right\} + KNG_j^{t+1} \left\{ \left(1-\tau^{KNG}\right) r^{KG}{}_j P_j^{INVKNGt-1} + \delta_j^{KNG} P_j^{INVKNGt} \right. \\
&\quad \left. - \left(P_j^{INVKNGt} - P_j^{INVKNGt-1}\right) + \tau^{PKNG} P_j^{INVKNGt-1} \right\} \Big] \cdot P_j^{SKG0}
\end{aligned} \tag{5.35}$$

Assuming that the current capital return rate $r^{KG}{}_j$ and the investment goods price P_j^{INVKGt} remain constant for the next period, the expected capital cost for the next period can be determined. Here, if the "trans-log" formulation is assumed for the long-term cost function, it becomes as follows:

$$\begin{aligned}
\ln C_j^{GRD*} &= \alpha_j^{GRD0} + \Sigma_k \alpha_j^{kGRD} \ln P_j^{k*} + \alpha_j^{XGRD} \ln X_j^{*} + \alpha_j^{tGRD} g\left(KNG^{t}{}_{\theta}\right) \\
&\quad + (1/2)\Sigma_k \Sigma_l \beta_j^{kl\,GRD} \ln P_j^{k*} \ln P_j^{l*} + \Sigma_k \beta_j^{kX\,GRD} \ln P_j^{k*} \ln X^{*}{}_j \\
&\quad + \Sigma_k \beta_j^{kt\,IT} \ln P_j^{k*} g\left(KNG^{t}{}_{\theta}\right) \\
&\quad \left(k = SKGRD, SKNGRD, LGRD, MDRD\right).
\end{aligned} \tag{5.36}$$

Given the planned production quantity, we can derive the share functions for all input factors from the cost minimizing behavior of the producer.

$$\begin{aligned}
v_j^{KGRD} &= \partial \ln C_j^{GRD*} / \partial \ln P_j^{SKGRD*} = a_j^{KPI*} + \sum_i \beta_j^{ki} \ln P_j^{i*} + \beta_j^{KGRDX} \ln X_j^{*} g\left(KNG^{t}{}_{\theta}\right) \\
v_j^{KGNRD} &= \partial \ln C_j^{GRD*} / \partial \ln P_j^{SKNGRD*} = a_j^{KNGRD*} + \sum_i \beta_j^{ki} \ln P_j^{i*} + \beta_j^{KNGRDX} \ln X_j^{*} + \beta_j^{KNGRDTX} g\left(KNG^{t}{}_{\theta}\right) \\
v_j^{LGRD} &= \partial \ln C_j^{GRD*} / \partial \ln P_j^{LGRD*} a_j^{LGRD} \sum_i \beta_j^{Li} \ln P_j^{i*} + \beta_j^{LGRDX} \ln Xj * + \beta KNGRD\, g\left(KNG^{t}{}_{\theta}\right) \\
v_j^{MGRD} &= \partial \ln C_j^{GRD*} / \partial \ln P_j^{MGRD*} = a_j^{MGRD} + \sum_i \beta_j^{MGRDX} \ln X_j^{*} + \beta_j^{MGRDT} g\left(KNG^{t}{}_{\theta}\right) \\
& i = SKGRD, SKNGRD, LGRD, MGRD
\end{aligned}$$

If the cost function is assumed to be a Douglas-type function, then the parameters β_j^{ki} (k = KGRD, KNGRD, LGRD, MGRD, i = KGRD, KNGRD, LGRD, MGRD) for the trans-log function become zero. Assuming that the share of each input element at the reference time $a_j^{k*}\left(k = KPI, KNPI, LPI, MPI\right)$ is given, the next period's input share for each production factor becomes a function of the planned production quantity and technological progress rate. Parameters β_j^{kX} and $\beta_j^{kT}\left(k = KGRD, KNGRD, LGRD, MGRD\right)$ are subject to the following constraint

$$\sum_k \beta_j^{kX} = 0 \text{ and } \sum_k \beta_j^{kT} = 0,$$

ensuring that the sum of input factor shares equals 1.0, which is a necessary and sufficient condition.

For the government's public R&D activities, the costs of tangible and intangible capital inputs can be directly calculated based on static assumptions about input prices, as per the scenario for the next period's R&D investment. If we denote the total cost of tangible and intangible capital as BS_j^{t+1}, as derived from the share functions, and provide the optimal shares v_j^{KGRD*} and v_j^{KNGRD*}, we can obtain a consistent long-term total cost for capital by satisfying the following equation:

$$v_j^{KGRD*} + v_j^{KNGRD*} = BS_j^{t+1} / C_j^{GRD*}.$$

The next period's planned quantities values for labor and intermediate inputs can be determined by multiplying the cost shares derived from the share function (5.36) by the total long-term cost.

$$LGRD_j^{t+1} = v_j^{LGRD**} \left(C_j^{GRD*} / P^{LGRD*}{}_j \right) \tag{5.37}$$

$$MGRD_j^{t+1} = v_j^{MGRD**} \left(C_j^{GRD*} / P^{LGRD*}{}_j \right) \tag{5.38}$$

The relationship with the labor market and determining intermediate input coefficients align with the procedures previously described for private industry R&D activities.

We formalize the behavior of these industries or companies for future technological choices under the hypothesis of bounded rationality, where firms aim to minimize production costs for the next period. The chosen technology will constrain the supply behavior of producers for the subsequent period and serve as a condition for determining short-term supply behavior. This iterative process of long-term technological choices and short-term supply actions is repeated sequentially, delineating a dynamic growth path.

Moreover, by utilizing the intermediate input costs obtained for each sector from the procedure, we can allocate them based on the share of domestic and imported intermediate inputs. This allocation enables the derivation of the cost of inputs by-product for domestic and imported goods and the nominal intermediate input coefficients for the subsequent period.

5.2 Measurement of Trans-Log Price Functions

5.2.1 Theoretical Constraints of Trans-Log Price Function

We estimate the parameters of the trans-log price function using the nonlinear three-stage least squares method, subject to the constraints of symmetry, additivity, homogeneity of degree one, monotonicity, and concavity (Christensen et al., 1973; Kuroda, 1988, 1993).[1] Generally, the price function is formulated in the following trans-log form:

$$\ln P_j^D = \alpha_j^0 + \Sigma_k \alpha_j^k \ln P_j^{k*} + \alpha_j^T T + (1/2)\Sigma_k \Sigma_l \ln \beta_j^{kl} \ln P_j^{k*} \ln P_j^{l*} + (1/2)\Sigma_k \beta_j^{kT} \ln P_j^{k*} T + (1/2)\beta_j^{TT} T^2 \quad (5.39)$$

The price function presumes homogeneity in technology. Therefore, in the cost function from the previous section, the total cost C_j^* is divided by the output scale X_j^*, resulting in the left-hand side representing the price P_j^D of the output. Additionally, the technological progress rate $g(KNG^t_\theta, KNRDE^t_j)$ is simplistically denoted as trend T. The right-hand side includes the generally assumed k types of input factors, such as the prices of tangible and intangible capital services, labor services, lows:

$$A = \begin{bmatrix} \alpha_j^1 \\ \alpha_j^2 \\ \\ \alpha_j^{k-1} \\ \alpha_j^k \end{bmatrix} \quad B^{kT} = \begin{bmatrix} \beta_j^{lT} \\ \beta_j^{2T} \\ \cdot \\ \beta^{k-1T} \\ \beta_j^{kT} \end{bmatrix}, \quad B^{kl} = \begin{bmatrix} \beta_j^{11} & \beta_j^{12} & \cdot\ \cdot & \beta_j^{1k-1} & \beta_j^{1k} \\ \beta_j^{21} & \beta_j^{22} & \cdot\ \cdot & \beta_j^{2k-1} & \beta_j^{2k} \\ \cdot & \cdot & \cdot\ \cdot & & \\ \beta_j^{k-11} & \beta_j^{k-12} & \cdot\ \cdot & \beta_j^{k-1k-1} & \beta_j^{k-1k} \\ \beta_j^{k1} & \beta_j^{k2} & \cdot\ \cdot & \beta_j^{kk-1} & \beta_j^{kk} \end{bmatrix}$$

In this model, by applying Shephard's lemma, the distribution rate w^{kt} for each production factor ($k = 1,\ldots,k$) and the negative technological progress rate w^{Tt} can be defined. These are derived from the shared functions in the preceding section as follows:

[1] For detailed information on parameter estimation, please refer to Kuroda's (1994) "Quantitative Analysis in General Equilibrium," pages 166–194.

$$\begin{bmatrix} w^{kt} \\ \cdot \\ w^{Tt} \end{bmatrix} = \begin{bmatrix} A \\ \cdot \\ \alpha^{T} \end{bmatrix} + \begin{bmatrix} B^{kk} & B^{kT} \\ & \\ B^{Tk} & \beta_j^{TT} \end{bmatrix} \begin{bmatrix} \ln P^{kt} \\ \\ T \end{bmatrix} \tag{5.40}$$

For this function to have meaning as a homogeneous price function, constraints such as symmetry, homogeneity, monotonicity, and quasi-concavity must be imposed on the parameters.

symmetry

$$\begin{aligned} \beta^{kl} &= \partial w^{kt} / \partial \ln P^{lt} = \partial^2 \ln P_j^{D} / \partial \ln P^{kt} \partial \ln P^{lt} \\ &= \partial^2 \ln P_j^{D} / \partial \ln P^{lt} \ln P^{kt} = \partial w^{lt} / \ln P^{kt} = \beta^{lk}\ , \end{aligned} \tag{5.41}$$

$$\beta^{Tk} = \beta^{kT}, \tag{5.42}$$

adding up and homogeneity,

$$A'i = 1,\ B^{kl} i = 0,\ B^{kT} i = 0. \tag{5.43}$$

monotonicity

$$w^{kt} > 0 \tag{5.44}$$

quasi-concavity

Suppose a homogeneous price function is formulated as a quadratic function of prices. In that case, the necessary and sufficient condition for concavity is that the Hessian matrix, comprising second-order derivatives, has a negative, definite sign. The Hessian matrix of the price function is given by

$$H = \partial^2 P_j^{D} / \partial P^{kt} P^{lt} = P_j^{D} N^{t*} \left[B^{kk} + w^{kt} \cdot w^{kt'} - \hat{W}^{t} \right] N^{t*} \tag{5.45}$$

where

$$N^{t*} = \begin{bmatrix} (1/P^{1t}) & 0 & 0 & 0 \\ 0 & (1/P^{2t}) & 0 & 0 \\ 0 & 0 & (1/P^{3t}) & 0 \\ 0 & 0 & 0 & 1/P^{kt}) \end{bmatrix} \quad \hat{W}^{t} = \begin{bmatrix} w^{1t} & 0 & 0 & 0 \\ 0 & w^{2t} & 0 & 0 \\ 0 & 0 & w^{3t} & 0 \\ 0 & 0 & 0 & w^{kt} \end{bmatrix}$$

In Eq. (5.45), the matrix [$wkt \cdot wkt' - \hat{W}t$], defined by the shares, is negative definite for all periods. Therefore, if parameter matrix *B* is also negative definite, it is a necessary and sufficient condition for the Hessian matrix to be negative definite.

Per the Cholesky factorization theorem, a definite matrix can be factored; indeed, it has been proven that its Cholesky factor is non-positive if it is negative definite, its Cholesky factor is non-positive:

Parameter matrix B can be factorized as follows:

$$B = \begin{bmatrix} 1 & 0 & 0 & 0 \\ \lambda_{12} & 1 & 0 & 0 \\ \lambda_{13} & \lambda_{23} & 1 & 0 \\ \lambda_{14} & \lambda_{24} & \lambda_{34} & 1 \end{bmatrix} \begin{bmatrix} \delta_1 & 0 & 0 & 0 \\ 0 & \delta_2 & 0 & 0 \\ 0 & 0 & \delta_3 & 0 \\ 0 & 0 & 0 & \delta_4 \end{bmatrix} \begin{bmatrix} 1 & \lambda_{12} & \lambda_{13} & \lambda_{1k} \\ 0 & 1 & \lambda_{23} & \lambda_{2k} \\ 0 & 0 & 1 & \lambda_{3k} \\ 0 & 0 & 0 & 1 \end{bmatrix}$$

It can be factored. As the numbers of parameters B_{kl} and Cholesky decomposition factors remain the same, by imposing constraint conditions on B_{kl}, Eq. (5.40) can be rewritten as the following share function:

To estimate the price function parameters, it is necessary to impose constraints (1) to (4) thereon. Once these constraints are applied, Eq. (5.40) can be rewritten to yield a share function, illustrated using four input elements: tangible fixed capital services (*K*), labor services (*L*), energy input (*E*), and intermediate input (*M*). As discussed in the preceding section, the types of input elements differ across various industrial and activity sectors, including main production activities, intra-firm ICT, intra-firm R&D, ICT industries, public and non-profit R&D, and industrial R&D. However, the subsequent development can be similarly applied across these sec

$$\begin{bmatrix} w^K \\ w^E \\ w^{Lt} \\ -w^T \end{bmatrix} = \begin{bmatrix} \alpha^K \\ \alpha^E \\ \alpha^L \\ \alpha^T \end{bmatrix} + \begin{bmatrix} \delta^K & \lambda^{KL}\delta^K & \lambda^{KE}\delta^K & \beta^{KT} \\ \lambda^{KE}\delta^K & \lambda^{KE}\lambda^{KL}\delta^K + \lambda^{LE}\delta^L & \lambda^{KE2}\delta^K + \lambda^{LE2}\delta^L + \delta^E & \beta^{ET} \\ \lambda^{KL}\delta^K & \lambda^{KL2}\delta^K + \delta^L & \lambda^{KE}\lambda^{KL}\delta^K + \lambda^{LE}\delta^L & \beta^{LT} \\ \beta^{KT} & \beta^{LT} & \beta^{ET} & \beta^{TT} \end{bmatrix}$$
$$\times \begin{bmatrix} \ln P^{Kt} - \ln P^{Mt} \\ \ln P^{Lt} - \ln P^{Mt} \\ \ln P^{Et} - \ln P^{Mt} \\ T \end{bmatrix} + \begin{bmatrix} u^{Kt} \\ u^{Lt} \\ u^{Et} \\ u^{Tt} \end{bmatrix} \tag{5.46}$$

Here, u^{it} ($i = K, L, E, T$) represents the disturbance terms introduced to shape the probability model, and it is assumed to follow the probability distribution described below. Regarding u^{it} ($i = K, L, E, M, t = 1, \ldots, n$), we assume the following conditions for its mean and variance: for $i = K, L, E$,

$$E(u_{kt}) = 0, E(u_{kt}^{\ 2}) = \sigma_k^2 \ (t = 1, \ldots, n), \ E(u_{kti} \cdot u_{ktj}) = 0 \quad (i,j = 1 \ldots n, \ i \neq j) \tag{5.47}$$

where negative technological progress $-w^T$ is an unobservable variable, and it is calculated as a discrete Divisa index from the following equation:

$$\begin{aligned} w^T &= -\left(\partial \ln P_j^D / \partial T\right)\left(dT/dt\right) \\ &= \Sigma\left(1/2\right)\left(w^{it} + w^{it-1}\right)\left(\ln P_{it} - \ln P_{it-1}\right) - \left(\ln P_j^{Dt} - \ln P_j^{Dt-1}\right). \end{aligned} \tag{5.48}$$

This is an approximate observed value of the technological progress rate provided during the parameter estimation of the model. It is more convenient to express other shared functions in a discreet way. Instead of (5.46), it is preferable to convert it into the following difference form:

$$\begin{bmatrix} \overline{w}^{Kt} \\ \overline{w}^{Lt} \\ \overline{w}^{E} \\ \overline{w}^{T} \end{bmatrix} = \begin{bmatrix} \alpha^K \\ \alpha^L \\ \alpha^E \\ \alpha^T \end{bmatrix} + \begin{bmatrix} \delta^K & \lambda^{KL}\delta^K & \lambda^{KE}\delta^K & \beta^{KT} \\ \lambda^{KL}\delta^K & \lambda^{KL2}\delta^K + \delta^L & \lambda^{KE}\lambda^{KL}\delta^K + \lambda^{LE}\delta & \beta^{LT} \\ \lambda^{KE}\delta^K & \lambda^{KE}\lambda^{KL}\delta^K + \lambda^{LE}\delta^L & \lambda^{KE2}\delta^K + \lambda^{LE2}\delta^L + \delta & \beta^{ET} \\ \beta^{KT} & \beta^{LT} & \beta^{ET} & \beta^{TT} \end{bmatrix} \times \begin{bmatrix} \ln\ P^{Kt} \\ \ln\ P^{Lt} \\ \ln\ P^{Et} \\ T^t \end{bmatrix} + \begin{bmatrix} \overline{u}^{Kt} \\ \overline{u}^{Lt} \\ \overline{u}^{Et} \\ \overline{u}^{Tt} \end{bmatrix} \tag{5.49}$$

where the variables are defined as

$$\begin{aligned} \overline{w}^{it} &= \left(1/2\right)\left(w^{it} + w^{it-1}\right)\left(i = K, L, E, T\right) \\ \overline{u}^{Kt} &= \left(1/2\right)\left(u^{it} + u^{it-1}\right)\left(i = K, L, E, T\right) \\ \ln P^{it} &= \left(1/2\right)\left(\ln P^{it} - \ln P^{it-1}\right) - \left(1/2\right)\left(\ln P^{Mt} - \ln P^{Mt-1}\right) \\ T &= \left(1/2\right)\left(T^t - T^{t-1}\right) \end{aligned}$$

In this development, because the disturbance term is newly defined as the difference between two points in time as (1/2) (u_{kt} + u_{kt-1}) (where k = K,L,E,T), under the assumption of the previously defined disturbance term, it inevitably possesses auto-correlation. Thus, it becomes necessary to correct for this during estimation.

Under the assumption of the previously mentioned disturbance term, when expanding the variance-covariance matrix of the disturbance terms, even if the disturbance terms observed at two points in time are independently distributed, first-order autocorrelation will arise.

Then,

$$u^{it} = \left(1/2\right)\left(u^{it} + u^{it-1}\right) \text{ and } u^{iT-1} = \left(1/2\right)\left(u^{it-1} + u^{it-2}\right).$$

Thus, a correlation exists between u^{it} and u^{it-1}.

Currently, the variance and covariance of u^{it} in the self-correlation are

$$Var\left(u^{it}\right)=E\left(u^{it2}\right)=(1/4)\left(2\sigma^{i2}\right)=(1/2)\sigma^{i2}$$
$$Var\left(u^{it}u^{it+1}\right)=E\left(u^{it}u^{it+1}\right)=(1/4)\sigma^{i2}\quad Var\left(u^{it}u^{it+j}\right)=0.$$

Thus, the covariance matrix described in Eq. (5.49) exhibits autocorrelation between periods t and $t+1$. In Eq. (5.49), the variance-covariance matrix of the disturbance term u^{it} for each KLEMT share function can be represented by a Laurent matrix as follows:

$$VAR\begin{bmatrix} u^{i2} \\ u^{i3} \\ u^{i4} \\ \cdot \\ u^{in} \end{bmatrix}=\sigma^{i2}\cdot\begin{bmatrix} 1/2 & 1/4 & 0 & \cdots & 0 \\ 1/4 & 1/2 & 1/4 & \cdots & 0 \\ 0 & 1/4 & 1/2 & 1/4 & 0 \\ \cdot\cdot & \cdot & \cdot & \cdot & \\ \cdot\cdot & \cdot & \cdot & 1/4 & 1/2 \end{bmatrix}=\sigma^{i2}\Omega$$

The variance-covariance matrix Ω is a positive definite matrix; Ω can be factorized using Cholesky decomposition, where Ω is diagonalized into an orthogonal matrix D and a lower triangular matrix L.

$$L'\Omega L=D \quad \text{or} \quad \Omega^{-1}=LDL'.$$

Here, D is a diagonal matrix with positive diagonal elements λ^{i}, eigenvalues of Ω. Let $\Lambda^{(1/2)}$ denote the diagonal matrix with diagonal elements expressed as $\lambda_i^{(1/2)}$, and let $\Lambda^{(-1/2)}$ denote its inverse. Thus, we can define the transformation matrix F as

$$F=\Lambda^{(-1/2)}L'$$

Using this transformation matrix, we can redefine each variable in Eq. (5.49)

$$Y^{i}=F\begin{bmatrix} w^{K2} \\ w^{K3} \\ w^{K4} \\ \\ \cdot \\ W^{Kn} \end{bmatrix},\ X=F\begin{bmatrix} 1 & \ln P^{k2} & \cdots\cdot & T^{2} \\ 1 & \ln P^{k3} & \cdots\cdot & T^{3} \\ 1 & \ln p^{k4} & \cdots\cdot & T^{4} \\ \cdot & \cdot & \cdots\cdot & \\ \\ 1 & \ln p^{kn} & \cdots\cdot & T^{n} \end{bmatrix},\ e^{k}=F\cdot\begin{bmatrix} u^{k2} \\ u^{k3} \\ u^{k4} \\ \cdot \\ \\ u^{kn} \end{bmatrix} \qquad (5.50)$$

Therefore, autocorrelation no longer exists.

Thus, the estimation involves parameter estimation for each variable transformed by F. Ultimately, λ^{ij} ($i, j = K, L, E, T$) and δ^{i} ($i = K, L, E$) are estimated using the nonlinear three-stage least squares method.

$$Y^i = XB^i + e^i \left(i = K, L, E, T\right) \tag{5.51}$$

The three distribution rate functions and one technological progress rate function are treated as a simultaneous system of equations. Using exogenous variables as instrumental variables, the parameters are estimated using a nonlinear three-stage least squares method.

The matrix B of parameters is related as follows;

$$B^K = \begin{bmatrix} \delta^K \\ \lambda^{KL}\delta^K \\ \lambda^{KE}\delta^K \\ \beta^{KT} \end{bmatrix} \quad B^L = \begin{bmatrix} \lambda^{KL}\delta^K \\ \lambda^{KL2}\delta^K + \delta^L \\ \lambda^{KE}\lambda^{KL}\delta^K + \lambda^{LE}\delta^L \\ \beta^{LT} \end{bmatrix} \quad B^E = \begin{bmatrix} \lambda^{KE}\delta^K \\ \lambda^{KE}\lambda^{KL}\delta^K + \lambda^{LE}\delta^L \\ \lambda^{KE2}\delta^K + \lambda^{LE2}\delta^L + \delta^E \\ \beta^{ET} \end{bmatrix}$$

$$B^T = \begin{bmatrix} \beta^{KT} \\ \beta^L \\ \beta^{ET} \\ \beta^{TT} \end{bmatrix} \tag{5.52}$$

The estimated parameters δ^i (i = K,L,E,T) and λij (i,j = K,L,E), including βij ($i,j = K,L,E,T$), are determined. Furthermore, given the homogeneity constraint, $\beta_M j$ (j = K,L,E,M,T) can also be determined. Additionally, αi (i = K,L,E,M,T) can be estimated using the definition formulas for the baseline distribution rates and technological progress rates, assuming a baseline price of 1.0.

In the nonlinear estimation, it is necessary to ensure that δi ($i = K,L,E,M,T$) meets non-positivity condition given the quasi-concavity constraint. If this condition is not met during the estimation stage, one of the δi parameters may be linearly set to zero to satisfy the quasi-concavity condition and recompute the estimation. By sequentially setting δi to zero and repeating the nonlinear estimation, parameters that satisfy the quasi-concavity condition can be found. However, if δi becomes zero for all input factors, the price function will be reduced to a linearly homogeneous Cobb-Douglas type function.

5.3 Measurement Results of Trans-Log Price Function

We utilize data from input-output tables (1995–2011) to measure trans-log price functions across various industry sectors and activity types. The input-output tables are aggregated into 93 sectors based on sector classification. These sectors are broadly divided into main production, intra-firm ICT, and intra-firm R&D activities. In certain sectors where intra-firm ICT or R&D activities are not separately listed, we combine them with the main production activities, considering them as a single sector. There are also sectors in which the activity classification is unspecified, and both main production and intra-firm ICT or R&D activities are not differentiated, hence treated as a single sector. The number of input elements in the industry's price function varies based on the classification of activities within the industry, and we categorize them accordingly.

Type 1: There are 23 industries, each divided into main production, intra-firm ICT, and intra-firm R&D activities. These industries include food manufacturing, textile manufacturing, pulp and paper manufacturing, chemical and materials manufacturing, general machinery, consumer electronics and electrical machinery, optical fiber and cable, semiconductor manufacturing equipment, communications machinery, electronic computers and related equipment, semiconductor devices and integrated circuits, other electronic components, heavy electrical machinery and transportation machinery, robotics, other precision machinery, petroleum and coal products, other manufacturing, energy, construction, transportation, communication and broadcasting, and commerce. This results in a total of 69 activities (23 × 3).

Type 2: There are six industries in which the main production and intra-firm ICT activities are aggregated, and the activity classification is divided into main production and intra-firm R&D. These industries include agriculture, forestry, and fisheries; mining and resources, software; ICT and provision services; Internet; and other services. This results in a total of 12 activities (6 × 2).

Type 3: There are industries where main production, intra-firm ICT, and intra-firm research activities are all aggregated into one activity classification. These include the healthcare and welfare services industry, education industry, public and non-profit R&D sectors (R&D life sciences, R&D information and communication, R&D materials and substances, R&D environment and energy, R&D others), and industrial R&D sectors (R&D life sciences, R&D information and communication, R&D materials and substances, R&D environment and energy, R&D others), totaling 12 activity sectors. This classification results that whole industry is divided into 93 sectors. Type I: Industry Classification I: Sector Divided into Three Activities: Main Production, Information and Communication Treating (ICT), and Research and Development (R&D) Activities.

5.3.1 Estimated Results

Type 1: Case of the industry which classified into Main, R&D, and ICT Activities

(a) **Estimated Results for Price Function of Main Activity**

The price function in Eq. (5.39) is presented by the trans-log function as follows:

$$\ln P_j^D = \alpha_j^0 + \Sigma_k \alpha_j^k \ln Pkj^* + \alpha_j^T T + (1/2)\Sigma_k \Sigma_l \ln \beta_j^{kl} \ln P_j^{k*} \ln P_j^{l*} + (1/2)\Sigma_k \beta_j^{kT} \ln P_j^k K^* T + (1/2)\beta_j^{TT} T^2 \tag{5.53}$$

where the input factors are separated into the following six inputs: information processing intangible capital service (**K_i**), intra-firm R&D intangible capital service (**K_r**), tangible capital service (**K**), labor input service (**L**), intermediate material input (**M**), and technical progress trend (**T**).

The parameters $\boldsymbol{\delta}^i,\boldsymbol{\lambda}^{ij}$ $(i,j = K_r,K_i,K,L,M)$ and β^{ij} $(i,j = K_r,K_i,K,L,T)$ for the Cholesky decomposition corresponding to Eqs. (5.51) and (5.52) with six input factors, including the technological progress trend (**T**), can be determined. Furthermore, β^{Mj} $(i,j = K,L,E,M,T)$ can be derived from the constraint of homogeneity of degree one.

Moreover, for $\boldsymbol{\alpha}_i$ $(i = K_r, K_i, K, L, M, T)$, the distribution rates at the base period and the definitions of technological progress rates allow the estimation of base period prices as 1.0. In nonlinear estimation, the Cholesky decomposition factors δ_i $(i = K, L, E, T)$ must be non-positive, a necessary and sufficient condition derived from the concavity condition.

If the conditions are not met at the estimation stage, an attempt is made to satisfy the condition of suspected concavity by a priori setting one or more of the $\boldsymbol{\delta}^i$ to zero and recalculating. By sequentially setting $\boldsymbol{\delta}^i$ to zero and repeating the nonlinear estimation, parameters that satisfy the condition of suspected concavity are sought. If $\boldsymbol{\delta}^i$ becomes zero for all input elements, the price function is reduced to a first-degree homogeneous Cobb-Douglas type function.

The correspondence between the Cholesky decomposition factors and the trans-log price function parameters is presented in Tables 5.1 and 5.2.

The estimation of price functions for the 23 main production sectors can be summarized as follows: the price function parameters are estimated under the theoretical constraints of symmetry, additivity, monotonicity, homogeneity, and quasi-convexity. There are 16 potential combinations of Cholesky factors (δ), including unconstrained cases, as depicted in the second column of Table 5.3. If all Cholesky factors are zero, the function simplifies to a Cobb-Douglas type. The results of the nonlinear three-stage least squares estimation are summarized in the third column. When multiple constraint conditions are satisfied, selecting a more significant functional form is determined by the fit of the share functions and significance testing of the parameters.

Table 5.1 Correspondence Table between Cholesky decomposition factors and the trans-log price function parameters

		1	2	3	4
		Kr	*Ki*	*K*	*L*
1	*Kr*	*1*	*R12 = λkrki*	*R13 = λkrk*	*R12 = λkrl*
2	*Ki*		*1*	*R23 = λklk*	*R24 = λlil*
3	*K*			*1*	*R34 = λkl*
4	*L*				*1*
		1	2	3	4
		Kr	*ki*	*K*	*L*
1	*Kr*	*D1 = δkr*			
2	*Ki*		*D2 = δki*		
3	*K*			*D3 = δk*	
4	*L*				*D1 = δl*

Table 5.2 The relationship between trans-log parameter and Cholesky decomposition factors

		1	2	3	4	5
		Kr	***Ki***	***K***	***L***	***M***
1	***Kr***	***D1 = βkrkr***	***R12*D1 = βkrki***	***R13*D1 = βkrk***	***R14*D1 = βkrl***	
2	***Ki***	***R12*D1 = βklkr***	***R12**2*D1 + D2 = βkiki***	***R13*R12*D1 + R23*D2 = βkik***	***R14*R12*D1 + R24*D2 = βkil***	
3	***K***	***R13*D1 = βkkr***	***R13*R12*D1 + R23*D2 = βkki***		***R14*R13*D1 + R24*R23*D2 + R23*D3 = βkl***	
4	***L***	***R14*D1 = βlkr***	***R14*R12*D1 + R24*D2 = βlki***		***R14**2*D1 + R24**2*D2 + R34**2*D3 + D4 = βll***	
	M	***Bmkr = 1 − Σβikr (i = kr, ki, k l)***	***Bmki = 1 − Σβiki (i = kr, ki, k l)***	***Bmk = 1 − Σβik (i = kr, ki, k l)***	***Bml = 1 − Σβil (i = kr, ki, k l)***	***Bmm = 1 − Σβmi (i = kr, ki, k l)***

Table 5.3 Constraints of the parameter of the price function in main production activity: Case in which the activities are divided into main, intra-firm ICT, and intra-firm R&D activities

	(0) No constraint	–
The main production is divided into three sectoral activities	(1) $\delta 1 = 0$	–
	(2) $\delta 2 = 0$	–
	(3) $\delta 3 = 0$	
	(4) $\delta 4 = 0$	
	(5) $\delta 1 = \delta 2 = 0$	–
	(6) $\delta 1 = \delta 3 = 0$	
	(7) $\delta 1 = \delta 4 = 0$	038: Semiconductor devices (Main products); 41: Electronic component (Main products)
	(8) $\delta 1 = 0$	
	(9) $\delta 2 = \delta 3 = 0$	Communication devices (Main products)
	(10) $\delta 2 = \delta 4 = 0$	
	(11) $\delta 1 = \delta 2 = \delta 3 = 0$	5: Food (Main products); 8: Textile, wooden product (Main products); 23: Electronic devices (Main products); 26: Fiber optic cable (Main products); 35: Computing equipment (Main products); 50: Precise machinery (Main products)
	(12) $\delta 1 = \delta 2 = \delta 4 = 0$	20: Machinery (Main products); 29: Semiconductor manufacturing (Main products)
	(13) $\delta 1 = \delta 3 = \delta 4 = 0$	
	(14) $\delta 2 = \delta 3 = \delta 4 = 0$	
	(15) $\delta 1 = \delta 2 = \delta 3 = \delta 4 = 0$	Pulp, paper (Main products); 14: Chemical (Main products); 17: Synthetic (Main products); 44: Heavy machinery, transportation equipment (Main products); 47: Robot (Main products); 53: Petroleum, coal (Main products); 56: Miscellaneous manufacturing (Main products); 59: Energy manufacturing (Main products); 62: Construction (Main products); 65: Transportation (Main products); 68: Communication (Main products); 71: Commerce (Main products)

Table 5.3 shows that in estimating price functions for main production activities in the 23 industrial sectors, divided into main production, intra-firm ICT, and intra-firm R&D activities, no industries fit the trans-log model without constraints. For three industries, one Cholesky factor had to be set to zero; for eight industries, two factors had to be constrained to zero; for nine industries, three factors had to be constrained to zero; and the remaining three industries followed the Cobb-Douglas function.

Tables 5.4 and 5.5 summarize the measurement results for the main production activities of the 23 industries classified into three activities. The input elements are divided into five input services: tangible fixed services (***K***), intellectual capital

Table 5.4 Estimated results of the trans-log price function

		Main activity which is divided into three activities					
		αkr	*αki*	*αK*	*αL*	*αM*	*αT*
05	Food products	5.38E−03	5.44E−04	0.11596	0.126157	7.52E−01	3.08E−03
	t-value	7.52481	6.77515	17.1615	6.88831		0.512398
08	Textile and wood manufacturing	7.56E−03	4.70E−04	0.08891	0.218758	6.84E−01	2.96E−03
	t-value	7.92093	6.13697	19.3115	10.2593		1.33779
11	Pulp and paper manufacturing	2.89E−03	5.20E−04	0.14293	0.185882	6.68E−01	−3.23E−03
	t-value	4.4803	4.90147	20.2255	18.6693		−0.617994
14	Chemical manufacturing	0.062847	1.47E−03	0.15837	0.147258	0.630053	3.83E−03
	t-value	9.47915	10.9697	17.5438	15.9781		0.749065
17	Iron and steel manufacturing	9.79E−03	1.18E−03	0.12413	0.225015	6.40E−01	−9.97E−04
	t-value	7.56853	6.44463	20.8715	18.4786		−0.30275
20	General machinery manufacturing	0.014471	3.11E−03	0.12567	0.231852	6.25E−01	3.66E−03
	t-value	0.031204	6.70E−03	0.13974	0.290331		7.86E−03
23	Electric machinery manufacturing	0.017731	3.03E−03	0.14185	0.13958	6.98E−01	−2.20E−03
	t-value	4.75174	3.62441	13.0781	4.72203		−0.283938
26	Optic-fiber cable manufacturing	0.09673	1.53E−03	0.14635	0.193295	5.62E−01	3.94E−03
	t-value	7.61002	7.20094	11.3388	17.1132		0.421878
29	Semiconductor and equipment products	0.078908	2.44E−03	0.12823	0.150663	6.40E−01	4.70E−02
	t-value	0.170177	5.27E−03	0.241	0.194207		0.100972
32	Communication machine	0.042145	4.00E−02	0.04712	0.044179	8.27E−01	4.56E−03
	t-value	5.83896	5.7806	4.65508	1.85705		0.348064

(continued)

Table 5.4 (continued)

		Main activity which is divided into three activities					
		αkr	*αki*	*αK*	*αL*	*αM*	*αT*
35	Electric computer manufacturing	0.053398	5.06E−03	0.08213	0.006438	8.53E−01	1.81E−02
	t-value	8.92858	4.27435	13.7576	0.37182		1.59948
38	Semiconductor and integrated circuit	0.211606	1.95E−03	0.10823	0.104618	5.74E−01	2.62E−02
	t-value	11.5318	1.57103	7.1725	2.448		1.21506
41	Other electric machinery	0.081075	3.46E−02	0.13683	0.147554	6.00E−01	−6.43E−03
	t-value	6.89328	5.21425	12.6825	2.7097		−0.428102
44	Heavy electric and transport machinery	0.010624	7.54E−04	0.11047	0.177542	7.01E−01	1.19E−03
	t-value	4.82675	8.22818	16.5389	16.4312		0.311706
47	Robot	0.059063	2.44E−03	0.16307	0.195236	5.80E−01	2.87E−02
	t-value	6.5885	8.09196	15.3079	21.8191		1.04345
50	Other precision machinery	0.040822	4.08E−03	0.12187	0.113511	7.20E−01	1.04E−02
	t-value	7.64804	7.64057	15.5114	3.95184		2.2223
53	Coal and oil products	0.010223	0.000339	0.09604	0.063613	0.82979	−0.007722
	t-value	3.97151	2.40334	9.22768	12.0496		−0.768108
56	Other manufacturing	7.30E−03	6.78E−04	0.11985	0.237943	6.34E−01	3.13E−03
	t-value	6.89333	4.5276	17.7777	18.3745		0.841604
59	Energy industry	5.68E−03	3.20E−04	0.40121	0.135045	4.58E−01	1.03E−03
	t-value	8.46391	3.6104	21.0243	16.0688		0.130173
62	Construction	4.15E−03	1.94 E−04	0.05811	0.360041	5.78E−01	8.32E−04
	t-value	8.84421	0.927821	16.8585	17.0858		0.239408
65	Transportation	4.66E−04	8.22E−04	0.13726	0.409265	4.52E−01	−2.72E−03
	t-value	6.49322	9.00956	20.6626	16.8216		−0.574791
68	Communication and broadcasting	−1.60E−06	−3.12E−03	0.30196	0.322881	3.78E−01	2.93E−02
	t-value	−7.21E−03	−1.72375	13.7707	16.9577		0.770616
71	Trade	−4.55E−05	1.76E−03	0.14092	0.512255	3.45E−01	4.42E−03
	t-value	−0.1206	4.28625	19.1392	17.1093		0.684382

services from R&D as intangible fixed assets ($\boldsymbol{K_r}$), ICT capital services as intangible fixed assets ($\boldsymbol{K_i}$), labor services ($\boldsymbol{L}$), intermediate goods services ($\boldsymbol{M}$), and technological trend ($\boldsymbol{T}$). The formulation of the trans-log price function is represented by Eq. (5.48), where the parameters are represented by $\boldsymbol{\alpha^i}$ ($\boldsymbol{i = K_r, K_i, K, L, M, T}$) and $\boldsymbol{\beta^{ij}}$ ($\boldsymbol{i, j = K_r, K_i, K, L, M, T}$). Parameter $\boldsymbol{\alpha i}$ ($\boldsymbol{i = K_r, K_i, K, L, M}$) represents the distribution ratios of each input element at the base annual time when the price of each input

Table 5.5 Estimated results of the trans-log price function

		Main activity which is divided into three activities					
		$\beta KrKr$	$\beta KrKi$	βKrK	βKrL	βKrM	βKrT
05	Food products	0	0	0	0	0	0.00028
	t-value						4.11851
08	Textile and wood manufacturing	0	0	0	0	0	−0.00013
	t-value						−1.43397
11	Pulp and paper manufacturing	0	0	0	0	0	0.00038
	t-value						6.18125
14	Chemical manufacturing	0	0	0	0	0	0.00058
	t-value						−0.41228
17	Iron and steel manufacturing	0	0	0	0	0	0.00035
	t-value						2.83452
20	General machinery manufacturing	0	0	0	0	0	0.00061
	t-value						0.01390
23	Electric machinery manufacturing	0	0	0	0	0	0.00267
	t-value						7.55866
26	Optic-fiber cable manufacturing	0	0	0	0	0	−0.00106
	t-value						−0.88367
29	Semiconductor and equipment products	0	0	0	0	0	0.00044
	t-value						0.00997
32	Communication machinery	−0.891	0.918142	0.072063	−0.0901	−0.0088826	0.00450
	t-value	−4.194					7.18148
35	Electric computer manufacturing.	0	0	0	0		−0.00010
	t-value						−0.17503
38	Semiconductor and integrated circuit	−0.069	0.0032	0.0680	0.0365	−3.87E−02	−0.01088
	t-value	−3.051					−6.17151
41	Other electric machinery	−0.138	0.168789	0.022257	−0.0406	−1.28E−02	0.00173
	t-value	−5.848					1.57372

(continued)

Table 5.5 (continued)

		Main activity which is divided into three activities					
		βKrKr	*βKrKi*	*βKrK*	*βKrL*	*βKrM*	*βKrT*
44	Heavy electric and transport machinery	0	0	0	0	0	0.00101
	t-value						4.82543
47	Robot	0	0	0	0	0	0.00165
	t-value						1.94201
50	Other precision machinery	0	0	0	0	0	0.00280
	t-value						5.54298
53	Coal and oil products	0	0	0	0	0	−0.00025604
	t-value						−1.04928
56	Other manufacturing	0	0	0	0	0	0.00051
	t-value						5.07108
59	Energy industry	0	0	0	0	0	0.00009
	t-value						1.33948
62	Construction	0	0	0	0	0	−0.00018
	t-value						−4.02807
65	Transportation	0	0	0	0	0	0.00003
	t-value						5.12474
68	Communication and broadcasting	0	0	0	0	0	0.00029
	t-value						13.71240
71	Trade	0	0	0	0	0	0.00020
	t-value						5.48114

element is set to $\boldsymbol{P}_j^k = 1.0$. Furthermore, parameter $\boldsymbol{\beta}^{ij}$ $(\boldsymbol{i}, \boldsymbol{j} = \boldsymbol{K}_r, \boldsymbol{K}_i, \boldsymbol{K}, \boldsymbol{L}, \boldsymbol{M})$ represents the substitutability or complementarity relationship between input elements. When $\boldsymbol{\beta}^{ij} > 0$, it indicates that production factors $\boldsymbol{i}$ and $\boldsymbol{j}$ are substitutable or competitive, whereas when $\boldsymbol{\beta}^{ij} < 0$, it suggests a complementary relationship between the elements.

Additionally, when $\boldsymbol{\beta}^{ij} = 0$, it implies a neutral relationship, indicating independence from other elements. Furthermore, parameter $\boldsymbol{\beta}^{iT}$ $(\boldsymbol{i} = \boldsymbol{K}_r, \boldsymbol{K}_i, \boldsymbol{K}, \boldsymbol{L}, \boldsymbol{M})$ represents the bias of technical change, where $\boldsymbol{\beta}^{iT} > 0$ indicates a biased technical change toward the use of production factor $\boldsymbol{i}$, and $\boldsymbol{\beta}^{iT} < 0$ indicates a biased technical change toward saving. Moreover, because of the symmetry of parameters, $\boldsymbol{\beta}^{iT} > 0$ implies $\boldsymbol{\beta}^{Ti} > 0$, indicating that an increase in the price of production factor $\boldsymbol{i}$ negatively affects technical progress. Conversely, $\boldsymbol{\beta}^{iT} < 0$ implies that an increase in the price of production factor $\boldsymbol{i}$ positively affects technical progress. Finally, $\boldsymbol{\beta}^{TT}$ represents the acceleration of the technical progress rate, indicating whether the rate of technical progress is increasing or decreasing.

Consider the estimation results in Tables 5.4 and 5.5, which present an estimation of the price function for main production activities. These activities are categorized into three types: main production, intra-firm R&D, and intra-firm ICT. The parameters were estimated in accordance with the theoretical constraints of the price function, including symmetry, additivity, monotonicity, homogeneity of degree one, and quasi-concavity, as outlined in Table 5.1.

Parameters $\boldsymbol{\alpha}^i$ ($\boldsymbol{i} = \boldsymbol{K_r}, \boldsymbol{K_i}, \boldsymbol{K}, \boldsymbol{L}, \boldsymbol{M}$) represent the distribution rates of each input element at the base time, with their estimated values reflecting these rates. Notably, $\boldsymbol{\alpha}^{Kr}$ denotes the distribution rate of R&D knowledge capital services, considered intangible fixed assets. This input is significant in IT-related industries such as chemical, optical fiber and cable, semiconductor equipment, communications machinery, semiconductor integrated circuits, other electronic components, robotics, and other precision machinery. Conversely, $\boldsymbol{\alpha}^{Ki}$ represents the distribution rate of ICT-related knowledge capital services, also considered intangible fixed assets. This input is significant in industries such as general machinery, consumer electronics, communications equipment, electronic equipment, precision machinery, and robotics. Furthermore, $\boldsymbol{\alpha}^K$ signifies the input of tangible fixed capital services, which is highly weighted in industries such as chemical, material manufacturing, general machinery, electrical machinery, energy, transportation, and communication/broadcasting; $\boldsymbol{\alpha}^L$ corresponds to the distribution rate of labor service inputs and is significant in labor-intensive industries such as the textile, general machinery, other manufacturing, construction, transportation, communication/broadcasting, and commerce. Finally, $\boldsymbol{\alpha}^T$ is an adjustment coefficient at the base time (Table 5.6).

Parameters $\boldsymbol{\beta}^{ij}$ ($\boldsymbol{i,j} = \boldsymbol{K_r}, \boldsymbol{K_i}, \boldsymbol{K}, \boldsymbol{L}, \boldsymbol{M}$) act as indicators of substitutability and complementarity between each pair of input elements. Among them, $\boldsymbol{\beta}^{ii}$ ($\boldsymbol{i} = \boldsymbol{K_r}, \boldsymbol{K_i}, \boldsymbol{K}, \boldsymbol{L}, \boldsymbol{M}$) is expected to take a negative value, as changes in the own-factor price of input elements yield a decrease in their input, suggesting elasticity to be negative or zero. Table 5.5 presents parameters $\boldsymbol{\beta}^{ii}$ ($\boldsymbol{i} = \boldsymbol{K_r}, \boldsymbol{K_i}, \boldsymbol{K}, \boldsymbol{L}, \boldsymbol{M}$) for main production activities. The relationship between intra-firm R&D knowledge capital services, intra-firm ICT capital services, and tangible fixed capital services is indicated by $\boldsymbol{\beta}^{ij}$ ($\boldsymbol{i,j} = \boldsymbol{K_r}, \boldsymbol{K_i}$) and $\boldsymbol{\beta}^{ij}$ ($\boldsymbol{i,j} = \boldsymbol{Kr}, \boldsymbol{K}$). In industries such as communications machinery, semiconductor devices/integrated circuits, and other electronic components, a substitutive relationship is observed, while a neutral relationship is found in other sectors. Additionally, the relationship between intra-firm R&D knowledge capital services and labor service inputs, $\boldsymbol{\beta}^{ij}$ ($\boldsymbol{i,j} = \boldsymbol{Kr}, \boldsymbol{L}$), is complementary in industries such as communications machinery and other electronic components. A similar complementary relationship is observed between intra-firm R&D knowledge capital services and intermediate inputs in industries such as communications machinery, semiconductor devices/integrated circuits, and other electronic components. In other industries, the relationship of $\boldsymbol{\beta}^{ij}$ ($\boldsymbol{i} = \boldsymbol{K_r}, \boldsymbol{j} = \boldsymbol{K_i}, \boldsymbol{K}, \boldsymbol{L}, \boldsymbol{M}$) is neutral.

Relationship $\boldsymbol{\beta}^{ij}$ ($\boldsymbol{i} = \boldsymbol{K_i}, \boldsymbol{j} = \boldsymbol{K}, \boldsymbol{L}, \boldsymbol{M}$) between intra-firm information processing capital services and other input factors exhibits different dynamics across industries. For telecommunications machinery, semiconductor devices, and integrated circuits, as well as other electronic components, industries, this relationship is complementary with tangible capital services. However, it remains neutral in other industries. A complementary relationship is observed in industries such as communications machinery, semiconductor devices, and integrated circuits when

Table 5.6 Estimated results of the trans-log price function

		Main activity which is divided into three activities														
		βKiKi	*βKiK*	*βKiL*	*βKiM*	*βKiT*	*βKK*	*βKL*	*βKM*	*βKT*	*βLL*	*βLM*	*βLT*	*βMM*	*βMT*	*βTT*
05	Food products	0	0	0	0	4.98E−06	0	0	0	0.00389	−0.45223	0.45223	0.00326	−0.45223	−0.01000	−0.00025
	t-value					6.54E−01				6.07959			2.35404			0.51240
08	Textile and wood manufacturing	0	0	0	0	−1.54E−05	0	0	0	−0.00033	−0.31924	0.31924	0.00552	−0.31924	0.01000	−0.00027
	t-value					−2.12E+00				−0.77705			3.20858			−1.31257
11	Pulp and paper manufacturing	0	0	0	0	2.18E−05	0	0	0	0.00051	0	0	0.00068	0	0.00000	0.00023
	t-value					2.17E+00				0.76530			0.71712			0.46487
14	Chemical manufacturing	0	0	0	0	−3.00E−05	0	0	0	−0.00035	0	0	−0.00140	0	0.00000	−0.00025
	t-value					−4.12E−01				−4.12E-01			−1.60375			−0.51139
17	Iron and steel manufacturing	0	0	0	0	7.03E−05	0	0	0	0	0	0	−0.00225	0	0.00000	0.00008
	t-value					4.04E+00				−0.7736			−1.94726			0.24718
20	General machinery manufacturing	0	0	0	0	2.83E−04	−0.00003	0.00143	−0.00135	0.00021	−0.06501	0.06363	0.00209	−0.06228	0.00000	−0.00032
	t-value					6.45E−03				0.00300			0.03238			−0.00726
23	Electric machinery manufacturing	0	0	0	0	1.25E−04	0	0	0	−0.00186	−0.16874	0.16874	0.00203	−0.16874	0.01000	0.00055
	t-value					1.57271				−1.8079			0.80999			0.756529

(continued)

Table 5.6 (continued)

		Main activity which is divided into three activities														
		βKiKi	βKiK	βKiL	βKiM	βKiT	βKK	βKL	βKM	βKT	βLL	βLM	βLT	βMM	βMT	βTT
26	Optic-fiber cable manufacturing	0	0	0	0	1.25E−04	0	0	0	−0.00443	−0.03929	0.03929	−0.00289	−0.03929	0.01000	−0.00063
	t-value					6.17E+00				−3.62497			−2.61719			−0.71803
29	Semiconductor and equipment products	0	0	0	0	2.39E−04	−0.00057	−0.01389	0.01446	0.00005	−0.33540	0.34928	0.00702	−0.36375	−0.01000	−0.00485
	t-value					5.44E−03				0.00005			0.10843			−0.11007
32	Communication machinery	−0.946	−0.07424	0.09284	0.00915	−1.98E−03	−0.03736	−0.08567	0.12521	0.00428	−0.28312	0.36607	0.00942	−0.49155	−0.01000	−0.00017
	t-value					−3.35E+00				4.47662			4.83231			−0.14000
35	Electric computer manufacturing	0	0	0	0	3.80E−04	0	0	0	0.00181	−0.24200	0.24200	0.01117	−0.242	0.00000	−0.00165
	t-value					3.39E+00				3.20005			7.78710			−1.53220
38	Semiconductor and integrated circuit	−1E−04	−0.00311	-0.00167	0.00177	1.02E−04	−0.06697	−0.03598	0.03806	0.00609	−0.30859	0.30971	0.00847	−0.31089	0.00000	−0.00251
	t-value					7.30E−01				3.61696			2.45017			−1.22814
41	Other electric machinery	−0.207	−0.02729	0.04978	0.01574	−2.00E−03	−0.00359	0.00656	0.00207	−0.00333	−0.25613	0.24037	0.00182	−0.24535	0.00000	0.00032
	t-value					−3.47E+00				−3.26815			0.38977			0.22822
44	Heavy electric and transport machinery	0	0	0	0	7.64E−05	0	0	0	−0.00140	0	0	−0.00080	0	0.00000	−0.00012
	t-value					8.79E+00				−2.21977			−0.78041			−0.34891

47	Robot	0	0	0	0	1.03E−04	0	0	0	−0.00326	0	0	0.00044	0	0.00000	−0.00280
	t-value					3.62E+00				−3.23286			0.52365			−1.07638
50	Other precision machinery	0	0	0	0	7.80E−04	0	0	0	−0.00063	−0.46289	0.46289	0.01164	−0.46289	0.00000	−0.00078
	t-value					1.54E+01				−0.85311			4.78517			−1.76583
53	Coal and oil products	0	0	0	0	1.91E−05	0	0	0	−0.0041	0	0	−0.0023	0	0.01	0.000518
	t-value					1.43E+00				−4.15362			−4.60435			0.54058
56	Other manufacturing	0	0	0	0	1.30E−40	0	0	0	0.00004	0	0	0.00263	0	0.00000	−0.00022
	t-value					9.18E+00				0.05623			2.14461			−0.62688
59	Energy industry	0	0	0	0	5.19E−05	0	0	0	−0.00710	0	0	−0.00043	0	0.01000	0.00014
	t-value					6.17E+00				−3.9301			−0.54203			0.19049
62	Construction	0	0	0	0	1.78E−04	0	0	0	0.00086	0	0	0.00460	0	−0.01000	−0.00027
	t-value					8.95E+00				2.63467			2.30467			−0.81575
65	Transportation	0	0	0	0	2.81E−05	0	0	0	0.00235	0	0	0.00299	0	−0.01000	0.00025
	t-value					3.25E+00				3.74058			1.29506			0.56669
68	Communication and broadcasting	0	0	0	0	2.71E−03	0	0	0	0.000274	0	0	−0.00076	0	0.00000	−0.00227
	t-value					1.58E+01				1.31749			−0.42209			−0.63110
71	Trade	0	0	0	0	5.89E−04	0	0	0	0.00570	0	0	0.00138	0	0.01000	−0.00032
	t-value					1.51E+01				8.16476			0.48635			−0.52377

considering tangible capital services. However, a neutral relationship is found in other sectors. In terms of labor service inputs and intermediate inputs, a complementary relationship is observed only in the semiconductor devices and integrated circuit industries. Conversely, either substitutive or neutral relationships are found in other industries. The relationship between tangible fixed capital services and labor service inputs also are observed the differences by industries. Substitutive relationships are typically found in the general machinery and other electronic components industries. Complementary relationships are observed in the semiconductor devices, communications machinery, and semiconductor devices/integrated circuits industries. Other industries typically exhibit neutral relationships. Lastly, concerning labor services and intermediate inputs, substitutive relationships are prevalent in sectors such as food manufacturing, textile/wood furniture manufacturing, general machinery, civilian electronic/electrical machinery, optical fiber/cables, semiconductor devices, communications machinery, electronic computers, semiconductor devices/integrated circuits, other electronic components, and other precision machinery. However, neutral relationships are found in other sectors.

Table 5.7 shows the estimated results of parameters $\boldsymbol{\beta}^{iT}(\boldsymbol{i} = \boldsymbol{Kr}, \boldsymbol{Ki}, \boldsymbol{K}, \boldsymbol{L}, \boldsymbol{M})$ and $\boldsymbol{\beta}^{TT}$, illustrating the characteristics of technological progress in main production activities. The former indicates whether technological progress has an aggregative nature in the use of each input element, with $\boldsymbol{\beta}^{iT} > 0$ (or if the increase in the price of that factor negatively affects technological progress), or whether it has a factor-saving nature with $\boldsymbol{\beta}^{iT} < 0$ (or if the increase in the price of that factor positively affects technological progress). Additionally, $\boldsymbol{\beta}^{TT}$ indicates the acceleration of technological progress, showing whether the rate of technological progress accelerates or decelerates over time.

Consider the effects of incorporating intra-firm R&D knowledge services ($\boldsymbol{K_r}$) as intangible fixed assets into the main production activities of various industries. It is observed that, with the advancement of technology, the utilization of intra-firm R&D elements becomes more prevalent in the main production activities of industries excluding textiles, optical fibers, electronic computers, semiconductor devices, integrated circuits, petroleum and coal, and construction ($\boldsymbol{\beta}^{KrT} > 0$). This suggests a bias toward higher distribution rates of R&D knowledge services due to technological progress, particularly in consumer electronics, electrical machinery, communications machinery, and other precision machinery. However, an increase in the cost of intra-firm R&D knowledge services is found to have a negative impact on technological progress in main production activities, leading to increased cost burden and decreased production efficiency. Conversely, in industries such as textiles, optical fibers, electronic computers, semiconductor devices, integrated circuits, petroleum and coal, and construction, where $\boldsymbol{\beta}^{KrT} < 0$, an increase in the price of intra-firm R&D tends to result in conservation of the input, thereby reducing the input distribution rate. The price increase of these elements stimulates technological progress, thereby enhancing the production efficiency of main production activities.

Additionally, in industries such as textiles, chemicals, communications machinery, and other electronic components, an increase in the price of intra-firm ICT

Table 5.7 Estimated parameters of trans-log price function: main activity by industry (technology bias and rate of technological progress)

		βKrT	*βKiT*	*βKT*	*βLT*	*βMT*	*βTT*
05	Food products	0.00028	4.98E−06	0.00389	0.00326	−0.00743	−0.00025
08	Textile manufacturing	−0.00013	−1.54E−05	−0.00034	0.00552	−0.00504	−0.00027
11	Paper and pulp manufacturing	0.00038	2.18E−05	0.00051	0.00068	−0.00159	0.00023
14	Chemical manufacturing	0.00058	−3.00E−05	−0.00035	−0.00140	0.00120	−0.00025
17	Iron and steel manufacturing	0.00035	7.03E−05	−0.00044	−0.00225	0.00226	0.00008
20	General machinery	0.00061	2.83E−04	0.00021	0.00209	−0.00320	−0.00032
23	Electric machinery	0.00267	1.25E−04	−0.00186	0.00203	−0.00297	0.00055
26	Fabricated metal	−0.00106	1.25E−04	−0.00443	−0.00289	0.00826	−0.00063
29	Semiconductor equipment	0.00044	2.39E−04	0.00005	0.00702	−0.00775	−0.00485
32	Communication machine	0.00450	−1.98E−03	0.00428	0.00942	−0.01622	−0.00017
35	Electric computer	−0.00010	3.80E−04	0.00181	0.01117	−0.01326	−0.00165
38	Semiconductor and integrated circuit	−0.01088	1.02E−04	0.00609	0.00847	−0.00379	−0.00251
41	Other electric parts	0.00173	−2.00E−03	−0.00334	0.00182	0.00179	0.00032
44	Heavy electric and transport machinery	0.00101	7.64E−05	−0.00140	−0.00080	0.00112	−0.00012
47	Robots	0.00165	1.03E−04	−0.00326	0.00044	0.00107	−0.00280
50	Other precision machinery	0.00280	7.80E−04	−0.00064	0.01164	−0.01459	−0.00078
53	Oil and coal products	−0.00026	1.91E−05	−0.00409	−0.00230	0.00663	0.00052
56	Other manufacturing	0.00051	1.30E−04	0.00004	0.00263	−0.00331	−0.00022
59	Energy industry	0.00009	5.19E−05	−0.00710	−0.00043	0.00740	0.00014
62	Construction	−0.00018	1.78E−04	0.00086	0.00460	−0.00546	−0.00027
65	Transportation	0.00003	2.81E−05	0.00235	0.00299	−0.00540	0.00025
68	Communication	0.00029	2.71E−03	0.00274	−0.00076	−0.00498	−0.00227
71	Trade	0.00020	5.89E−04	0.00570	0.00138	−0.00786	−0.00032

capital services contributes to economizing on this element and positively affects the rate of technological progress. Conversely, in other sectors, a bias toward using intra-firm ICT services is observed, negatively impacting technological progress by constraining it. In the above industries, an increase in the price of elements stimulates technological progress and contributes to the efficiency of main production activities. However, their influence is less than the impacts by the R&D, as indicated by comparing their parameter magnitudes.

The bias of technological progress toward the use of tangible fixed assets β^{KT} is further examined. It is observed that capital-intensive industries, such as textiles, chemical, material manufacturing, consumer electronics and electrical machinery, optical fibers, other electronic components, heavy electrical machinery, robotics, other precision machinery, petroleum and coal, and energy, tend to exhibit a saving bias. An increase in the price of tangible fixed asset services in these industries stimulates technological progress. Additionally, these capital-intensive industries show a high tendency for labor-saving technological progress bias regarding labor services. Many industries exhibit a saving bias in technological progress in intermediate inputs. These industries are incentivized to enhance the rate of technological progress when labor service prices and intermediate input prices increase alongside the price increase of tangible fixed asset services. These characteristics indicate the direction of factor savings and changes in the rate of technological progress based on fluctuations in each factor's price. However, it is crucial to note that actual outcomes may vary depending on the relationship with the fluctuations in factor prices.

In the subsequent section, the selection of technologies based on actual fluctuations in factor prices is examined. Focusing on the impacts by the indicator represented by β^{TT}, it is evident that the acceleration of technological progress in main production activities is negative in many industries. This observation aligns with the general economic situation, as the observation period used for parameter estimation spanned that of Japan's transition to low growth from 1995 to 2011. During this period, total factor productivity growth stagnated in many industrial sectors, consistent with the results of the price function parameter measurement. Finally, we will examine the measurement results of the price function concerning main production activities in industrial sectors, which are divided into two activities.

Table 5.8 delineates the results of the trans-log price function measurements for industries where distinguishing between main production activities and intra-firm ICT and R&D activities is challenging. Industries such as agriculture, forestry, fisheries, mining, and other services encounter difficulties differentiating between main production and intra-firm ICT activities because of statistical limitations, hence their integration. However, these industries are categorized as where intra-firm R&D activities are identifiable. For industries such as software, ICT and provision services, and Internet services, main production activities are inherently ICT activities, complicating the distinction between main production and intra-firm ICT activities. Nevertheless, intra-firm R&D activities are differentiated. The healthcare, welfare services, and education sectors are perceived as industries where neither intra-firm ICT nor intra-firm R&D activities can be differentiated from main production activities.

In the estimation results, all industries, excluding software and the Internet, exhibit Cobb-Douglas functions. In the software industry, the relationship between R&D knowledge capital services and other production factors is neutral, while tangible fixed capital and labor are complementary. The relationship between intermediate input and labor is substitutive. Conversely, in the Internet industry, R&D knowledge capital services, and tangible fixed capital services are neutral with respect to other input factors, while the relationship between labor services and intermediate input is substitutive.

Table 5.8 Estimated results of the trans-log price function: main activity in which cannot divided into three activities

		αkr	αK	αL	αM	αT	βKrKr	βKrK	βKrL	βKrM	βKrT	βKK	βKL	βKM	βKT	βLL	βLM	βLT	βMM	βMT	βTT
01	Agriculture (Main products, ICT activity)	0.0004	0.2233	0.2936	0.4801	0.0033	0	0	0	0	−7.0E−06	0	0	0	0.0014	0	0	0.0013	0	−0.0027	−0.0005
	t-value	10.4570	21.8651	13.9167		0.5700					−2.0E+00				1.4036			0.6502			−0.8293
03	Mining (Main products, ICT activity)	0.0055	0.2424	0.2379	0.5112	0.0075	0	0	0	0	3.8E−05	0	0	0	−0.0036	0	0	0.0025	0	0.0011	−0.0005
	t-value	8.5801	14.3317	11.1563		0.4517					6.3E−01				−2.2632			1.2537			−0.3223
74	Software (Main products)	0.0023	0.0673	0.1885	0.7423	0.0396	0	0	0	0	3.4E−03	−0.0909	−0.1734	0.2643	0.0110	−0.3307	0.5041	0.0129	−0.7684	−0.0273	−0.0032
	t-value	0.8486	−3.9433	2.2895		1.4179					2.3E+00				4.7792			1.8464			−1.1915
76	Information Management (Main products)	−0.0004	0.0633	0.6530	0.2881	0.0019	0	0	0	0	2.4E−03	0	0	0	0.0040	0	0	−0.0044	0	−0.0019	−0.0001
	t-value	−0.1949	8.1671	12.6025		0.1281					1.3E+01				5.4223			−0.9046			−0.0774
78	Internet (Main products)	−0.0107	0.1080	−0.1479	1.0551	0.0384	0	0	0	0	1.7E−03	0	0	0	0.0053	−2.3512	2.3512	0.0338	−2.3512	−0.0408	−0.0047
	t-value	−3.5274	2.9787	−4.1600		0.7415					8.6E+00				2.1991			12.3288			−1.3583
80	Medical, welfare services (Main products)	0.0000	0.0760	0.4386	0.4800	−0.0008	0	0	0	0	1.4E−10	0	0	0	0.0006	0	0	0.0100	0	−0.0106	0.0001
	t-value	0.0000	12.6958	18.6027		−0.1186					4.0E−05				12.6958			4.4662			0.2074
82	R&D Life science (Public, non-profit)	0.0000	0.0947	0.6808	0.2200	0.0007	0	0	0	0	1.3E−09	0	0	0	0.0039	0	0	0.0081	0	−0.0120	−0.0001
	t-value	−0.0004	19.3016	19.2515		0.2788					4.8E−04				8.3538			2.4287			−0.3356
94	Miscellaneous service (Main products)	0.0007	0.3083	0.3014	0.3800	0.0009	0	0	0	0	8.7E−05	0	0	0	0.0054	0	0	0.0026	0	−0.0081	0.0000
	t-value	2.7631	18.9928	17.2231		0.1556					3.7E+00				3.5140			1.5693			−0.0258

Regarding the technological bias of technological progress, all industries exhibit such biases, except for the agricultural, forestry, and fisheries industries, where internal R&D knowledge capital has a saving bias. In these industrial sectors, a bias toward using factors is observed, wherein an increase in factor prices leads to more intensive use of these factors and enhances their distribution rates. This suggests a tendency for factor price increases to dampen technological progress. Lastly, concerning the acceleration indicator of technological progress rates, except for healthcare and other services, where both negative and positive values are shown, the values are minimal. This indicates that the rate of increase in technological progress during this period has been minimal.

Parameter $\boldsymbol{\alpha}_L$ typically assumes negative or minuscule positive values, except in the energy, construction, transportation, telecommunications and broadcasting, and commerce industries. This trait is indicative of an internal ICT department. Industries with negative or small positive values are deemed to have statistically insignificant t-values, approaching zero. In terms of intra-factor substitution, it alternates between tangible fixed capital and labor in the optical fiber and telecommunications equipment industry, complements in the petroleum and coal industry, and remains neutral primarily for tangible fixed capital and intermediate input. However, it alternates in the optical fiber, petroleum, and coal industries and complements in the telecommunications equipment industry.

Technological progress bias often exhibits usage-based bias for tangible fixed capital and labor and intermediate input. The expansion of intra-firm ICT activities within companies suggests that tangible capital, labor, and intermediate goods display a usage-based nature. This implies that increasing the price of ICT capital services, coupled with technological innovation that enhances production efficiency, could potentially inhibit production efficiency as other factors also become usage-based. Labor and intermediate inputs demonstrate complementary relationships in all industries, excluding energy, construction, transportation, telecommunications, broadcasting, and commerce. Moreover, many industries exhibit negative values for accelerating technological progress rates, indicating a diminishing rate of increase in production efficiency owing to technological progress.

Table 5.10 encapsulates the estimated results of the trans-log price function parameters for intra-firm R&D activities across industries categorized into main production, intra-firm R&D, and intra-firm ICT activities.

The input factors for intra-firm R&D encompass tangible capital services, labor services, and intermediate inputs. Intercepts $\boldsymbol{\alpha}_i$ ($\boldsymbol{i} = \boldsymbol{K}, \boldsymbol{L}, \boldsymbol{M}$) are statistically significant in most industry sectors, particularly labor services, signifying a consistent input share, a feature that distinguishes them from the intra-firm ICT activities outlined in Table 5.9.

Regarding technological progress bias, tangible capital services tend to be savings-oriented in industry sectors, excluding agriculture, forestry, fishing, mining, food manufacturing, pulp and paper manufacturing, telecommunications, broadcasting, commerce, software, ICT services, Internet, and other services. Conversely, labor input tends to be usage-oriented in all industries except software and ICT services.

Table 5.9 Estimated parameters of trans-log price function: Intra-firm information processing activity

		αK	αL	αM	αT	βKK	βKL	βKM	βLL	βLM	βMM	βKT	βLT	βMT	βTT
06	Food manufacturing	0.31	−0.06	0.75	0.00	0.00	0.00	0.00	−2.72	2.72	−2.72	0.01	0.02	−0.04	0.00
	t-value	11.81	−0.61		−0.19							5.61	3.08		0.15
09	Textile and wooden furniture manufacturing	0.28	−0.05	0.78	0.03	0.00	0.00	0.00	−2.77	2.77	−2.77	0.00	0.03	−0.03	0.00
	t-value	10.67	−0.53		1.83							1.98	3.20		−1.81
12	Pulp and paper manufacturing	0.22	0.13	0.66	0.00	0.00	0.00	0.00	−1.63	1.63	−1.63	0.00	0.01	−0.02	0.00
	t-value	13.76	1.27		0.37							1.54	1.86		−0.42
15	Chemical industry	0.26	−0.04	0.78	0.02	0.00	0.00	0.00	−2.73	2.73	−2.73	0.00	0.03	−0.03	0.00
	t-value	13.82	−0.38		1.97							1.92	2.97		−1.73
18	Material and metal manufacturing industry	0.25	−0.04	0.79	0.00	0.00	0.00	0.00	−2.74	2.74	−2.74	0.00	0.03	−0.03	0.00
	t-value	20.69	−0.36		−0.39							−0.36	3.21		0.49
21	General machinery	0.28	−0.04	0.75	0.02	0.00	0.00	0.00	−2.60	2.60	−2.60	0.00	0.03	−0.03	0.00
	t-value	14.50	−0.34		2.01							0.52	3.02		−1.75
24	Consumer electronics and electrical machinery	0.28	0.05	0.67	0.01	0.00	0.00	0.00	−1.97	1.97	−1.97	0.00	0.02	−0.02	0.00
	t-value	16.36	0.46		0.42							1.37	2.48		−0.43
27	Fiber optic cables	0.18	−0.01	0.83	0.02	−0.23	0.06	0.17	−2.73	2.67	−2.84	0.00	0.03	−0.03	0.00
	t-value	7.62	−0.09		0.83							0.89	2.62		−0.82
30	Semiconductor manufacturing and equipment products	0.30	−0.03	0.72	0.08	0.00	0.00	0.00	−2.45	2.45	−2.45	0.00	0.02	−0.03	−0.01
	t-value	12.96	−0.24		1.18							0.87	2.89		−1.19
33	Telecommunication machinery	0.28	−0.08	0.80	0.02	−0.10	0.23	−0.13	−2.65	2.42	−2.29	0.00	0.03	−0.03	0.00
	t-value	9.10	−0.80		1.18							1.06	3.55		−0.96
36	Computers and accessories	0.23	−0.06	0.83	0.11	0.00	0.00	0.00	−2.77	2.77	−2.77	0.01	0.03	−0.04	−0.01
	t-value	11.77	−0.50		1.03							2.78	3.23		−0.99

(continued)

Table 5.9 (continued)

		αK	αL	αM	αT	βKK	βKL	βKM	βLL	βLM	βMM	βKT	βLT	βMT	βTT
39	Semiconductor devices and integrated circuits	0.07	0.03	0.89	0.05	0.00	0.00	0.00	−2.81	2.81	−2.81	0.01	0.02	−0.04	0.00
	t-value	4.61	0.28		0.65							8.37	2.42		−0.65
42	Electronic component (intra-firm ICT activity)	0.18	−0.03	0.86	0.04	0.00	0.00	0.00	−3.01	3.01	−3.01	0.00	0.03	−0.03	0.00
	t-value	12.40	12.40		0.92							2.93	3.46		−1.13
45	Heavy electrical machinery and transportation equipment	0.24	0.01	0.76	0.00	0.00	0.00	0.00	−2.43	2.43	−2.43	0.01	0.02	−0.03	0.00
	t-value	18.06	0.06		0.10							4.51	2.58		−0.10
48	Robots	0.22	0.02	0.77	0.06	0.00	0.00	0.00	−2.46	2.46	−2.46	0.00	0.02	−0.02	−0.01
	t-value	19.72	0.15		1.09							0.51	2.67		−1.18
51	Other precision machinery	0.28	−0.01	0.73	0.04	0.00	0.00	0.00	−2.40	2.40	−2.40	0.00	0.03	−0.03	0.00
	t-value	15.95	−0.13		1.20							0.87	2.93		−1.11
54	Petroleum and coal products	0.18	−0.03	0.86	−0.04	−0.01	−0.03	0.04	−2.80	2.83	−2.88	0.01	0.03	−0.03	0.01
	t-value	11.16	−0.31		−0.40	−0.64						5.74	2.96		0.57
57	Other manufacturing industry	0.28	0.00	0.72	0.02	0.00	0.00	0.00	−2.37	2.37	−2.37	0.00	0.02	−0.03	0.00
	t-value	15.00	−0.04		1.42							0.80	2.75		−1.34
60	Energy industries	0.21	0.35	0.44	0.02	0.00	0.00	0.00	0.00	0.00	0.00	0.00	0.01	−0.02	0.00
	t-value	14.13	0.92		1.01							3.01	1.04		−0.86
63	Construction	0.16	0.38	0.46	0.02	0.00	0.00	0.00	0.00	0.00	0.00	0.00	0.00	0.00	0.00
	t-value	15.30	10.44		0.97							2.29	−1.00		−1.09
66	Transportation	0.21	0.35	0.44	0.01	0.00	0.00	0.00	0.00	0.00	0.00	0.00	0.00	0.00	0.00
	t-value	11.97	9.65		0.96							1.84	−0.87		−0.72
69	Telecommunications and broadcasting	0.19	0.39	0.41	−0.01	0.00	0.00	0.00	0.00	0.00	0.00	0.00	0.00	−0.01	0.00
	t-value	17.46	14.04		−0.37							3.33	1.79		0.45
72	Commerce	0.29	0.31	0.40	0.00	0.00	0.00	0.00	0.00	0.00	0.00	0.01	0.00	−0.01	0.00
	t-value	17.80	10.33		−0.09							6.97	−1.68		0.30

Table 5.10 Estimated parameters of trans-log price function: intra-firm R&D activity

		αK	αL	αM	αT	βKK	βKL	βKM	βLL	βLM	βMM	βKT	βLT	βMT	βTT
02	Intra-firm R&D (agriculture, forestry, and fisheries)	0.14	0.46	0.40	0.01	0.00	0.00	0.00	0.00	0.00	0.00	0.00	0.00	0.00	0.00
	t-value	16.24	17.30		2.84							0.61	1.58	−2.19	−3.10
04	Intra-firm R&D (mineral resources)	0.30	0.28	0.42	0.00	0.00	0.00	0.00	−0.91	0.91	−0.91	−0.01	0.02	−0.01	0.00
	t-value	13.88	6.49		0.65							−4.85	4.88	−0.04	−0.11
07	Food manufacturing	0.20	0.23	0.56	0.00	0.00	0.00	0.00	−1.71	1.71	−1.71	0.00	0.02	−0.02	0.00
	t-value	18.75	3.44		1.08							2.76	3.26	−6.02	−1.10
10	Textile and wooden furniture manufacturing	0.20	0.32	0.48	0.01	0.00	0.00	0.00	−0.95	0.95	−0.95	0.00	0.01	−0.01	0.00
	t-value	12.16	5.77		1.54							−2.93	3.35	−0.42	−1.25
13	Pulp and paper manufacturing	0.21	0.31	0.48	0.00	0.00	0.00	0.00	−1.05	1.05	−1.05	0.00	0.01	−0.01	0.00
	t-value	13.33	5.84		0.01							0.10	0.02	−0.12	0.00
16	Chemical industry	0.20	0.30	0.50	0.01	0.00	0.00	0.00	−1.16	1.16	−1.16	0.00	0.02	−0.01	0.00
	t-value	15.63	4.50		2.31							−2.18	3.01	−0.82	−2.00
19	Materials manufacturing division	0.22	0.29	0.48	0.00	0.00	0.00	0.00	−1.08	1.08	−1.08	0.00	0.01	−0.01	0.00
	t-value	20.21	5.46		0.40							−2.31	3.49	−1.18	0.00
22	General machinery	0.19	0.29	0.53	0.01	0.00	0.00	0.00	−1.34	1.34	−1.34	0.00	0.02	−0.01	0.00
	t-value	12.88	6.50		2.07							−2.72	5.02	−2.30	−1.50
25	Consumer electronics and electrical machinery	0.11	0.36	0.53	0.00	0.00	0.00	0.00	0.00	0.00	0.00	0.00	0.01	−0.01	0.00
	t-value	12.79	4.68		0.55							−0.68	2.27	−1.58	−0.31
28	Fiber optic cables	0.19	0.29	0.52	0.00	0.00	0.00	0.00	−1.29	1.29	−1.29	0.00	0.02	−0.01	0.00
	t-value	14.30	4.23		−0.38							−2.59	3.20	−0.61	0.50

(continued)

Table 5.10 (continued)

		αK	αL	αM	αT	βKK	βKL	βKM	βLL	βLM	βMM	βKT	βLT	βMT	βTT
31	Semiconductor manufacturing equipment	0.24	0.26	0.50	0.01	0.00	0.00	0.00	−1.27	1.27	−1.27	−0.01	0.02	−0.01	0.00
	t-value	13.76	5.29		2.06							−3.83	4.82	−0.99	−1.72
34	Telecommunication machinery	0.19	0.29	0.52	0.00	0.00	0.00	0.00	−1.27	1.27	−1.27	−0.01	0.02	−0.01	0.00
	t-value	12.17	4.26		1.57							−4.13	3.36	0.77	−1.25
37	Computers and accessories	0.19	0.43	0.38	0.01	0.00	0.00	0.00	0.00	0.00	0.00	0.00	0.01	0.00	0.00
	t-value	11.45	18.40		1.95							18.40	3.02	−21.42	−1.62
40	Semiconductor devices and integrated circuits	0.25	0.28	0.47	0.01	0.00	0.00	0.00	0.00	0.00	0.00	−0.01	0.02	−0.01	0.00
	t-value	13.08	4.33		1.35							−5.70	3.70	2.00	−1.19
43	Other electronic components	0.18	0.44	0.38	0.00	0.00	0.00	0.00	0.00	0.00	0.00	0.00	0.01	0.00	0.00
	t-value	13.22	18.10		−0.16							−3.02	2.64	0.38	−0.14
46	Heavy electrical machinery and transportation equipment	0.16	0.32	0.52	0.00	0.00	0.00	0.00	−1.20	1.20	−1.20	0.00	0.02	−0.01	0.00
	t-value	13.58	13.58		1.90							−1.89	3.41	−1.52	−1.78
49	Robots	0.16	0.31	0.53	0.00	0.00	0.00	0.00	−1.27	1.27	−1.27	0.00	0.02	−0.01	0.00
	t-value	12.18	5.10		1.77							−2.79	3.50	−0.71	−1.44
52	Other precision machinery	0.14	0.33	0.53	0.00	0.00	0.00	0.00	0.00	0.00	0.00	0.00	0.02	−0.01	0.00
	t-value	13.28	4.31		1.76							−2.28	2.52	−0.23	−1.48
55	Petroleum and coal products	0.15	0.46	0.39	0.00	0.00	0.00	0.00	0.00	0.00	0.00	0.00	0.01	0.00	0.00
	t-value	10.23	20.21		0.53							−2.61	2.72	−0.11	−0.39

58	Other manufacturing industry	0.17	0.29	0.53	0.01	0.00	0.00	0.00	−1.37	1.37	−1.37	0.00	0.02	−0.01	0.00
	t-value	14.37	4.70		1.72							−2.45	3.49	−1.04	−1.53
61	Energy industries	0.13	0.48	0.40	0.00	0.00	0.00	0.00	0.00	0.00	0.00	0.00	0.00	0.00	0.00
	t-value	18.67	17.29		0.85							−3.36	1.86	1.50	−0.41
64	Construction	0.34	0.35	0.32	0.01	0.00	0.00	0.00	0.00	0.00	0.00	0.00	0.01	0.00	0.00
	t-value	15.17	16.91		1.62							−2.36	3.81	−1.45	−1.35
67	Transportation	0.13	0.47	0.40	0.00	0.00	0.00	0.00	0.00	0.00	0.00	0.00	0.00	0.00	0.00
	t-value	18.31	17.99		0.98							−1.51	1.94	−0.44	−0.64
70	Telecommunications and broadcasting	0.11	0.48	0.41	0.00	0.00	0.00	0.00	0.00	0.00	0.00	0.00	0.00	−0.01	0.00
	t-value	12.03	17.05		0.75							3.93	0.77	−4.70	−0.54
73	Commerce	0.26	0.39	0.35	0.01	0.00	0.00	0.00	0.00	0.00	0.00	0.00	0.00	−0.01	0.00
	t-value	16.69	16.19		0.52							2.11	1.29	−3.40	−0.43
75	Software industry	0.08	0.49	0.42	0.04	0.00	0.00	0.00	0.00	0.00	0.00	0.02	0.00	−0.01	0.00
	t-value	7.51	16.55		1.65							15.41	−1.49	−13.93	−1.38
77	Information processing and provision services	0.14	0.46	0.40	0.04	0.00	0.00	0.00	0.00	0.00	0.00	0.02	0.00	−0.01	0.00
	t-value	10.92	16.45		1.37							13.25	−1.66	−11.59	−1.16
79	Internet industry	−0.16	0.27	0.89	−0.06	0.00	0.00	0.00	0.00	0.00	0.00	0.02	0.02	−0.04	0.00
	t-value	−2.04	2.83		−0.43							3.66	2.47	−6.13	0.28
95	Other services	0.13	0.33	0.54	0.00	−0.11	−0.20	0.31	−0.37	0.56	−0.87	0.01	0.01	−0.02	0.00
	t-value	6.84	7.30		0.99							5.36	2.71	−8.07	−0.86

An increase in the price of tangible capital services in intra-firm R&D generally triggers technological progress, an increase in the price of labor services impedes technological progress in industries outside of software and ICT services. The acceleration indicator of technological progress rates is predominantly negative across industry sectors, suggesting a dearth of factors accelerating technological innovation in intra-firm R&D departments during this period. The increase in production efficiency due to technological progress is decelerating. Only a handful of industries, such as fiber optics and Internet, exhibit a trend of technological advancement, albeit with extremely small values (Table 5.10).

Regarding factor substitution, the relationship between tangible fixed capital, labor, and intermediate input is neutral, excluding other service industries. However, a substitutive relationship exists between labor and intermediate inputs.

Turning our attention to the R&D-specialized sector's price function measurements, as depicted in Table 5.11, this sector comprises departments conducting R&D activities in both profit-oriented industrial sectors and academic institutions, including national and private universities. This sector is divided into five research fields: life sciences, information and communication, materials science, environmental and energy, and others. The input factors under consideration are tangible fixed assets, labor, and intermediate input. Labor services and intermediate inputs are relatively high in terms of input, while tangible fixed asset input is relatively low. This is a well-established trend in parameter α, with statistical significance being nearly certain. The functional form is Cobb-Douglas for all sectors, suggesting neutral substitution between factors across all fields.

Regarding technological progress bias, capital usage bias is observed in all fields for national R&D and is statistically significant. Conversely, in industrial sectors, a capital-saving bias is observed in the life sciences, information and communication, and other fields, with a capital usage bias in the remaining sectors, although these are not statistically significant.

In the national sector, a labor usage bias is present in the life sciences and materials fields, while a labor-saving bias is observed in the information and communication, environmental and energy, and other fields, albeit with low statistical significance. In the industrial sector, a labor-saving bias is prevalent across all fields. In the R&D-specialized sector, an increase in input factor prices leading to factor usage implies that the efficiency of knowledge stock production does not contribute positively. If this trend persists, it poses a challenge to the country's science and technology policy, as it may indicate that the expansion of knowledge stock through the country's scientific and technological progress is either not occurring or not being realized. As previously mentioned, this depends on the actual trend of input factor price changes, a topic that I intend to explore in detail in the subsequent section.

Lastly, concerning the acceleration of technological progress indicators, only the materials and other fields in the national sector and materials field in the industrial sector show positive values. In other fields, both in the national and industrial sectors, the indicator displays negative values, indicating stagnation in technological progress during this period.

Table 5.11 Estimated parameters of trans-log price function: public and industrial R&D sector

		αK	αL	αM	αT	βKT	βLT	βMT	βTT
84	R&D life (national, public, non-profit)	0.06	0.74	0.20	0.00	0.00	0.00	0.00	0.00
	t-value	12.82	17.67		0.20	5.69	0.15		−0.03
85	R&D information and communication (national,public,non-	0.06	0.76	0.18	0.00	0.00	0.00	0.00	0.00
	t-value	11.14	17.41		0.17	3.46	−0.52		−0.04
86	R&D materials and substances (national, public, non-profit)	0.02	0.64	0.34	0.00	0.00	0.01	−0.01	0.00
	t-value	2.48	6.50		−0.39	8.17	0.73		0.59
87	R&D environment and energy (national, public, non-profit)	0.03	0.77	0.20	0.00	0.00	0.00	0.00	0.00
	t-value	6.18	15.82		0.64	6.15	−0.98		−0.24
88	R&D and other R&D (national, public, non-profit)	0.05	0.74	0.21	0.00	0.00	0.00	0.00	0.00
	t-value	13.66	17.54		−0.80	6.81	−0.14		0.74
89	R&D life (industry)	0.00	0.72	0.28	0.14	0.00	−0.01	0.01	−0.01
	t-value	0.62	0.62		0.35	−0.12	−2.67		−0.40
90	R&D information and communication (industry)	0.01	0.72	0.28	0.09	0.00	−0.01	0.01	−0.01
	t-value	3.91	14.58		0.43	−1.44	−2.67		−0.48
91	R&D substances and materials (industry)	0.00	0.62	0.38	−0.05	0.00	−0.01	0.01	0.00
	t-value	0.71	5.59		−0.16	0.40	−0.67		0.05
92	R&D environment and energy (industry)	0.00	0.72	0.28	0.07	0.00	−0.01	0.01	−0.01
	t-value	1.85	14.58		0.32	0.02	−2.67		−0.33
93	R&D and other R&D (industry)	0.01	0.70	0.29	0.06	0.00	−0.01	0.01	−0.01
	t-value	2.97	14.67		0.32	−0.57	−2.48		−0.41

5.4 Variations in Relative Price and Technology Selection

In this section, we aim to elucidate the implications of industrial-technological choices, inferred from the parameter estimation results of the trans-log price function discussed previously, considering the observed changes in relative element prices during the observation period. The industrial classification for the price function measurement is compiled in Table 5.12.

Table 5.12 Industry and Activity Category Correspondence

	Main products		Intra-firm ICT		Intra-firm R&D
1	Agriculture (Main products, ICT activity)			2	Agriculture, forestry, and fisheries
3	Mining (Main products, ICT activity)			4	Mineral resources
05	Food manufacturing	6	Food manufacturing	7	Food manufacturing
08	Textile and wooden furniture manufacturing	9	Textile and wooden furniture manufacturing	10	Textile and wooden furniture manufacturing
11	Pulp and paper manufacturing	12	Pulp and paper manufacturing	13	Pulp and paper manufacturing
14	Chemical industry	15	Chemical industry	16	Chemical industry
17	Materials manufacturing	18	Materials manufacturing	19	Materials manufacturing division
20	General machinery	21	General machinery	22	General machinery
23	Consumer electronics and electrical machinery	24	Consumer electronics and electrical machinery	25	Consumer electronics and electrical machinery
26	Fiber optic cables	27	Fiber optic cables	28	Fiber optic cables
29	Semiconductor manufacturing equipment	30	Semiconductor manufacturing equipment	31	Semiconductor manufacturing equipment
32	Telecommunication machinery	33	Telecommunication machinery	34	Telecommunication machinery
35	Computers and accessories	36	Computers and accessories	37	Computers and accessories
38	Semiconductor devices and integrated circuits	39	Semiconductor devices and integrated circuits	40	Semiconductor devices and integrated circuits
41	Other electronic components	42	Other electronic components	43	Other electronic components
44	Heavy electrical machinery and transportation equipment	45	Heavy electrical machinery and transportation equipment	46	Heavy electrical machinery and transportation equipment
47	Robots	48	Robots	49	Robots
50	Other precision machinery	51	Other precision machinery	52	Other precision machinery
53	Petroleum and coal products	54	Petroleum and coal products	55	Petroleum and coal products
56	Other manufacturing industry	57	Other manufacturing industry	58	Other manufacturing industry
59	Energy industries	60	Energy industries	61	Energy industries
62	Construction	63	Construction	64	Construction
65	Transportation	66	Transportation	67	Transportation

(continued)

Table 5.12 (continued)

	Main products		Intra-firm ICT		Intra-firm R&D
68	Telecommunications and broadcasting	69	Telecommunications and broadcasting	70	Telecommunications and broadcasting
71	Commerce	72	Commerce	73	Commerce
74	Software (Main products)			75	Software industry
76	Information management (Main products)			77	Information processing and provision services
78	Internet (Main products)			79	Internet industry
80	Medical and social services (main production)			81	N/A
82	Education (Main production division)			83	N/A
84	R&D life (national, public, non-profit)				
85	R&D information and communication (national, public, non-profit)				
86	R&D materials and substances (national, public, non-profit)				
87	R&D environment and energy (national, public, non-profit)				
88	R&D and other R&D (national, public, non-profit)				
89	R&D life (industry)				
90	R&D information and communication (industry)				
91	R&D substances and materials (industry)				
92	R&D environment and energy (industry)				
93	R&D and other R&D (industry)				
94	Other services (Main products)			95	Other services (R&D)

The industrial sectors are categorized into 41, ranging from agriculture, forestry, fisheries, and aquaculture to other services. Of these, 23 industries, from Food Manufacturing to Commerce, can be subdivided into three areas: main production, intra-firm information processing, and intra-firm R&D activities. With the addition of Other Services, this makes 24 industries. Furthermore, six industries, including agriculture, forestry, fisheries, aquaculture, mining, software industry, information processing services, Internet services, and Other services, are divided into two activities: main production and intra-firm R&D.

Among these six industries, data acquisition is challenging in agriculture, forestry, fisheries, aquaculture, mining, other services, and intra-firm information processing cannot be classified. Industries 74–77 specialize in information processing services, where distinguishing between main production and intra-firm information processing activities is not meaningful. Therefore, these activities are integrated, and only main production and intra-firm R&D are used as activity classifications. For the medical welfare services and education industry, intra-firm R&D and intra-firm ICT activities are currently integrated (industry numbers 81 and 83 are placeholders, with no data developed for them).

Lastly, industries 84–93 specialize in R&D activities. Industries 84–88 represent the activities of national, public, and non-profit universities and research institutions, divided into five scientific fields (life, information and communication, materials, environment and energy, and others). Similarly, sectors 89–93 distinguish for-profit institutions specializing in research as an industry, also divided into the five scientific fields. This distinction underscores profit-driven R&D activities, separate from the non-profit R&D activities of national and public institutions. In total, there are 41 industries and 95 sectors when considering both industry and activity classifications (with the numbers 81 and 82 missing, effectively making it 93 sectors).

Regarding the measurement results of the trans-log cost function discussed in the previous section; we seek to reconfirm the characteristics of the technology from the perspective of technological progress bias.

1. Industrial sectors are categorized into main production, intra-firm information processing (ICT), and intra-firm R&D.

(a) Main Production Activities: Set the price function for elements as knowledge intangible capital services of R&D Kr, knowledge intangible capital services of information processing Ki, tangible fixed capital services K, labor services L, and intermediate raw material services M, capturing parameter shifts due to technological advancement with trend T.
(b) Intra-firm Information Processing (ICT) Activities: Set the price function for elements as tangible fixed capital services K associated with intra-firm information processing activities, labor services L for intra-firm information processing activities, and intermediate raw material services M, capturing parameter shifts because of technological advancement with trend T.
(c) Intra-firm R&D Activities: The elements include tangible fixed capital services K, labor services L, and intermediate raw material services M for intra-firm R&D. The price functions for these elements are established by capturing parameter shifts given technological advancement through rend T.

2. Sectoral activity is divided into main production activities, including intra-firm information processing (ICT) and intra-firm R&D activities.

(a) Main Production Activities, including Intra-firm Information Processing: These activities are characterized by the input of tangible fixed capital services $K + Ki$ that include capital services from intra-firm information processing activities, intangible knowledge capital services Kr from intra-firm R&D, labor services L, and intermediate raw material services M, all explained through the trend of technological advancement T.
(b) Intra-firm R&D Activities: These involve setting the price functions for tangible fixed capital services K associated with intra-firm R&D activities, labor services L for intra-firm R&D, and intermediate raw material services M. The aim is to capture parameter shifts owing to technological progress through trend T.

3. One sector views main production as solely R&D within the scientific field.

To organize and compare the technological advancement bias in each of the aforementioned cases, we summarize the key points of each activity.

Figure 5.1 delineates the technological advancement bias in main production activities, categorizing such into three distinct sectors: main production, intra-firm information processing, and intra-firm R&D. This categorization likely underscores the unique impact of technological advancements on each sector, demonstrating the direction of technology's influence on the activities. This passage suggests reorganizing the technological characteristics within the main production activities of various industrial sectors in Japan from 1995 to 2011. This period follows the

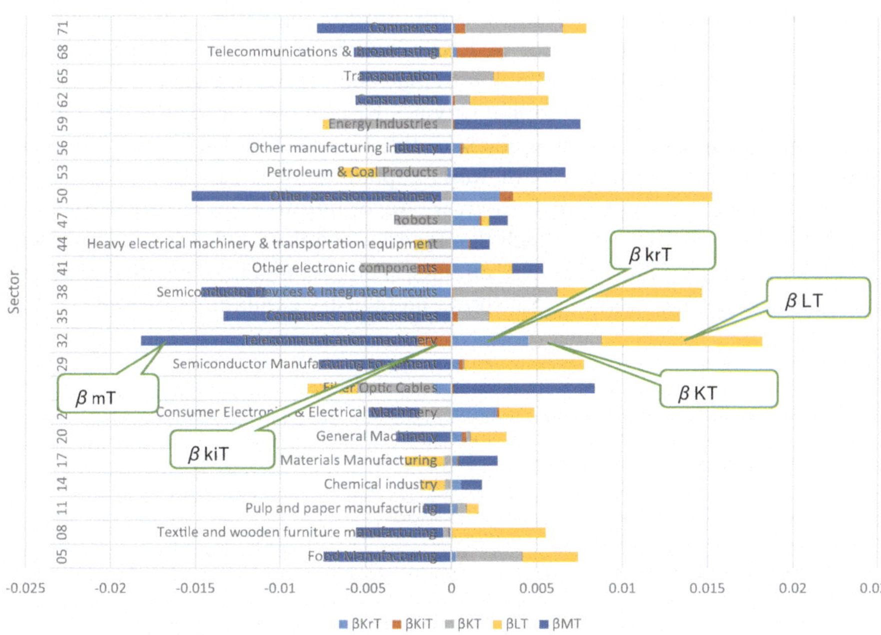

Fig. 5.1 Comparison of bias of technical progress: main activity

collapse of the early 1990s' bubble economy. The reorganization is proposed from the perspective of input structures—specifically capital, labor, and intermediate materials—impacted by technological progress. The primary aim is to analyze the influence of technological advancements on the input structures of these sectors during the specified period. This analysis highlights the transition and adaptation following the economic downturn experienced after the collapse of the bubble economy. The characteristics of technological progress are described as follows:

(a) $\beta_{krT} > 0$: The cost of input for R&D capital services (Kr) rises relative to the total input costs. This indicates an increase in the proportion of this specific production factor in the total input. This characteristic could suggest a reluctance toward technological progress because of a usage bias of this factor or because the conditions denoted by $\beta_{krT} > 0$ result in an increase in the cost of R&D capital services. As shown in Fig. 5.1, this $\beta_{krT} > 0$ characteristic is observed in the main production activities of various sectors, including Food, Paper and pulp, Chemical, Materials, General machinery, Electrical machinery, Semiconductor devices, Communication machinery, Other electronic components, Heavy electric and transport machinery, Robots, Other precision machinery, Other manufacturing, Energy, Transportation, Communications and broadcasting, and Commerce.

(b) Moreover, Fig. 5.2 shows a trend in which technological progress increases the use of R&D capital services in sectors where corporate information processing cannot be differentiated as the main production activity. This trend, coupled with an increase in the cost of this intangible capital, leads to a widespread reluctance toward technological advancement across several industries. This reluctance is influenced by the rising prices of R&D capital, suggesting that the increased cost of incorporating new technologies may deter technological progress.

(c) $\beta_{kiT} < 0$: The input of R&D capital services Kr becomes more efficient, indicating an economization or saving bias for this factor. This characteristic suggests that an increase in the prices of R&D capital services can lead to a positive attitude toward technological progress. According to Fig. 5.1, industries that exhibit this characteristic include Textiles, Optics and fiber, Computers, Semiconductors and integrated circuits, Petroleum and coal products, and Construction. However, Fig. 5.2 shows no industry exhibiting this trait.

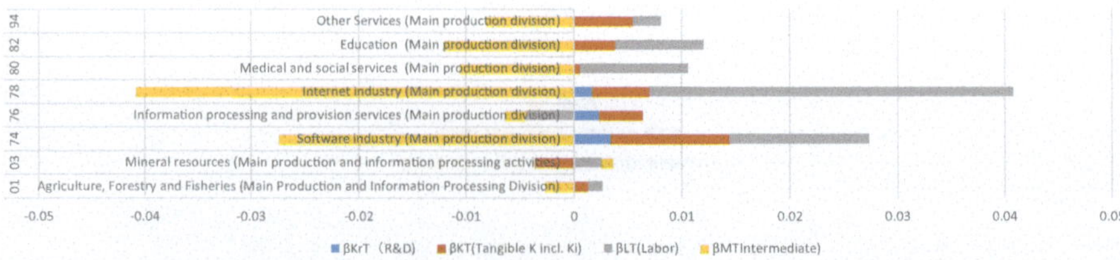

Fig. 5.2 Comparison of technology bias in the main production activity

(d) $\beta_{kiT} > 0$: The use of information processing capital services *Ki* increases, and a rise in the price of these services leads to a negative attitude toward technological progress. According to Fig. 5.1, industries exhibiting this characteristic include Food, Paper and pulp, Materials, General machinery, Electrical machinery, Optics and fiber, Semiconductor devices, Computers, Semiconductors and integrated circuits, Heavy electric and transport machinery, Robots, Other precision machinery, Petroleum and coal products, Other manufacturing, Energy, Construction, Transportation, Communications and broadcasting, and Commerce. An increase in the price of information processing services can hinder technological efficiency improvement in these main industries.

(e) $\beta_{kiT} < 0$: **There is an economical use of information processing capital services *Ki***. This characteristic suggests that an increase in the prices of information processing capital services can lead to a positive attitude toward technological advancement, implying that industries are encouraged to adopt new technologies despite the rising costs of these services. As shown in Fig. 5.1, industries demonstrating this characteristic include Textiles, Chemicals, Communication machinery, and Other electronic components. However, only a few industries possess this trait.

(f) $\beta_{kT} > 0$: **The utility of tangible fixed capital services and an increase in the price of tangible fixed capital services can lead to a reluctance toward technological progress**. In Fig. 5.1, this applies to industries such as Food, Paper and pulp, General machinery, Semiconductor equipment, Communication equipment, Electronic computers, Semiconductors and integrated circuits, Other manufacturing, Construction, Transportation, Communications and broadcasting, and Commerce. In Table 5.10, the trend of $\beta_{kT} > 0$ remains unchanged in industries other than mining, indicating that many industries still exhibit this characteristic despite a decrease in the price of tangible fixed capital services.

(g) $\beta_{kT} < 0$: **Tangible fixed capital services *k* becomes more efficient or the price of tangible fixed capital services increases, which positively contributes to technological advancement**. In Fig. 5.1, the industries include Textiles, Chemicals, Materials, Electrical machinery, Fiber optics, Other electronic parts, Heavy machinery and transportation equipment, Robots, Other precision machinery, Petroleum and coal products, and Energy. In Fig. 5.2, only the mining industry displays this trait. The fact that many industries favor technological progress despite the rising prices of tangible capital is largely due to numerous industries continuing their growth pattern centered around tangible fixed capital investments from the period of high economic growth. This characteristic contradicts the recent trend where investments in intangible capital have become significant.

(h) $\beta^{LT} > 0$: Labor becomes more useful, and an increase in the price of labor leads to a reluctance toward technological progress. In Fig. 5.1, from Food sector to Other manufacturing sectors, all manufacturing sectors exhibit $\boldsymbol{\beta^{LT} > 0}$, and in the service industry, Communications and broadcasting sector also shows $\boldsymbol{\beta^{LT} > 0}$, indicating that the increase in labor service prices leads to a reluctance toward technological advancement in corporate information processing technology.

(i) $\beta^{LT} < 0$: **Labor services become more efficient, or as the price of labor services increases, technology becomes more favorable**. As shown in Fig. 5.3, labor-intensive industries like Energy, Constructions, Transportation, and Commerce are included.

(j) $\beta^{MT} > 0$: **Intermediate raw materials become more intensive, or as the price of intermediate raw materials increases, technology progresses less favorably**. In Fig. 5.1, the industries include Chemicals, Materials, Fiber optics, Other electronic parts, Heavy machinery and transportation equipment, Robots, Petroleum and coal products, and Energy. As shown in Fig. 5.2, only the mining industry is included.

(k) $\beta^{MT} < 0$: **As intermediate materials are saved, or as the price of intermediate materials rises**. According to Fig. 5.3, all sectors except for the Construction exhibit $\beta^{MT} < 0$.

With regard to intermediate raw materials, technological advancement tends to be resource-saving.

5.4.1 *Technology Biases in Corporate Information Processing Activities in Sectors Where Industrial Activities Are Divided into Two or Three Categories*

Figures 5.3 and 5.4 present the estimated parameter results for biases in technological progress, specifically within corporate information processing and R&D activities. These estimates are derived when industrial activities are segmented into two

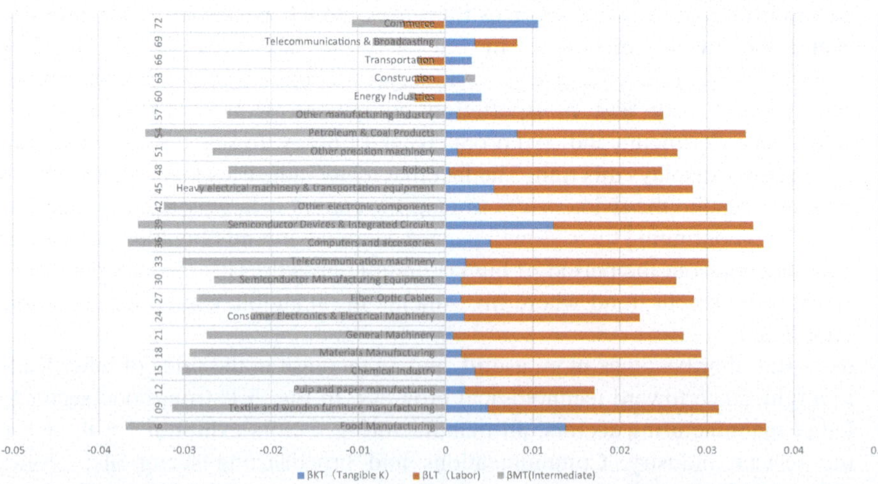

Fig. 5.3 Technological progress bias of information processing activities by industry (three activity areas) $\beta_i^T (i = K,L,M)$

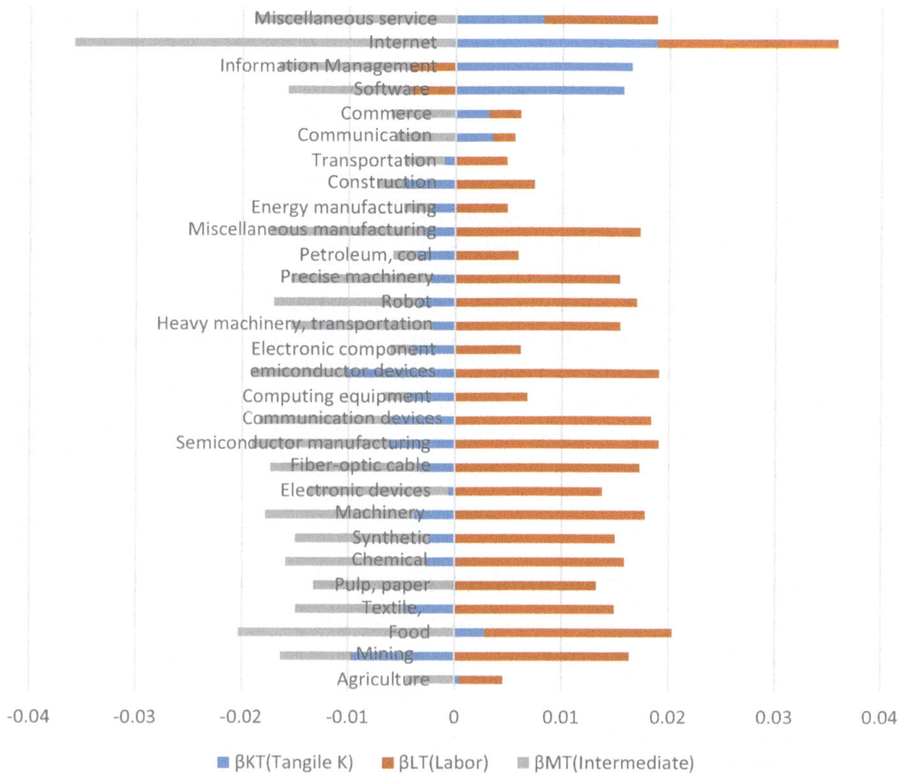

Fig. 5.4 Technological progress bias in research activities within companies by industry (three activity areas) Bias $\beta_i^T(i = K,L,M)$

or three sectors. Notably, Fig. 5.4 encompasses the industrial sectors, which are binding into two distinct areas for corporate R&D activities.

(l) $\beta^{KT} > 0$: **The use of tangible fixed capital services related to corporate information processing activities becomes more efficient, and the price of tangible fixed capital services increases, leading to a reluctance toward technological advancement in corporate information processing**. According to Fig. 5.3, all sectors exhibit $\beta_{KT} > 0$. In particular, significant values are shown in the sectors of Food, Semiconductors and integrated circuits, Petroleum and coal products, and Commerce.

(m) $\beta^{LT} > 0$: **Labor services related to corporate information processing activities become more efficient, and as the price of labor services involved in those activities increases, it leads to a reluctance toward technological advancement**. According to Fig. 5.3, from sector Food to Other manufacturing, all manufacturing sectors exhibit $\beta^{LT} > 0$, and in the service industry, Communications and broadcasting also shows $\beta^{LT} > 0$.

(n) $\beta^{LT} < 0$: **Labor services related to information processing activities become more efficient, and as the price of labor services increases, it leads to a positive attitude toward technological advancement**. According to Fig. 5.3, this includes the service sectors of Energy, Construction, Transportation, and Commerce, and this tendency is observed in labor-intensive industries.
(o) $\beta^{MT} > 0$: **Intermediate raw material services related to corporate information processing activities become more efficient, and as the price of intermediate raw materials increases, it leads to a reluctance toward technological advancement**. This includes only the Construction industry.
(p) $\beta^{MT} < 0$: **Intermediate raw material services related to corporate information processing activities become resource-saving, and as the price of intermediate raw materials increases, it leads to a positive attitude toward technological advancement**. According to Fig. 5.3, all sectors except for Construction exhibit $\beta^{MT} < 0$. Regarding the input of intermediate raw materials, technological advancement tends to be resource-saving.

5.4.2 Property of the Technology Bias in the Intra-firm R&D Activities in which the Industry Classified into Several Intra-firm Activities

Figure 5.4 depicts the bias in technological progress within corporate R&D. Notably, the characteristics of technological progress associated with corporate research, including R&D activities in industrial sectors segmented into two parts (main production activities and intra-firm R&D activities), display a bias in technological progress that contrasts with that of corporate information processing. Specifically, the bias in technological progress for tangible fixed capital services demonstrates contrasting characteristics in the context of corporate information processing and R&D (See Kuroda and Nomura, 2004).

(q) $\beta^{KT} > 0$: As the price of tangible fixed capital services increases, services related to corporate R&D activities become more efficient, and technological progress related to corporate R&D becomes reluctant. Contrary to corporate information processing, a bias toward using tangible fixed capital is evident only in food manufacturing within the manufacturing sector. However, in non-manufacturing sectors, almost all industries exhibit a bias toward using tangible fixed capital.
(r) $\beta^{KT} < 0$: As the price of tangible fixed capital services increases, services related to corporate R&D activities become resource-saving, leading to a positive attitude toward technological progress. This is observed in all manufacturing sectors except for the food manufacturing industry.
(s) $\beta^{LT} > 0$: **Labor services related to corporate R&D activities become more efficient, and as the price of labor services increases, it leads to a reluctance toward technological advancement**. In all industrial sectors, except for Software and Information processing services, an increase in labor service

prices during corporate R&D activities exhibits a bias toward a reluctance regarding technological advancement.

(t) $\beta^{MT} < 0$: **Intermediate input services related to corporate R&D activities become resource-saving, and as the price of intermediate input services increases, it yields a positive attitude toward technological advancement**. Regarding corporate R&D activities across all industries, the increase in the price of intermediate input services leads to a positive stance toward technological advancement.

5.4.3 Technological Advancement Bias in R&D Activities in Sectors Where All Industrial Activities Are Aggregated into One Activity

In Fig. 5.5, we examine the bias in technological advancement in sectors solely dedicated to R&D. These sectors are divided into two categories: national/public and non-profit private universities research institutions and industrial research institutions. Each category is further divided into five scientific fields: Life/Biosciences, Information Science, Materials Science, Energy/Environmental Science; and Other Natural Sciences and Humanities.

Technological advancements in national/public and nonprofit research institutions reveal a utilization bias in tangible capital services across all scientific fields. This suggests that technological progress in these fields enhances the efficiency of tangible fixed capital for R&D. Conversely, this finding implies that an increase in the price of such capital services can hinder technological progress. In terms of labor services, a utilization bias is observed in the life sciences and materials science fields, while information and communications, environmental and energy, and other fields exhibit resource-saving characteristics. Regarding intermediate raw materials, the life sciences, materials science, and other fields are resource-saving,

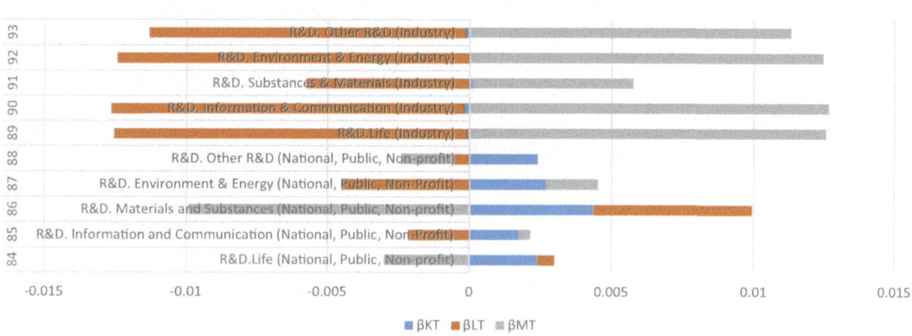

Fig. 5.5 Technology bias in sectors engaged in R&D

whereas the information and communications and environmental and energy fields show a utilization bias. R&D in national/public institutions, to some extent, accumulates public goods-like intellectual assets, and investment in R&D is likely to be influenced by government policy decisions. Therefore, when making policy decisions, it is crucial to consider the characteristics of technological advancement bias. In contrast, the technological characteristics of R&D in private research institutions (industries), engaged in for-profit activities, show that the bias toward technological advancement in tangible fixed assets is minimal and primarily neutral. However, every scientific field has a labor-saving characteristic for labor services, and an increase in the price of labor services significantly stimulates technological development. For raw materials, a utilization bias is evident in every field.

References

Christensen, L. R., Jorgenson, F. W., & Lau, L. J. (1973). Transcendental logarithmic utility function. *American Economic Review, 65*(3), 367–383.

Haskel, J., & Westlake, S. (2018). *Capitalism without capital: The rise of of the intangible economy*. Prinston University Press.

Jorgenson, D. W. (1987). Japan-US Economic Growth Industry -level Productivity Comparison (Postwar U.S. economic growth) (Vol. 1). MIT Press.

Jorgenson, D. W. (1995). *Productivity* (International comparisons of economic growth) (Vol. 1). MIT Press.

Jorgenson, D. W., Kuroda, M., & Nishimizu, M. (1987). Japan-U.S. industry level productivity comparison, 1960–1970. *Journal of the Japanese and International Economies, 1*(1), 1–30.

Kuroda, M. (1988). A method of estimation for updating transaction matrix in the input-output relationships. In K. Uno & S. Shishido (Eds.), *Statistical data bank systems. Socio-economic database and model building in Japan*. North Holland.

Kuroda, M. (1993). Price and goods control in the Japanese postwar inflationary periods. Chapter 2. In J. Teranishi & U. Kosai (Eds.), *The Japanese experience of economic reforms*. Macmillan Press and St. Martin's Press.

Kuroda, M., & Nomura, K. (2004). Technological change and accumulated capital: A dynamic decomposition of Japanese economic growth. Chapter 15. In E. Dietzenbacher & M. L. Lahr (Eds.), *Wassily Leontief and input-output economics*. Cambridge University Press.

Chapter 6
Results of Expanded Investment in the Intangible Assets in Japan: The Decline of Technological Efficiency in the Twenty-First Century and the Analysis of the Factors

Masahiro Kuroda and Michael C. Huang

Abstract Since 1995, Japan's economy has shifted its investment focus from tangible to intangible capital in response to demographic changes affecting the labor force. While tangible fixed capital investment declined, macro-level trends indicate a significant expansion in intangible capital investments. We interpret these changes by focusing on shifts in industrial technology and investment patterns, as well as their impact on economic growth. However, there remain many issues regarding the expansion of intangible capital investments, which now surpasses tangible capital investments, affects industrial efficiency, shapes the future economic structure, and can be effectively guided through policy measures. These include understanding the implications of this shift and addressing the challenges it presents. This chapter tackles the task of evaluations of the merits and demerits of intangible fixed capital investments, which differ from the technological efficiency of tangible capital, and linking these investments to economic growth. In the 1990s, Japan began to show signs of a new era, shifting away from a growth pattern that had continued since the end of World War II—one characterized by the expansion of heavy and the industries through capital I 1995, investment, which leveraged an increasing labor force and economies of scale to enhance efficiency and production capacity. Instead, there emerged a move toward building a new economic structure centered on the expansion of investments in intangible assets, often referred to as intellectual assets, and the development of an industrial structure that capitalizes on them. As is evident from Table 6-1, a key feature of economic growth since 1995 has been a sharp decline in the labor force, reflecting

M. Kuroda (✉)
Professor Emeritus, Keio University, Tokyo, Japan

M. C. Huang
Ocean Policy Research Institute, The Sasakawa Peace Foundation, Tokyo, Japan
e-mail: michael-huang@spf.or.jp

M. Kuroda (ed.), *Intangible Capital Investment and Economic Growth*, Translational Systems Sciences 45,
https://doi.org/10.1007/978-981-96-7675-0_6

demographic changes, and a shift in capital accumulation from tangible to intangible assets. In Chapter 6, we will examine the outcomes of this expansion in intangible capital investment and consider its impact on the economy of the 21st century. The rise of intangible capital investments over tangible ones raises questions about industrial efficiency, future economic structures, and policy guidance. This shift's implications and challenges need to be understood and addressed. This shift is evident in the focus on intellectual investments, such as research and development (R&D) and advancements in information and communication technology (ICT). These sectors have redefined economic efficiency and innovation. The rise of intangible capital has altered investment strategies and reshaped industrial dynamics, requiring a deeper understanding of the implications. This transformation calls for a reevaluation of how these investments contribute to productivity and growth, highlighting their role in developing a knowledge-driven economy.

This chapter must tackle the task of evaluating the merits and demerits of intangible fixed capital investments, which differ from the technological efficiency of tangible capital, and linking these investments to economic growth. By assessing whether intellectual assets such as intra-firm R&D and ICT activities have enhanced economic efficiency and contributed to achieving policy goals related to the expansion of intangible capital, we clarify the significance and impact of the growing investments in intangible fixed assets.

Keywords Tangible and Intangible capital · STI investment · R&D activity · ICT activity · Spillover effect · Trans-log price function

6.1 The Core Issue

According to observations of the Japanese economy from 1995 to 2011, it is indicated that the economic growth pattern following the collapse of the so-called bubble in the early 1990s has deviated from previous patterns and the property differs significantly from the economic characteristics that supported growth in the twentieth century. During this period, the growth pattern and the quality of the macroeconomic growth was quite different from the high economic growth era, and the average annual growth rate of TFP (Total Factor Productivity) remained at −0.05% in the macro economy, which is explicitly lower than the growth rate during the high economic growth period (Table 6.1).

I had the opportunity to encounter the work *Capitalism without Capital: The Rise of the Intangible Economy published in 1995* by two British authors, Jonathan Haskel and Stian Westlake. Their insights are incredibly thought-provoking and remarkably capture the features of contemporary investments.

Quoting on page 19 from their work:

> "The starting point of this book is observation. The nature of investment has been gradually changing, but significantly, over the past few decades. This change is not solely about information technology. New investment does not take the form of robots, computers, or silicon chips. However, all of these are supporting actors in the story. The inexorable rise in investment is in intangibles—investments in ideas, knowledge, aesthetic sense, soft brands, networks, and relationships."

Table 6.1 Annual growth rate of Japanese economy: 1955–2020

	1955–1960	1960–1965	1965–1970	1970–1975	1975–1980	1980–1985	1985–1990	1990–1995	1995–2000	2000–2005	2005–2010	2010–2015	2015–2020
Nominal GDP	12.8831	14.3606	16.1884	13.9805	9.6999	5.7739	5.8867	2.3558	0.5224	−0.1087	−1.0401	1.2462	0.0046
Real GDP	8.4103	10.1260	11.7910	4.3363	4.6236	3.3021	4.5496	1.5731	1.0353	1.1802	−0.0483	1.0438	−0.3715
Total population	0.9589	0.9686	1.0723	1.5127	0.9623	0.6922	0.4566	0.3276	0.2666	0.1322	0.0452	−0.1508	−0.2808
Total labor	2.1915	1.7750	1.8620	0.5576	0.9367	0.7964	1.0310	0.7390	−0.0341	−0.1776	−0.2869	0.3244	0.8413
Employee	5.3189	4.1013	2.9864	2.0080	1.4839	1.9875	1.5849	1.5534	0.3503	0.1377	0.3929	0.5841	1.0659
Capital input	n.a	10.2740	10.1900	7.9580	4.6470	4.9900	n.	n.a	3.6391	2.0855	−3.9371		
Tangible capital									3.4016	−2.2113	−4.1256		
Intangible capital									7.5714	−0.2918	−1.2728		
Total factor productivity	n.a	2.9790	5.4820	1.0160	0.7040	0.8940			−0.2509	−0.1584	0.1931		

Their notes showed how the nature of investment over the past 20–25 years has significantly diverged from that of the twentieth century.

They highlight how these intangible investments are hard to transform the economic and social fabric and discuss how they should continue to evolve. According to the authors, investments in intangible assets possess several characteristics: (1) they involve significant sunk costs, making their valuation challenging and their resale being difficult; (2) however, their utility value comes with an easily expandable scalability effect; (3) this expandability leads to substantial spillover effects into other uses; (4) when combined with other tangible or intangible investments, they generate synergy effects. Consequently, due to these characteristics, the construction of networks yields more significant effects. In essence, the expansion of "knowledge" as an intangible asset has characteristics allowing the assessment of its economic effects through the functions of synergy, scalability, spillover, and network, which distinctly differ from the scale effects of tangible fixed assets that exhibit economies of scale.

This chapter focuses on examining the effectiveness of measures to improve technological efficiency in industries. By measuring the parameters of technological characteristics using a trans-log price function, we aim to determine whether these technological characteristics have played an effective role in enhancing technological efficiency in line with the observed trends in factor price changes during the period. The goal is to assess whether the choice of these technological characteristics has effectively contributed to the observed trends in factor prices and, consequently, led to improvements in technological efficiency across industries.

Particularly, we intend to investigate whether intra-firm research and development (R&D) activities and intra-firm information and communication technology (ICT) activities in each industry during this period align with the achievement of science and technology policy goals through the expansion of intangible capital.

In Tables 6.2 and 6.3, we tried to compare the average annual growth rates of tangible and intangible capital investments in intra-firm research and development (R&D) and information processing activities over 5-year intervals from 1995 to 2011. It shows that during this period, the growth rate of investment in intra-firm R&D and information processing related to intangible capital services significantly exceeded that of investment in tangible assets. To understand what this implies and how it has impacted on the improvement of technological efficiency in Japanese industries, it is essential to analyze this structure.

By focusing on these facts, the aim is to understand the structural dynamics and explore how they could be linked to improvements in technological efficiency. Therefore, the challenge of this chapter is to re-examine the actual trends, and analyze the observed reality, to examine whether the Japanese economy has fully realized the benefits of expanding intangible capital investments and seek policy measures for the improvement.

Based on these results, we will examine whether intra-firm Research and Development (R&D) activity and intra-firm information and communication technological (ICT) activity in each industry align with the policy guidelines of science and technology policy aimed at expanding intangible capital.

Table 6.2 Average annual growth rate of tangible and intangible capital services by Industry: main R&D and ICT activities during 1995–2011

	Growth rate 1995–2000		
	Tangible capital services	Intangible capital services R&D	Intangible capital services ICT
1. Agriculture, forestry, and fisheries (main production, intra-firm R&D, intra-firm ICT activities)	0.0956	0.0414	–
3. Mining (main production, intra-firm R&D, intra-firm ICT activities)	−0.8942	−0.0156	–
5. Food manufacturing	−0.0819	0.0258	0.0446
8. Textile and wood furniture manufacturing	−0.0256	−0.0949	−0.0684
11. Pulp and paper manufacturing	0.0256	0.0293	0.1901
14. Chemical manufacturing	0.0638	0.0214	0.0894
17. Metallurgical manufacturing	−0.0268	0.0191	0.1849
20. General machinery manufacturing	0.0677	0.0321	0.1397
23. Consumer electronics and electrical machinery manufacturing	−0.0075	−0.0028	0.0755
26. Optical fiber and cable manufacturing	−0.1749	0.0049	−0.0382
29. Semiconductor equipment manufacturing	0.2293	0.2038	0.3024
32. Communication equipment semiconductor device manufacturing	0.0708	0.1141	0.1471
35. Electronic computers and related accessories manufacturing	0.0689	−0.0327	0.6539
38. Semiconductor element and integrated circuit device manufacturing	0.0949	0.0494	0.0972
41. Other electronic parts manufacturing	0.0331	0.0634	0.8243
44. Heavy electrical equipment and transport machinery manufacturing	0.0641	0.0416	0.1197
47. Robot, heavy electrical equipment, and transportation machinery manufacturing	0.0866	0.0904	0.1899
50. Other precision machinery, heavy electrical equipment, and transport machinery manufacturing	0.0923	0.0582	0.1699
53. Petroleum and coal products, heavy electrical equipment, and transport machinery manufacturing	−0.3186	−0.0582	−0.0036
56. Other manufacturing	0.0756	0.0332	0.0786
59. Energy industry	0.0579	0.0971	0.0869
62. Construction	−0.0175	−0.0171	0.2779
65. Transportation	0.0265	0.0826	0.0848
68. Communication and broadcasting services	0.0841	0.1933	0.2848

(continued)

Table 6.2 (continued)

	Growth rate 1995–2000		
	Tangible capital services	Intangible capital services R&D	Intangible capital services ICT
71. Commerce	0.0524	0.0849	0.1423
74. Software, communication, and broadcasting services	0.0089	0.2434	–
76. Information processing. Services provided: communication and broadcasting services	−0.0437	0.1966	–
78. Internet, communication, and broadcasting services	–	–	–
87. R&D life (industry/main productive activities)	0.0813	–	–
88. R&D information and communications (industry/main production)	0.1213	–	–
89. R&D substances and materials (industry/main production)	–	–	–
90. R&D environment and energy (industry/main production)	0.4741	0.4784	–
91. R&D other research and development (industry/major production)	0.0751	0.0868	–
92. Other service	0.0058	0.1407	–

We begin to examine how R&D investment and ICT investment affect the production efficiency of each industry through the accumulation of knowledge stock. This analysis focuses on the investment trends of each industry based on past performance, as shown in Tables 6.2 and 6.3. By examining the intangible capital investment behaviors in intra-firm R&D and intra-firm ICT activities, we aim to understand how these investments link to improvements in the technological efficiency of main production activities. We will organize these technological characteristics based on the actual conditions of past tangible and intangible capital investments.

Through the measurement of the trans-log price function, we capture the technological characteristics of each industry as a production technology structure that integrates main production activities with intra-firm activities such as R&D and ICT. This approach helps clarify the overall structural technological characteristics of industry.

Based on this, Sects. 6.4–6.5 will explore the nature of the productivity spillover effects between industries brought about by investments in intangible capital. We will consider the optimal technological characteristics and industrial structures that efficiently realize these effects, based on the reality of these spillover effects.

Furthermore, we attempt to explore the significance of intangible investments as the accumulation of knowledge stock. To confirm the economic meaning of the investment effects of intangible fixed assets, we aim to examine the overall investment effect, including the accumulation effect of tangible fixed assets, and

Table 6.3 Main products activity by industry during the year 2000–2011

Growth rate 2000–2005			Growth rate 2000–2010			Growth rate 1995–2011		
Tangible capital services	Intangible capital services R&D	Intangible capital services ICT	Tangible capital services	Intangible capital services R&D	Intangible capital services ICT	Tangible capital services	Intangible capital services R&D	Intangible capital services ICT
0.0065	−0.0967	–	−0.0085	−0.0339	–	−0.0087	−0.0334	–
−0.0741	0.0408	–	−0.0181	−0.0821	–	−0.0566	−0.0142	–
0.0242	0.0276	−0.1941	−0.0652	0.0214	0.1351	0.0253	0.0236	0.0005
−0.0497	−0.1139	−0.3782	−0.1087	−0.0131	0.1585	−0.0591	−0.0572	−0.0812
−0.0922	0.0025	−0.0879	−0.1129	−0.0255	−0.1258	−0.0549	0.0022	−0.0219
0.0544	0.0309	−0.1521	−0.0535	0.0247	0.0355	0.0242	0.0345	−0.0159
−0.0513	−0.0135	−0.1536	−0.0134	−0.0061	0.0431	−0.0232	0.0039	0.0021
0.0087	0.0675	0.0418	−0.0353	0.0013	−0.0091	0.0087	0.0399	0.0547
−0.1271	−0.0232	0.0447	−0.1128	0.0377	−0.0556	−0.0679	−0.0131	0.0153
−0.1107	−0.1181	0.0069	−0.0263	0.0381	0.05135	−0.0742	−0.0263	−0.0089
−0.0165	0.0346	0.0279	0.0049	0.0109	0.08473	0.0734	0.0832	0.1326
0.0463	−0.0399	0.0845	−0.0959	−0.0576	−0.0047	0.0232	0.0081	0.0769
0.0346	−0.0563	0.1263	−0.1071	−0.0216	−0.0245	−0.0194	−0.0569	0.0464
0.1704	−0.0696	−0.1245	−0.1457	−0.1811	0.0068	0.0089	−0.0742	−0.0185
−0.1244	−0.0695	−0.1477	−0.2119	−0.1998	−0.1064	−0.1007	−0.0605	−0.0529
0.0133	0.6951	0.0178	−0.1351	0.0411	0.0255	−0.0188	0.0454	0.0385
−0.0081	0.0409	0.0016	−0.0861	0.0327	0.0186	−0.0196	0.0598	0.0575
0.0119	0.0224	0.0369	−0.0402	0.0696	0.0491	0.0225	0.0587	0.0817
−0.0376	−0.1251	0.0471	0.1307	0.0227	0.6741	−0.0841	−0.0305	0.0279
−0.0155	0.0456	0.0451	−0.0981	0.0015	0.0132	−0.0051	0.0341	0.0384

(continued)

Table 6.3 (Continued)

Growth rate 2000–2005			Growth rate 2000–2010			Growth rate 1995–2011		
−0.0511	0.0487	0.0985	0.0258	−0.0224	0.0961	0.0044	0.0395	0.0869
0.0019	−0.2595	−0.1032	0.00E+00	−0.0908	0.1374	−0.0194	−0.1157	0.1017
0.0622	0.0716	−0.0194	0.0025	0.0187	0.0177	0.0219	0.0629	0.0252
−0.1406	0.1971	0.0479	−0.0697	−0.0447	0.0637	−0.0461	0.4754	0.1231
0.0078	0.2161	−0.0201	−0.041	−0.0045	0.0588	0.0017	0.7723	0.0619
−0.0482	−0.0484	–	−0.1228	−0.1228	–	−0.0476	0.0254	–
0.7062	0.0588	–	−0.1022	−0.1032	–	−0.0214	0.0449	–
–	–	–	−0.1131	0.0407	–	–	–	–
−0.4726	−0.4617	–	−0.0774	0.1189	–	−0.1087	−0.0511	–
0.0216	0.0856	–	−0.1465	0.0215	–	0.0013	0.0809	–
–	–	–	−0.2204	−0.0083	–	–	–	–
−0.2317	−0.2559	–	−0.1526	0.0057	–	0.0201	0.0578	–
−0.2331	−0.2289	–	−0.1506	0.0264	–	−0.1041	−0.0495	–
−0.0454	0.0025	–	−0.0525	−0.1272	–	−0.0238	0.0173	–

understand the implications of economic fluctuations in the Japanese economy since the early twenty-first century. As mentioned in Chap. 5, the effects of these intangible investments, as captured by the technological characteristics expressed through price functions, vary significantly with the movements in the prices of factors such as capital, labor, and intermediate materials.

Therefore, in this chapter, by capturing the characteristics of these production factor markets and the impact of fluctuations in the prices of each production factor, we aim to scrutinize the future direction of science and technology policy in Japan's ongoing structural reforms. This analysis will contribute to the final chapter's discussion on shaping science and technology policies that align with the current economic structure and support sustainable growth.

6.2 Trend of the Factor's Relative Prices and the Choice of Technology

6.2.1 Synergy Effect of Technological Efficiency

In the analysis up to the previous chapter, we have accumulated observations from time-series data of input-output tables. Additionally, we have seen changes in external factors such as population structure shifts, scientific and technological advancements, and corresponding variations in people's preferences. Taking these external changes into account, we aim to observe more precisely how the technological efficiency of industries in the Japanese economy has undergone significant transformations during this period.

1. Shifting from an economic growth pattern driven by the vigorous expansion of investment demand, particularly in tangible fixed assets, to an era where intangible *capital investment, such as the accumulation of knowledge stocks and the diffusion* of information in both quality and quantity due to advances in information processing technology, leads to the new growth pattern of the economy.
2. This is also evidence we could observe when we were examining indicators of the macro-level total productivity as the measure during the recent periods. However, observations of the Japanese economy from 1995 to 2011 reveal that, although there are signs of structural change, a completion of the socio-economic structure to fully reap the benefits has not yet been achieved.
3. Amid these rapid changes, by the end of the 20th century, the globalized economic society was also undergoing major structural transformations, bringing about shifts in the global geopolitical landscape. However, our observations indicate that although there are signs of structural transformation, the Japanese economy has not yet transitioned into a socio-economic structure that can fully reap these benefits..
4. These points can also be inferred from the observations of economic efficiency by the TFP measure in the macro level since 2000. Furthermore, as shown in Table 6-1, this is consistent with the observed fact that the growth rate of tangible capital investment has fallen below that of intangible capital investment.

The activities within the industry are assumed to be interconnected with one another.

In the sector j, considering the integration of main production activities, intra-firm R&D activities, and intra-firm ICT activities, it is assumed that actions are taken to minimize long-term costs in the technological choices of the industry. In our analytical framework, the long-term cost function is assumed to be formulated as a trans-log cost function. Based on the cost minimization principle with input share functions for tangible capital services, intangible capital services for intra-firm R&D, intangible capital services for intra-firm ICT processing, labor services, and intermediate goods input and technology improvement rate coefficient $-v_j^T$, given the prices of each production factor, the input levels of production factors for long-term cost minimization to achieve the assumed planned production quantity and the technology chosen to realize it will be selected.

It will be assumed the cost-minimizing actions for the tangible capital services, labor services, and intermediate inputs both in the intra-firm R&D and ICT activities. These long-term cost-minimizing technology choices, under the interdependent relationships and technological conditions of main production activities, intra R&D, and intra information processing activities, are determined depending on the given prices of each production factor in the factor markets. The determination of these production factor prices is considered to have a general interdependent relationship through the actions of all participants in the market-demand decision-making entities.

5. The wage as labor service price and input prices for intermediate materials can be directly observed in labor and material markets. However, except in cases like rental markets for cars and houses, the service price is not directly observable. Our explanation which was discussed in Chap. 5, Sect. 5.2, is how to measure capital service prices under such circumstances. How is the rental fee for a rental car determined? Depending on the intricacies of the rental car market, fundamentally, the rental fee is likely set to an amount that ensures the total income from renting out a car, purchased and invested by the rental car company, for the rental period, does not fall below the total maintenance cost of that car during the rental period. In other words, it would be rational to set the rental price so that the present discounted value of the total income over the rental period is equal to the total cost during that period. The estimated discount rate might be close to the market interest rate if the market is perfectly competitive and there is no misinformation. The pricing of capital services proposed in Chap. 4 was inspired by this analogy with rental markets. Of course, as in the case of the rental car market, where directly measuring the fee is challenging, the assumption is made that the capital service price, for both tangible and intangible (Intra research and Intra information processing) assets, is set such that the discounted present value of the total investment cost (purchase cost of the investment goods, including maintenance costs) is equal. This was derived as shown in Sect. 5.3 of Chap. 5.

In the capital service price function, $r^{Kit}{}_j$ refers to discount rate (capital rate of return), when the present value of each tangible and intangible asset's income is determined. Since the investment amounts for tangible and intangible assets, as well as the purchase prices of these investment goods can be directly observed, tangible and intangible capital stocks are estimated, and it is assumed that the stock quantity is proportional to the service quantity provided by the stock. On the other hand, the capital cost for each industry is determined as the sum of the operating surplus and the capital depreciation allowance for tangible and intangible capital, within the value-added amount in the input-output table, serving as the total cost of all capital services. With this idea, if the discount rates for calculating the asset values of each capital service are equalized among assets, the discount rates (capital rates of return) are calculated. We tried to estimate the prices and quantities of tangible and intangible capital services.

Thus, it is possible to obtain estimates for the quantities and prices of tangible and intangible capital services, as well as the capital rate of return (discount rate), corresponding to the accumulation of investment that realizes tangible and intangible capital service quantities in the form of capital stock. The estimated annual average growth rates of tangible and intangible capital service quantities by industry are summarized in Tables 6.2 and 6.3. The calculated capital rate of return also influences the level of expected future planned output and is involved in the determination of factor input quantities to realize cost-optimized technology choices.

The main subject of this chapter is how to construct the theoretical framework to be able to explain the empirical evidence consistently.

6.3 Measurement of Synergy Effects by Trans-Log Price Function

A schematic representation of the relationships in the production activities of the three activities within an industry (main production, corporate R&D, corporate information processing ICT can be depicted as shown in Fig. 6.1).

Let us focus on the three activities: The figure illustrates the Synergy effect between these three activities, captured by the technological characteristics formulated by the trans-log price functions. The technological characteristics of the production activity are identified by the sign conditions of the parameters β^{TT} and β_j^{iT} ($i = K, Ki, Kr, L, M$). In the formulation with trans-log price function, the technical progress in the production activity is expressed as follows,

$$v_j^T = -\ln P_j / \partial \boldsymbol{T} = -\left(\alpha_j^T + \sum_i \beta_j^{iT} \ln P_j^i + \beta_j^{TT} T\right). \tag{6.1}$$

where the rate of technical progress v_j^T is represented by the function of technical trend T with the parameters β_j^{TT} and input price of each factor P_j^i with the parameters β_j^{iT}.

In the capital service price function, $r^{Kit}{}_j$ refers to discount rate (capital rate of return), when the present value of each tangible and intangible asset's income is determined. Since the investment amounts for tangible and intangible assets, as well as the purchase

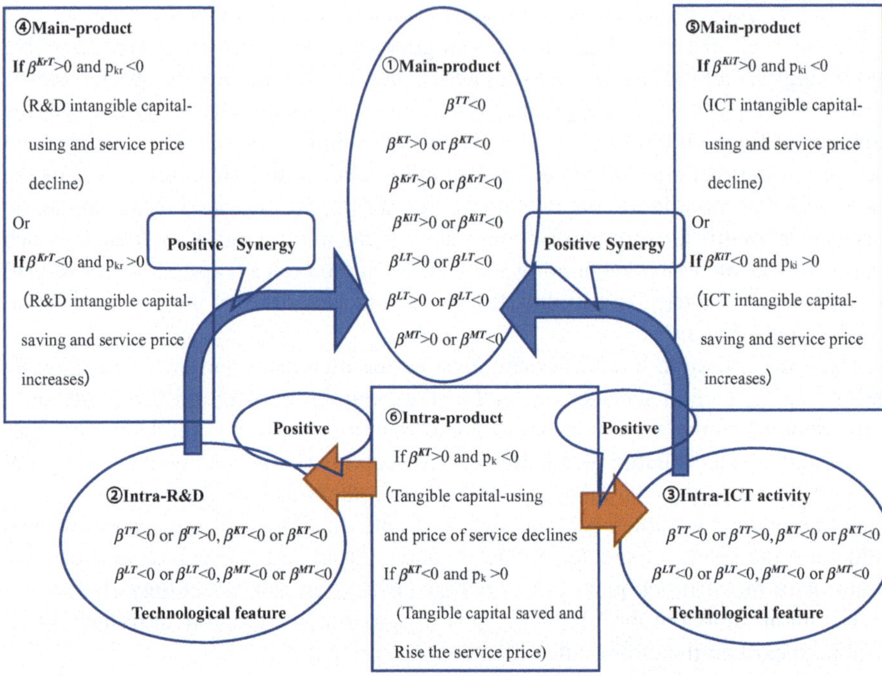

Fig. 6.1 Framework of the relative relationships among three industrial activity

prices of these investment goods can be directly observed, tangible and intangible capital stocks are estimated, and it is assumed that the stock quantity is proportional to the service quantity provided by the stock. On the other hand, the capital cost for each industry is determined as the sum of the operating surplus and the capital depreciation allowance for tangible and intangible capital, within the value-added amount in the input-output table, serving as the total cost of all capital services. With this idea, if the discount rates for calculating the asset values of each capital service are equalized among assets, the discount rates (capital rates of return) are calculated. We tried to estimate the prices and quantities of tangible and intangible capital services.

Thus, it is possible to obtain estimates for the quantities and prices of tangible and intangible capital services, as well as the capital rate of return (discount rate), corresponding to the accumulation of investment that realizes tangible and intangible capital service quantities in the form of capital stock. The estimated annual average growth rates of tangible and intangible capital service quantities by industry are summarized in Tables 6.2 and 6.3. The calculated capital rate of return also influences the level of expected future planned output and is involved in the determination of factor input quantities to realize cost-optimized technology choices.

Regarding the input factors in main production, technological progress has a characteristic that determines the direction of these input factors—whether they become more utilitarian or more frugal. This characteristic applies to tangible fixed capital (K), intangible capital from corporate R&D (K_r), intangible capital from information processing (Ki), labor services (L), and intermediate inputs (M). In other words, when β^{i} ($i = K, Ki, Kr, L, M$) is negative, it indicates a technologically progressive bias toward becoming more utilitarian, while a negative value indicates a bias toward becoming more frugal. This characteristic is also defined by the sign of the parameter β^{iT} ($i = K, L, M$) in the activity.

In our model, the function is assumed as a first-degree homogeneous function, if the technological efficiency does not change with the output scale for a given technology at a specific point in time. Instead, technological efficiency shifts over time according to the trend of technological progress (β^{TT}), which represents technological advancement. As will be explained later, technological characteristics such as synergy can be expressed through the magnitude of the parameters in the trend of technological progress in major production and intra-firm activities. The production activity is linked to two others: intra-firm R&D and intra-firm ICT activities. **We call it as "Internal Synergy Effect**."

If the accumulation of knowledge embodied in intangible fixed assets reflects the aggregation of knowledge stock in an industry, improvements in the technological efficiency of knowledge stock accumulation in that industry will lead to a reduction in the prices of the industry's main production outputs. If the ripple effects of this efficiency improvement function as an intangible knowledge stock effect on other industries, it may influence changes in the technological efficiency of those industries through the market transaction. This represents the synergy and scalability inherent in intangible knowledge stock. **We call it as "External Synergy Effect**."

Rate of the technological progress in Eq. (6.1) depends on the parameters of the progress trend (β^{TT}) and technological biases (β^{iT}), serving as functions of production factor's prices P_j^i and time-trend T. Similar technological progress is derived from the technical progress in the intra-firm R&D and the intra-firm ICT activities. The parameters of these intra-firm price functions for both intra-firm activities reflect the technological characteristics of the industry. The parameter β^{TT} of the technological progress trend for the three activities, including main production, intra R&D and ICT activities, reflects the degree of shift in the price function due to technological progress by time trend, assuming a first-degree homogeneous function. If the parameter β^{TT} is negative, where the larger its absolute value, the higher the shift rate of the function—indicating a higher technological progress rate and a decrease in the product price due to technological progress. Our formulation assumes a trans-log price function capturing a first-degree homogeneous relationship, where we are assuming no change in technological efficiency with output scale expansion for a given technology at a specific point in time. Instead, we assume that technological efficiency shifts between points in time through the technological progress trend β^{TT}, signifying technological progress. We believe that the effects of Synergy and Scalability as technological characteristics can be expressed through the magnitude of the parameters in the technological progress trends of main production, which is linked by two intra-firm R&D, and infra-firm ICT activities. Particularly, if the accumulation of knowledge embodied in intangible fixed assets reflects the aggregation of knowledge stocks in an industry, an increase in the technological efficiency of knowledge stock accumulation in that industry may lead to a decrease in the output price of the industry's main production activity. If this efficiency's ripple effect serves as an intangible knowledge stock effect in other industries, it could influence changes in the technological efficiency of other industries—a representation of synergy and scalability inherent in intangible knowledge stocks.

Regarding the input factors of main production, there is a characteristic of technological progress determining the direction of input factors—whether they become more utilitarian or more frugal. This characteristic applies to tangible fixed capital (*K*), intangible capital from corporate R&D (*Kr*), intangible capital from information processing (*Ki*), labor services (*L*), and intermediate inputs (*M*). In other words, β^{iT} ($i = K, Ki, Kr, L, M$) being positive implies a technologically progressive bias toward being more utilitarian, while a negative value implies a technologically progressive bias toward being more frugal. This characteristic is also characterized by the sign of the parameter β^{iT} ($i = K, L, M$) for both corporate activities.

The price of intangible fixed capital services in the main production activity is constrained by the price dynamics of intangible capital services within the industry, determined by the technological efficiency of corporate R&D and information processing ICT activities. This is illustrated in Fig. 6.1 with points ② and ③. In ②, concerning the input of R&D intangible fixed capital services in main production, when the technological bias (β^{KrT}) is positive, indicating that the technology progress makes this factor more utilitarian, the price of intangible fixed capital services from corporate R&D decreases, leading to more efficient technological progress in main production. Conversely, when the technological bias is negative, indicating frugality, an increase in the price of intangible fixed capital services inhibits the rise in technological efficiency in main production.

We refer to interdependence among the three activities within industries that are well-versed in the technological characteristics of these industries as the **Internal Synergy Effect** of the main production and intra-firm activities of each industry. Therefore, whether the synergy effect functions effectively depend on the technological bias characteristics of the main production activities and both intra-firm activities, as well as the trends in factor prices.

Regarding the price of tangible capital services, it not only depends on the main production activities and intra-firm activities of the industry but also on the trends in the demand for tangible capital services from other industries through the domestic market for investment goods that determine the investment prices of tangible capital. Thus, it is influenced by the supply and demand structure for tangible capital investment goods from other industries in the investment goods market.

Furthermore, this applies not only to capital service prices but also reflects the fluctuations in factor prices through the supply and demand structure in the market for labor services and intermediate inputs, considering the factor demands among industries in the domestic market. Here too, as characteristics of the labor and intermediate goods markets, price fluctuations in the market influenced by the actions of other industries will have an impact.

The effects of these influences, reflecting the interdependence among industrial sectors within a national economy, are referred to as the **External Synergy Effects**, distinguishing them from the Internal Synergy Effects. Observing the internal and external synergy effects of an industry as indicators to evaluate the impact of science and technology policies is crucial for assessing the effectiveness of these policies.

Similarly, as indicated in ③, the technological bias of intangible fixed capital services for information processing in the main production activity and the price trends of information processing services from corporate information processing activities influence the technological efficiency of the main production activity. On the other hand, these two corporate activities are influenced by the price of tangible fixed capital services in the main production activity and the technological bias parameters (β^{KT}) for tangible fixed capital services in both corporate activities. If the technological bias parameters for both corporate activities are positive (negative), the efficiency change direction will depend on whether the price of tangible capital services in the main production activity decreases (increases).

We refer to the interdependence relationship between the three activities of an industry through the technological characteristics of the industry as the **Internal Synergy Effect** between the main production and intra-firm activities. Therefore, whether the Synergy Effect works effectively depends on the technological bias characteristics of both main production activities and intra-firm activities, as well as the factor prices. Regarding the price of tangible capital services, it is not only influenced by the main production activities and both intra-firm activities of the industry but also depends on the demand trends for tangible capital services in other industries through the domestic market for investment goods, determining the investment prices for tangible capital. This influence extends not only to the price of capital services but also to labor services and intermediate inputs. The fluctuation in factor prices through the demand and supply structure of tangible capital investment goods in the investment goods market reflects the inter-industry factor demand within the domestic market. Similarly, for labor and intermediate goods markets, the impact of the trend of price changes in the market reflecting the behavior of other industries will be reflected.

The prices of tangible and intangible capital services, as formulated in Chap. 5, depend on the investment goods prices for tangible and intangible assets, their capital rates of return, as well as corporate income tax rates, fixed asset tax rates, economic capital depreciation rates, and so on. As mentioned earlier, we assume here that the capital rates of return are equalized for tangible and intangible capital, and their determination level is influenced by factors such as the operating surplus of each activity, current corporate profit amounts, and their taxation, tangible, and intangible investment goods prices. Although tax rates and capital depreciation rates may be exogenously given as policy parameters, the capital rate of return reflects on the demand and supply relationships in all markets for goods and services, depending on the fluctuation in investment goods prices in the market. As a result of all these interrelated changes, the rational decision-making for long-term cost minimization in each industry leads to the selection and determination of the technological characteristics of factor inputs for the next period. While the overall interdependence of the economy is a complex structural issue, in this context, we assume that the parameters of the price functions reflect the technological choices resulting from the long-term decisions of each industry. The transition of tangible and intangible fixed capital investment is expected to have a similar impact on the technological efficiency of the industry. It enhances synergy between the main production and two intra-firm activities, influencing the technological efficiency of the industry. Whether policies inducing trends in factor prices related to each activity can guide the industry's technological efficiency in a direction that enhances it is considered a crucial challenge in science and technology policy.

The exogenous synergy effect suggests that the impact of shifts in science and technology policies in a specific industry on other industries varies based on the strength of the relationship between the specific industry and other industries. It is conceivable that the influence and propagation of the impact on inter-industry spillovers differ depending on the intensity of the relationship between a specific industry and others. This variation is expected to be influenced by the scale of changes in science and technology policies within a particular industry, resulting in differences in the magnitude and extent of the impact. It is believed that this effect varies based on the extent and scale of the interdependence of technologies between industries. To distinguish this effect from the endogenous synergy effect mentioned above, it is referred to as the external scalability effect and external spillover effect. When we try to evaluate the impact of science and technology policies, it is crucial to pay attention to these indicators of synergy, scalability, and spillover effects.

Here we revisit the conceptualization and estimation methodology of our capital rate of return. In a general context where industry operations are categorized into three distinct sectors, namely "main production activities," "intra-firm R&D activities," and "intra-firm information processing activities," the aggregate capital cost of the industry can be expressed as follows:

$$
\begin{aligned}
BS^t{}_j = {} & [K^t{}_j P^{KSt}{}_j + KITE^t{}_j P^{KSITEt}{}_j + KNITE^t{}_j P^{KSNITEt}{}_j + KRDE^t{}_j P^{KSNRDEt}{}_j \\
& + KNRDE^t{}_j P^{KSNRDEt}{}_j] \cdot P^{SK0} \\
= {} & [K^t{}_j \left\{ \left(1-\tau^K\right) r^{Kt}{}_j P^{INVKt-1}{}_j + \delta^K{}_j P^{INVKt}{}_j - \left(P^{INVKt}{}_j - P^{INVKt-1}{}_j\right) + \tau^P P^{INVKt-1}{}_j \right\} \\
& + KITE^t{}_j \{ \left(1-\tau^{KITE}\right) r^{Kt}{}_j P^{INVKIETt-1}{}_j + \delta^{KITE}{}_{j+1} P^{INVKITEt}{}_j \\
& - \left(P^{INVKITEt}{}_j - P^{INVKITEt-1}{}_j\right) + \tau^{PKITE} P^{INVKITEt-1}{}_j \} \\
& + KNITE^t{}_j \left(1-\tau^{KNITE}\right) r^{Kt}{}_j P^{INVKNITEt-1}{}_j + \delta^{KNITE}{}_{j+1} P^{INVKNITEt}{}_j \\
& - \left(P^{INVKNITEt}{}_j - P^{INVKNITEt-1}{}_j\right) + \tau^{PKNITE} P^{INVKNITEt-1}{}_j \} \\
& + KRDE^t{}_j \{ \left(1-\tau^{KRDE}\right) r^{Kt}{}_j P^{INVKRDEt-1}{}_j + \delta^{KRDE}{}_j P^{INVKRDEt}{}_j \\
& - \left(P^{INVKRDEt}{}_j - P^{INVKRDEt-1}{}_j\right) + \tau^{PKRDE} P^{INVKRDEt-1}{}_j \} \\
& + KNRDE^t{}_j \{ \left(1-\tau^{KNRDE}\right) r^{Kt}{}_j P^{INVKNRDEt-1}{}_j + \delta^{KNRDE}{}_j P^{INVKNRDEt}{}_j \\
& - \left(P^{INVKNRDEt}{}_j - P^{INVKNRDEt-1}{}_j\right) + \tau^{PKNRDE} P^{INVKNRDEt-1}{}_j \}] \cdot P^{SKO}{}_j
\end{aligned}
\tag{6.2}
$$

$BS^t{}_j$ represents the total capital cost of the three activities, serving as the sum of operating surplus and the total of capital depreciation allowances, which are observable from the input-output table. Additionally, $K^t{}_j$, $P^{KSt}{}_j$, $KITE^t{}_j$ $P^{KSITEt}j$, $KNITE^t{}_j$, $P^{KSNITEt}{}_j$, $KRDE^t{}_j$ $P^{KSRDEt}{}_j$ are, respectively, the tangible fixed capital service quantity and its service price for main production activities ($K^t{}_j$ $P^{KSt}{}_j$), the tangible fixed capital service quantity and its service price for intra-firm information processing activities ($KITE^t_j$ $P^{KSITEt}{}_j$), the intangible fixed capital service quantity and its service price for intra-firm information processing activities ($KNITE^t_j$ $P^{KSNITEt}{}_j$), the tangible fixed capital service quantity and its service price for intra-firm R&D activities ($KRDE^t_j$ $P^{KSRDEt}{}_j$), and the intangible fixed capital service quantity and its service price for intra-firm R&D activities ($KNRDE^t_j$ $P^{KSNRDEt}{}_j$). $P^{KS0}{}_j$ is the adjustment coefficient at the reference time.

We can induce the capital service price for the tangible and intangible asset.

$$P^{kSt}{}_j = \left(1-\tau^k\right) r^{kt}{}_j P^{INVkt-1}{}_j + \delta^K{}_j P^{INVKk\ t}{}_j - \left(P^{INVk\ t}{}_j - P^{INVkt-1}{}_j\right) + \tau^P P^{INVkt-1}{}_j$$
$$k = K, KITE, KNITE, KRDE, KNRDE \tag{6.3}$$

Here, r^{kt}, $\delta^k{}_j$, τ^k and τ^P represent the rate of return on tangible and intangible fixed capital services, the economic depreciation rate of capital, the capital income tax rate, and the property tax rate, respectively. P^{INVktj} and P^{INVNkt} are the price of investment goods for tangible and intangible capital.

r^{kt}, δ^k, τ^k, τ^P (k = $K,KITE,KNITE,KRDE,KNRDE$) represent the capital rate of return, the capital depreciation rate, the capital gains tax rate for tangible and intangible capital, and the fixed asset tax rate for tangible capital. It introduces P^{INVkt} as the investment goods price for tangible and intangible capital. Assuming that the capital rate of return r^{kt} for industry j is equal for tangible and intangible capital services, the derived capital profit rate is obtained, as defined in the Eq. (5.11) from the previous chapter. Therefore, the equalized capital rate of return is defined as follows.

$$r^{*t}{}_j = F(BS^t{}_j, P^{INVkt}, P^{INVkt-1} \left(k = K,KITE,KNITE,KRDE,KNRDE\right); \delta^k{}_j, \tau^k, \tau^P) \tag{6.4}$$

The current capital cost is a function of the investment goods prices or tangible and intangible fixed capital in the current and previous periods. It functions as a function of policy variables, including the capital depreciation rate, capital gains tax rate, and fixed capital tax rate. Since capital costs and the investment goods prices are endogenously determined because of the supply and demand equilibrium in the current factor market, the equilibrium levels of capital costs and capital investment goods prices for the current period are assumed to statically continue into the next period when determining the minimal cost for next period's investment. In this scenario, the capital depreciation rate, profit tax rate, and fixed tax rate are considered by exogenous policy variables, and policies can influence the determination of the capital profit rate and the prices of tangible and intangible capital services. Furthermore, the capital profit rate is believed to influence not only the determination of capital service prices mentioned earlier but also the decision on the planned production quantity for the next period. A higher capital profit rate can impact future expected profit rates, influencing the levels of each production factor derived from the minimal cost principle and consequently affecting the technological choices for the next period.

Keeping in mind the determination mechanism of production factors from long-term equilibrium, let's examine the structure of the technological characteristics obtained from the trans-log price function measured in the previous chapter. Applying the trans-log price function to observations from 1995 to 2011, and using these measured price function values, we believe we effectively capture the observed facts during this period. First, we confirm how well the trans-log function fits the data. Tables 6.4 and 6.5 show the checking results of the goodness of fit of the price functions concerning main production activities in each industry sector, intra-firm R&D activities, and intra-firm ICT activities, using correlation coefficients between observed and theoretical values for the distribution rates of each production factor and the rate of technological progress. This allows us to ascertain to what extent the model reproduces the movements of observed data regarding the distribution rates of each production factor and the rate of technological progress.

The values of the measured correlation coefficients may not necessarily indicate a good fit from a statistical perspective. However, the observed input shares and technological progress rates for each factor contain disturbances that cannot completely explain by the limited variables captured explicitly in the trans-log functions. At this stage, the importance is not to delve into these disruptive variations, but rather to capture the trend of observed changes in input shares and technological progress rates as trends driven by changes in dominant variables.

As an example, the fitting of input shares and technological progress rates for each factor in the main production activity of the textile manufacturing industry are illustrated in Fig. 6.2 (1)–(6). While the theoretical values may not capture the detailed variations at each time point completely, it is important that they broadly follow the trend of the time-series variations.

For both intra-firm activities, regarding the time-series changes in input share and technological progress rate, it is considered that the theoretical values almost reflect the trends observed in the empirical data. Regarding the correlation coefficients seen in Fig. 6.3 (1)–(8), the fitness is not good in several industries. However, as seen in the plots for the textile manufacturing industry mentioned above, it is judged that the theoretical values roughly follow the trends of observed time variations. From the technological characteristics captured by the trans-log price function, it is believed that it is possible to explore the structure of changes in technological efficiency for each industry during this period.

Keeping in mind the industrial classification based on the technological characteristics shown in the previous chapter, let's consider its structural implications. In this context, we would like to classify industries based on observations related to the capital return rates and capital service prices of the main production activities during the observation period of 1995–2011. We will explore whether there is a downward trend, an upward trend, or no discernible trend in these time-series observations. Additionally, we will focus on the sign conditions of the trend coefficients β^{TT} and

Table 6.4 Correlation coefficients between observed and theoretical values for distribution rates of production factors and rate of technical progress

	Main activity						
Industry classification	Technology bias	skr	ski	sk	sl	sm	st
a-1 BKT < 0							
a-1-1 BTT < 0,BKrT<0							
8–10 Textile industry	BkiT < 0,BLT > 0,BMT < 0	0.6963	0.7188	0.8602	−0.4501	−0.6914	0.3765
a-1-2 BTT < 0,BkrT>0							
14–15–16 Chemical industry	BkiT < 0,BLT < 0,BMT > 0	−0.2421	0.8031	0.8907	0.9453	0.9091	0.4136
44–45–46 Heavy electrical equipment and transport machinery manufacturing.	BkiT > 0,BLT < 0,BMT > 0	0.6892	0.8481	0.9208	0.8348	0.8994	0.6056
47–48–49 Robot industry	BkiT > 0,BLT > 0,BMT > 0	0.1511	0.4642	0.8391	−0.7164	0.8512	−0.05
50–51–51 Other precision machinery	BkiT > 0,BLT > 0,BMT < 0	0.7386	0.9398	0.8151	−0.5079	−0.4935	0.4138
a-2 BKT > 0							
a-2-1 BTT < 0,BkrT < 0							
35–36–37 Electronic computers	BkiT > 0,BLT > 0,BMT < 0	0.4368	0.5319	0.2632	−0.7008	−0.5042	0.5991
38–39–40 Semi-conductor industry	BkiT > 0,BLT > 0,BMT < 0	0.9071	0.7504	0.1718	−0.2314	0.3342	0.3167
62–63–64 Construction	BkiT > 0,BLT > 0,BMT < 0	0.8199	0.8862	−0.0427	−0.3936	−0.4139	0.5941
a-2-2 BTT < 0,BkrT>0							
20–21–22 General machinery	BkiT > 0,BLT > 0,BMT < 0	0.4621	0.7518	−0.6595	−0.7634	−0.7779	0.5924
29–30–31 Semi-conductor equipment	BkiT > 0,BLT > 0,BMT < 0	−0.2571	0.7797	0.0111	0.3216	−0.0969	−0.0137
32–33–34 Communication machinery	BkiT < 0,BLT > 0,BMT < 0	0.3284	0.1331	0.5115	−0.2419	0.0687	−0.0803
b-1 BKT < 0							
b-1-1 BTT > 0,BKrT > 0							
59–60–61 Energy industry	BkiT > 0,BLT < 0,BMT > 0	0.4621	0.8809	0.9128	0.8384	0.9204	0.0352
b-2 BKT > 0							
b-2-1 BTT < 0,BkrT>0							
65–66–67 Transportation service	BkiT > 0,BLT > 0,BMT < 0	0.6988	0.3848	0.2209	−0.8589	−0.8219	−0.1261
c-1 BKT < 0							
c-1-1 BTT < 0,BkrT < 0							
26–27 Fabricated metal	BkiT > 0,BLT < 0,BMT > 0	0.4015	0.7288	0.8645	0.7782	0.8611	−0.0581
c-2 BKT > 0							

(continued)

Table 6.4 (Continued)

Industry classification	Main activity						
	Technology bias	skr	ski	sk	sl	sm	st
c-2-2 BTT < 0,BkrT > 0							
68–69–70 Communication broadcast	BkiT > 0,BLT < 0,BMT < 0	0.9517	0.9639	−0.3845	0.9071	−0.0936	−0.0938
71–72–73 Trade	BkiT > 0,BLT > 0,BMT < 0	0.7751	0.9535	0.8313	−0.8689	−0.5798	−0.5798
74–75 Software industry	BLT > 0,BMT < 0	0.9439		0.2997	−0.0926	0.2521	0.2521
76–77 Information and communication	BLT < 0,BMT < 0	0.9501		0.6842	0.7775	−0.4848	−0.4848
93–94 Other service	BLT > 0,BMT < 0	0.5965		0.4163	−0.8487	−0.1189	−0.1189
d-1 BKT < 0							
d-1-1 BTT < 0,BKrT > 0							
17–18–19 Iron and steel industry	BkiT > 0,BLT < 0,BMT > 0	0.3558	0.5439	0.8897	0.8995	0.1779	0.1779
23–24–25 Electric machinery	BkiT > 0,BLT > 0,BMT < 0	0.8401	0.2727	0.8922	0.7604	−0.0797	−0.0797
41–42–43 Other electric parts	BkiT < 0,BLT > 0,BMT < 0	−0.0174	0.3961	0.7884	−0.3611	−0.1885	−0.1885
d-2 BKT > 0							
d-2-2 BTT < 0,BkrT > 0							
11–12–13 Pulp and paper industry	BkiT > 0,BLT > 0,BMT < 0	0.7846	0.3372	−0.7995	−0.7767	−0.8351	−0.1624
e-1 BKT < 0							
e-1-2 BTT < 0,BkrT > 0							
3–4 Mining	BLT > 0,BMT > 0	−0.1931		0.8433	−0.3193	0.8001	−0.4254
e-2 BKT > 0							
e-2-1 BTT < 0,BkrT < 0							
1–2 Agriculture, forestry, and fishery	BLT > 0,BMT < 0	0.9027		−0.4698	−0.7357	0.6997	0.6997
e-2-2 BTT < 0,BkrT > 0							
5–7 Food products industry	BkiT > 0,BLT > 0,BMT < 0	0.5548	−0.1987	0.6571	−0.8621	0.2247	0.2247
56+57–58 Other manufacturing	BkiT > 0,BLT > 0,BMT < 0	0.6975	0.8793	−0.8132	−0.6152	0.5458	0.5458
f-1 BTT > 0, BKT < 0							
f-1-1 Bkr < 0							
53–54–55 Oil and coal products	BkiT > 0,BLT < 0,BMT > 0	0.3561	0.2779	0.8901	0.9253	0.2977	0.2977

Table 6.5 Correlation coefficients between observed and theoretical values for distribution rates of production factors and rate of technical progress

Intra-firm research and development activity (RDA)					Intra-firm information processing activity				
Technology bias	sk	sl	sm	st	Technology bias	sk	sl	sm	st
BKT < 0,BLT > 0,BMT < 0	0.9008	−0.5333	−0.6973	−0.0349	BKT > 0,BLT < 0,BMT < 0	0.1059	0.6251	−0.7404	0.6234
BKT < 0,BLT > 0,BMT < 0	0.8874	−0.6677	−0.7634	0.1577	BKT > 0,BLT > 0,BMT < 0	−0.0508	0.6491	−0.8033	0.6093
BKT < 0,BLT > 0,BMT < 0	0.8611	−0.8101	−0.7501	0.2382	BKT > 0,BLT > 0,BMT < 0	0.3961	0.7123	−0.7196	−0.1012
BKT < 0,BLT > 0,BMT < 0	0.8462	−0.6655	−0.7581	−0.2324	BKT > 0,BLT > 0,BMT < 0	−0.4347	0.6252	0.8399	−0.0033
BKT < 0,BLT > 0,BMT < 0	0.8291	−0.4008	−0.9019	−0.1937	BKT > 0,BLT > 0,BMT < 0	−0.3204	0.5735	−0.7809	0.4095
BKT < 0,BLT > 0,BMT < 0	0.8736	−0.0005	−0.3332	0.2219	BKT > 0,BLT > 0,BMT < 0	0.2259	−0.3551	−0.3352	0.3064
BKT < 0,BLT > 0,BMT < 0	0.9371	0.6074	−0.9286	0.2008	BKT > 0,BLT > 0,BMT < 0	0.8721	0.8127	−0.3204	0.0182
BKT < 0,BLT > 0,BMT > 0	0.8279	0.4467	−0.8596	0.4441	BKT > 0,BLT < 0,BMT < 0	−0.0097	0.7014	0.7371	0.3231
BKT < 0,BLT > 0,BMT < 0	0.9097	−0.6489	−0.7219	−0.1371	BKT > 0,BLT > 0,BMT < 0	−0.3316	0.5815	0.8546	0.4211
BKT < 0,BLT > 0,BMT < 0	0.9293	−0.2151	−0.6924	−0.0616	BKT > 0,BLT > 0,BMT < 0	−0.3277	0.5664	−0.8525	0.0286
BKT < 0,BLT > 0,BMT < 0	0.9101	−0.4424	−0.6551	−0.1553	BKT > 0,BLT > 0,BMT < 0	0.3757	0.6995	−0.6975	0.3919
BKT < 0,BLT > 0,BMT < 0	0.7866	−0.6063	−0.8369	−0.3541	BKT > 0,BLT > 0,BMT < 0	0.2088	0.6783	−0.7846	0.0386
BKT < 0,BLT > 0,BMT < 0	0.8755	−0.6324	−0.8538	−0.1118	BKT > 0,BLT < 0,BMT < 0	0.0278	0.6524	0.8069	−0.0435
BKT < 0,BLT > 0,BMT < 0	0.8312	−0.5505	−0.7747	−0.3072	BKT > 0,BLT > 0,BMT < 0	0.9445	0.1168	0.2876	0.1486
BKT < 0,BLT > 0,BMT < 0	0.4489	−0.7742	−0.7533	−0.5736	BKT > 0,BLT > 0,BMT < 0	0.1717	−0.2963	−0.3203	0.0262
BKT < 0,BLT > 0,BMT < 0	−0.1489	−0.6196	−0.7725	−0.4141	BKT > 0,BLT < 0,BMT < 0	07768	0.7631	−0.5482	−0.0121

(continued)

Table 6.5 (Continued)

Intra-firm research and development activity (RDA)					Intra-firm information processing activity				
Technology bias	sk	sl	sm	st	Technology bias	sk	sl	sm	st
BKT > 0,BLT < 0,BMT < 0	0.9525	0.9214	0.7581	−0.0332					
BKT > 0,BLT > 0,BMT < 0	0.9773	0.9268	0.7102	0.1192					
BKT > 0,BLT > 0,BMT < 0	−0.6909	−0.6009	0.4487	−0.0093					
BKT < 0,BLT > 0,BMT < 0	0.8676	−0.5451	−0.8352	0.6148	BKT > 0,BLT > 0,BMT < 0	−0.2318	0.6556	0.8325	0.1347
BKT < 0,BLT > 0,BMT < 0	0.5938	−0.6224	−0.8582	0.2241	BKT > 0,BLT > 0,BMT < 0	−0.3951	0.61 86	−0.8122	0.1624
BKT < 0,BLT > 0,BMT < 0	0.8613	−0.1572	−0.8914	0.5704	BKT > 0,BLT > 0,BMT < 0	0.2311	0.7184	−0.6981	0.3763
BKT > 0,BLT > 0,BMT < 0	−0.5811	−0.4726	−0.5838	0.4389	BKT > 0,BLT < 0,BMT > 0	−0.3521	0.7045	0.7293	−0.2136
BKT < 0,BLT > 0,BM < 0	0.9054	0.4135	−0.6416	0.5841					
BKT > 0,BLT > 0,BMT < 0	−0.4925	−0.6461	−0.8425	0.7193					
BKT > 0,BLT > 0,BMT < 0	0.0517	−0.6617	−0.8093	0.4356	BKT > 0,BLT < 0,BMT < 0	0.6342	0.7785	−0.3174	0.2156
BKT < 0,BLT > 0,BMT < 0	0.8492	−0.7425	−0.7421	0.1768	BKT > 0,BLT > 0,BMT < 0	−0.4439	0.6161	0.8301	0.3038
BKT < 0,BLT > 0,BMT < 0	0.8524	−0.2746	−0.8715	0.5901	BKT > 0,BLT > 0,BMT < 0	0.6171	0.7208	−0.6904	−0.2755

technological bias coefficients β^{KT}, β^{KiT}, β^{LT}, β^{MT} in the price functions to classify industries and investigate their characters.

6.4 Property of Industrial Technology and the Internal Synergy

From the parameters of the measured price function, I would like to consider the structural implications of the endogenous synergy effects due to the technological characteristics in industries. By focusing on the sign conditions of the technological progress trend coefficient β^{TT} and the technological bias coefficients β^{KT}, β^{KiT}, β^{LT}, β^{MT} in the price function, I try to classify industries and explore the meanings of each characteristic.

6.4.1 Internal Synergy Effect

1. **Industries with the declining trend of technological progress.**
   (a) **In the main production activities, industries with $\beta^{TT} > 0$:**
   11. Pulp and Paper Manufacturing, 17. Iron and Steel Product Manufacturing, 23. Electric Machinery Products, 41. Other Electric Parts, 53. Oil and Coal Products, 59. Energy Industry, 65. Transportation Service Industry.
   $\beta^{TT} < 0$ implies that the rate of technical progress in this sector is expected to be increasing gradually. In the industry indicated as $\beta^{TT} > 0$, the trend of technical progress is gradually decreasing. In the group with $\beta^{TT} > 0$, there are important sectors which were a leading sector of the Japanese economy in the twentieth century, such as 11. Pulp and Paper, 17. Iron and Steel Product and 23. Electric Machinery Products, which they had led the growth of Japan's economy in the twentieth century through expansion of the tangible capital due to the equipment investment. Additionally, industries such as 53. Oil and Coal Products and 59. Energy Industry, which are not blessed with abundant natural resources and where industrial competitiveness has further declined, especially since the global rise in oil prices in the 1970s, are also included.
   (b) **In intra-firm research and development (R&D) activities, industries with $\beta^{TT} > 0$**
   The analysis will delve into the period between 1995 and 2011, which includes significant economic transitions impacting industrial structures. This time frame witnessed the aftermath of the economic bubble burst, influencing technological investments and efficiency across various industries.
   (c) **In intra-firm information processing activities, industries with $\beta^{TT} > 0$:**

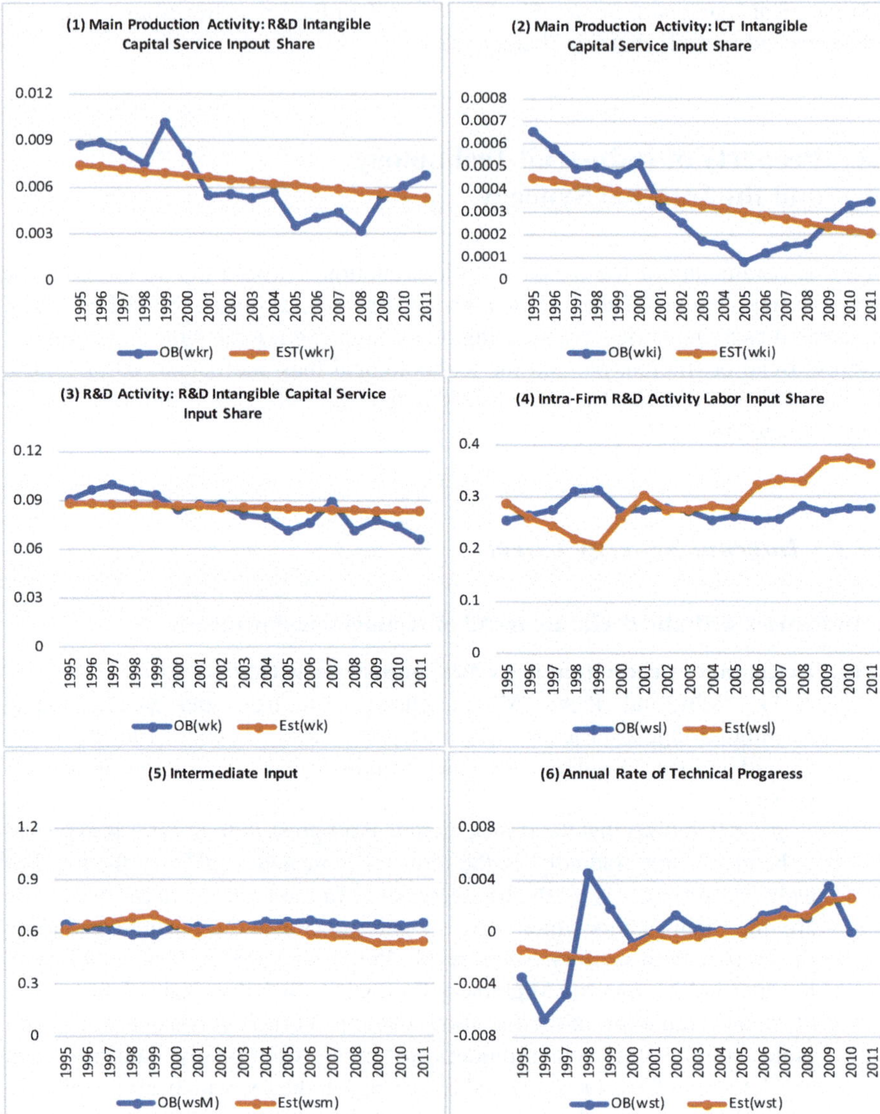

Fig. 6.2 (1)–(6) Comparison between Observed and Theoretical Values of Factor Inputs in Main Production Activity: Textile Manufacturing (1995-2011)

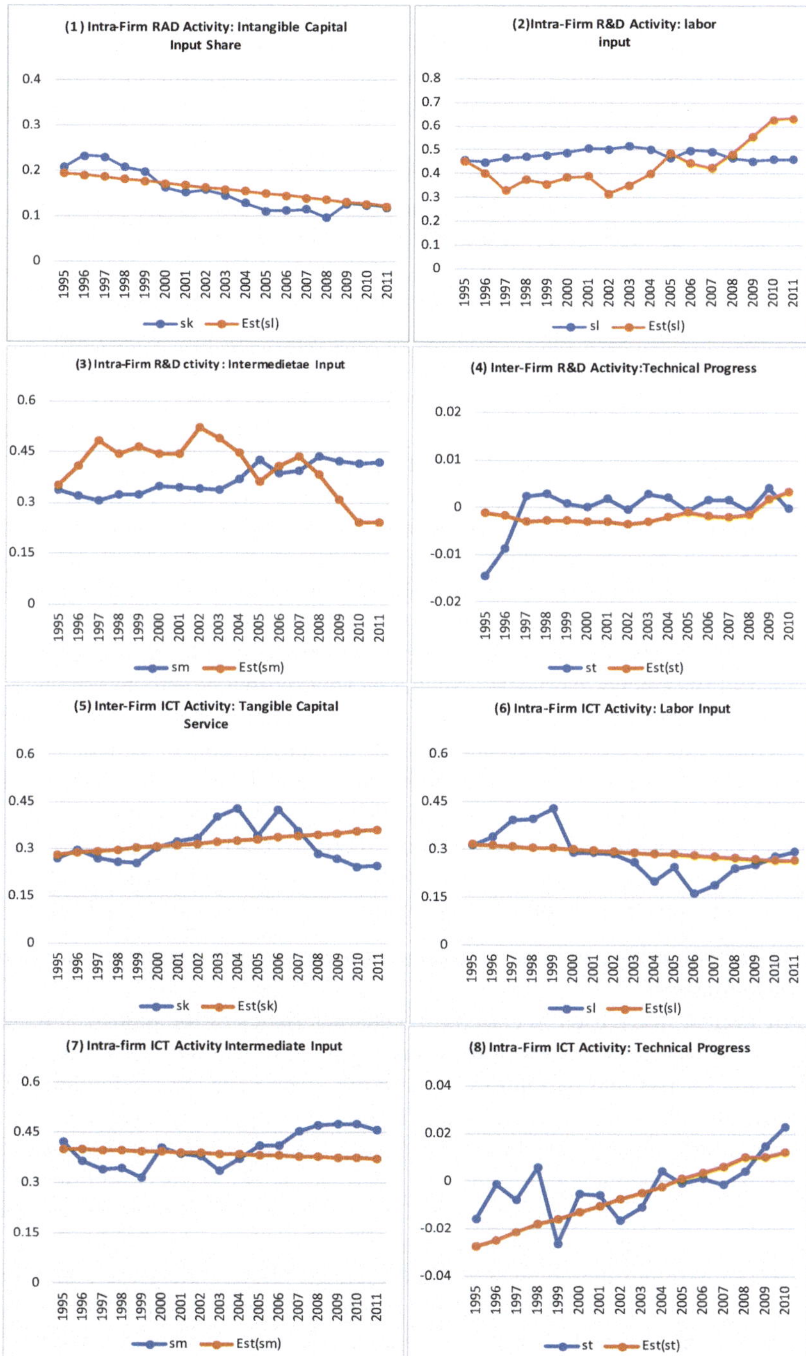

Fig. 6.3 Comparison between Observed and Theoretical Values of Factor Inputs in R&D and ICT Activity in Textile Manufacturing (1995-2011)

06. Food Manufacturing, 18. Iron and Steel Products Manufacturing, 54. Petroleum and Coal Products, 69. Communication and Broadcasting, 72. Trade. In these industry groups, the delay in internalizing information processing activities through self-development has hindered the efficiency of corporate information processing, which in turn weakens the technological characteristics of the main production activities of these industries, thereby becoming one factor that diminishes the improvement of productivity in the main production sectors. Particularly notable is the inclusion of 18. Iron and Steel Products, which was a central industry in the Japanese economy in the first half of the twentieth century. It is believed that the failure of these industries to keep up with information processing technology has been a factor in delaying their technological advancements in the new era.

The observation period mentioned here includes the industrial transition period spanning from 1995 to 2011, which encompassed the economic bubble. The industries listed with $\beta^{TT} > 0$ during this time were mostly those that continued to rely on tangible capital investments from the high growth period and lagged in adopting and expanding new research and development and information processing technologies. These industries were unable to advantageously harness the endogenous technological synergy effects within the industry through research and development and information processing. The primary premise for the technological progress effect in industries is that the technological advancements in primary activities and both corporate activities should progress in balance, and these activities should generate endogenous technological synergy effects, leading to interdependent improvements in technological efficiency and diffusion. However, during this period, the industries mentioned here were insufficient in making this transition into technological characteristics.

Based on the above classification, there were no industrial sectors where the technology progress trend $\beta^{TT} > 0$ across all main production and the two intra-company activities during the entire period from 1995 to 2011.

2. **Industries that we can observe technical progress ($\beta^{TT} < 0$) by the structural transformation**

In Table 6.6, we tried to classify the characteristics of the industrial technology by the estimated results of the trans-log price function in each primary production activities and two intra-firm activities. The table shows the main of the classification by the properties of the technology through the parameter β^{TT} and the technological progress bias β^{iT} (where $i = K, K^r, K^i, L, M$). The header of Table 6.6 lists parameters β^{TT} and β^{kT} ($k = K, K^r, K^i, L, M$) as conditions for the time-series trend that shows improvement in technological efficiency and a shift toward factor-saving technological biases. Each parameter (β^{TT} and β^{kT} ($k = K, K^r, K^i, L, M$)) and the changes in their element prices satisfy the condition of being negative, that is, it is considered that technological progress occurs due to technological bias and changes in element prices, marked with a circle (◯), while industries that do not meet this condition are marked with a circle (X).

(a) **Industries with technological progress in all activities ($\beta^{TT} < 0$)**

In terms of main production, intra-firm research and development and intra-firm information processing activities, the industries where $\beta^{TT} < 0$ for all activities are the following 12 industries: 08. Textile Manufacturing, 15. Chemical Industry, 20. General Machinery Manufacturing, 29. Semiconductor Equipment Industry, 32. Communication Machinery, 35. Electronic Computers and Equipment, 38. Semiconductor and Integrated Circuit, 44. Heavy Electrical Equipment and Transportation Machinery, 47. Robots, 50. Other Precision Machinery, 56. Other Manufacturing Industries, and 62. Construction Industry. These industries will be classified into categories A–C as follows.

Group A: Among the above 12 industries, the following four industries are notable for their main production activities where **technological progress ($\beta^{TT} < 0$) has led to a decrease in output prices, as well as a saving in tangible capital services as a production factor ($\beta^{KT} < 0$), which in turn suggests further technological progress due to the reduction in the price of these factors**: 08. Textile Manufacturing, 14. Chemical Industry, 26. Fiber Optics 44. Heavy Electrical Equipment and Transportation Machinery, and 47. Robots and 50. Other Precision Machinery.

A-1: **Industries where the decrease in the price of tangible capital services enhances the technological efficiency of research and development activities ($\beta^{KT} < 0$ in intra-firm R&D)**: 08. Textile Manufacturing, 14. Chemical Industry, 44. Heavy Electrical Equipment and Transportation Machinery, 47. Robot and 50. Other Precision Machinery.

A-2: In the Intra-firm ICT activity, the tangible capital services are economized ($\beta^{KT} < 0$ in intra-firm ICT) and leaded to be enhanced the technological efficiency due to the decrease in the price of tangible capital services. However, during this observation period, there were no industries that met this criterion. This may reflect a lag in intra-firm ICT activities in Japan.

Based on the above conditions, the synergy effect between main production activities and intra-firm R&D activities indicates that the technological efficiency of both activities can be improved by the reduction in the price of tangible capital services. However, the only industry that demonstrates endogenous synergy effect, where the promotion of intra-firm research activities further enhances the technological efficiency of main production activities, is 08. Textile Manufacturing.

Nevertheless, the influence of main production on intra-firm R&D activities suggests that the formation of tangible capital in the main production sector can contribute to the accumulation of tangible capital in intra-firm research. If this leads to a decrease in the market price of tangible capital services, the effects on intra-firm research activities could be possible in most industries. However, this effect is dependent on the trend of factor prices in the market for tangible capital services, suggesting that we should consider not only the endogenous effects but also the exogenous effects.

Table 6.6 Industrial classification by technology trends and bias parameters

	Main activity: positive condition of the parameters						Intra-firm RDA: positive condition of the parameters				Intra-firm IPA: positive condition of the parameters			
	$\beta^{TT}<0$	$\beta^{kT}<0$	$\beta^{krT}<0$	$\beta^{kiT}<0$	$\beta^{LT}<0$	$\beta^{MT}<0$	$\beta^{TT}<0$	$\beta^{kT}<0$	$\beta^{LT}<0$	$\beta^{MT}<0$	$\beta^{TT}<0$	$\beta^{kT}<0$	$\beta^{LT}<0$	$\beta^{MT}<0$
05. Food products manufacturing	○	X	X	X	X	○	○	X	X	○	X	X	X	X
08. Textile manufacturing	○	○	○	○	X	○	○	○	X	○	○	X	X	X
11. Paper and pulp manufacturing	X	X	X	X	X	○	X	X	X	○	○	X	X	X
15. Chemical manufacturing	○	○	X	○	○	X	○	○	X	○	○	X	X	X
17. Iron and steel manufacturing	X	○	X	X	○	X	○	○	X	○	X	X	X	X
20. General machinery manufacturing	○	X	X	X	X	○	○	○	X	○	○	X	X	X
23. Electric machinery products	X	○	X	X	X	○	○	○	X	○	○	X	X	X
26. Fabricated metal products	○	○	X	X	○	X	X	○	X	○	○	X	X	X
29. Semi-conductor equipment products	○	X	X	X	X	○	○	○	X	○	○	X	X	X
32. Communication machinery	○	X	X	○	X	○	○	○	X	○	○	X	X	X
35. Electric computer and equipment	○	X	○	X	X	○	○	○	X	○	○	X	X	X
38. Semiconductor and integrated circuit	○	X	X	X	X	○	○	○	X	○	○	X	X	X
41. Other electric parts	X	X	X	X	X	X	○	○	X	○	○	X	X	X

Table 6.6 (continued)

	Main activity: positive condition of the parameters						Intra-firm RDA: positive condition of the parameters				Intra-firm IPA: positive condition of the parameters			
	$\beta^{TT} < 0$	$\beta^{kT} < 0$	$\beta^{krT} < 0$	$\beta^{kiT} < 0$	$\beta^{LT} < 0$	$\beta^{MT} < 0$	$\beta^{TT} < 0$	$\beta^{kT} < 0$	$\beta^{LT} < 0$	$\beta^{MT} < 0$	$\beta^{TT} < 0$	$\beta^{kT} < 0$	$\beta^{LT} < 0$	$\beta^{MT} < 0$
44. Heavy and transport machinery	O	O	X	X	O	X	O	O	X	O	O	X	X	X
47. Robot	O	O	X	X	X	X	O	O	X	O	O	X	X	X
50. Other precision machinery	O	O	X	X	X	O	O	O	X	O	O	X	X	X
53. Oil and coal products	X	O	O	X	O	X	O	O	X	O	X	X	X	X
56. Other manufacturing	O	X	X	X	X	O	O	O	X	O	O	X	X	X
59. Energy industry	X	O	X	X	X	X	O	O	X	O	O	X	X	
62. Construction	O	X	X	X	X	O	O	X	O	X	O	X	O	
65. Transportation service	X	X	X	X	X	O	O	O	X	O	O	X	O	
68. Communication and broadcasting	O	X	X	X	O	O	O	X	X	O	X	X	X	
71. Trade	O	X	X	X	X	O	O	X	X	O	X	X	O	

On the other hand, synergy effects related to tangible capital services between main activities and intra-firm information processing activities have been barely observed. This is the evidence realized by the fact that $\beta^{KiT} < 0$ in main production activities and $\beta^{KT} < 0$ in intra-firm information processing activities have not been observed in most industries. This may reflect a delay in the adoption of information technology in Japanese industries during the observation period.

Group B: Technical progress in the intra-firm R&D activities will expectedly decrease the price of intangible capital services (K_r) as the products by the activities in the following industries: 05. Food Manufacturing, 08. Textile Manufacturing, 14. Chemical Products Manufacturing, 17. Iron and Steel Manufacturing, 20. General Machinery Industry, 23. Consumer Electronics and Electrical Machinery, 29. Semiconductor Equipment, 32. Communication Machinery, 35. Electronic Computers and Components, 38. Semiconductor Devices and Integrated Circuit and 41. Other Electronic Components, 44. Heavy Electrical Equipment and Transportation Machinery, 47. Robots, 50. Other Precision Machinery, 56. Other Manufacturing Industries, 59. Energy Industry, 62. Construction, 65. Transportation.

B-1: If there are the productivity increase in the main activity and the price reduction of the main products which are used in the intra-firm activity by the intangible fixed asset services from intra-firm R&D are factor-intensive in main production activities ($\beta^{KrT} > 0$), even if technological innovation in intra-firm R&D progresses and the price of intangible fixed assets decreases, the technological efficiency of the main production sector may not be greatly improved.

According to Table 6.6, in the main production activities of the following sectors, $\beta^{KrT} > 0$ is observed: 08. Textile Manufacturing, 26. Optical Fiber and Cable, 35. Electronic Computers and Components, 38. Semiconductor Devices and Integrated Circuits, 53. Petroleum and Coal Products, 62. Construction Industry.

In these sectors, if the price of intangible capital developed through intra-firm R&D decreases, technological progress in the main production activities is expected to advance further. However, for intra-firm R&D, if the price of tangible capital services decreases, the industries showing a usage bias ($\beta^{KT} > 0$) suggest that the production efficiency of intangible fixed capital may not increase significantly. In industries where synergy effects operate between main production activities and R&D activities, a decrease in the price of intangible fixed capital services from R&D activities will lead to technological progress in main production activities. However, the price trend of tangible fixed capital services represents the overall trend in the capital services market, and while the external synergy effect trend of the tangible fixed capital services market may indirectly affect it, there is no recognized effect that tangible capital investment for R&D in main production activities promotes intra-firm R&D.

Group C:

In the case of intra-firm information processing activities, like the intra-firm R&D, many industries show $\beta^{KiT} > 0$ in the main production sector, indicating a usage bias of intangible fixed capital services from intra-firm information processing activities. This suggests that if technological progress in intra-firm information processing leads to a decrease in the price of intangible fixed capital services, it has the potential to bring about technological progress in main production activities.

On the other hand, in most industries, intra-firm information processing activities show a usage bias of tangible fixed asset services ($\beta^{KT} > 0$), indicating that a decrease in the price of tangible capital services is crucial for technological progress in intra-firm information processing. This is particularly important for the progress of intra-firm information processing technology in Japan, in which it is said to be less than the level of the Western developed countries in adopting information processing technology. It implies that expanding tangible fixed capital to promote information processing technology is essential.

This effect is also related to the exogenous synergy effect of the price trend of tangible fixed capital services in the capital services market. The decrease in the price of tangible fixed capital services, both endogenous and exogenous, promotes technological progress in the intra-firm information processing department. Result in which the price of the output from information processing decreases, further enhancing the technological efficiency of main production activities.

The technological progress of intra-firm information processing activities (ICT) benefits from exogenous synergy effects accelerates the endogenous synergy effects, and finally it creates an important technical characteristic that leads to further synergy effects.

6.4.2 Inter-dependent Effects Among Main Activity and Two Subsidiary Activities

Regarding main production, intra-firm R&D, and intra-firm ICT activities, there is technological progress as a trend ($\beta^{TT} < 0$) with $\beta^{KT} > 0$, $\beta^{KrT} > 0$, $\beta^{KiT} > 0$, and $\beta^{LT} > 0$, there is technological progress in all activities with element usage, and if element prices decrease, the technological efficiency in these activities will be maintained compared to when prices rise. Therefore, if technological progress in intra-firm research and development and information processing activities leads to a decrease in the price of intangible capital, it will positively affect main production. If this further reduces the price of tangible capital services in main production activities, it will create a positive feedback loop among the three activities, leading to a further decrease in the price of intangible capital services and generating a synergy effect that expands the technological efficiency of main production activities, intra-firm R&D, and intra-firm ICT activities.

Additionally, for main production, intra-firm R&D, and intra-firm ICT activities, if $\beta^{KT} < 0$, $\beta^{KrT} < 0$, $\beta^{KiT} < 0$, and $\beta^{LT} < 0$ indicate element-saving technological characteristics, the technological efficiency in these activities will increase if the element prices rise compared to when they remain unchanged. Therefore, **if technological progress in intra-firm research and development and information processing activities reduces the price of intangible capital, positively affecting main production, and further lowers the price of tangible capital services, it will create a positive feedback loop that further improves the technological efficiency of main production activities. This creates a positive cycle of synergy effects.**

6.5 Rationality in the Selection of the Technological Property

Let's review the economic situation from 1995 to the collapse of the so-called IT bubble in 2011. As I mentioned before, from 1995 to the collapse of the so-called IT bubble in 2005, the capital rate of return had been decreasing. Subsequently, from 2005 to just before the Lehman Shock in 2007, the returns remained relatively stable. However, although there was a sharp decline in 2008 due to the Lehman Shock, it has been followed by a recovery. **After 2010, the capital rate of return and capital service price gradually increased, where we can find the following industries in the growing industries with positive technological progress**.

08. Textile Manufacturing, 14. Chemical Products, 20. General Machinery, 26. Fabricated Metal Products, 29. Semiconductor Equipment, 32. Communication Machinery, 35. Electric Computer, 38. Semiconductor and Integrated Circuit, 44. Heavy and Transportation Machinery, 47. Robot, 50. Other Precision Machinery, 56. Other Manufacturing, 62. Construction, 68. Communication and Broadcasting, 71. Trade.

When we select the industry with the property of the positive technical progress ($\beta^{TT} < 0$) and the saving property of the tangible capital service is negative ($\beta^{KT} < 0$). Industry with such property is limited to the following six industries:

08. Textile Manufacturing, 14. Chemical Products, 26. Fabricated Metal Products, 44. Heavy Electric and Transportation Machinery, 47. Robot and 50. Other Precision Machinery.

Moreover, in these industries they have capital saving property, which has a property in which the tangible capital is expected gradually with technical progress.

Here, we would like to focus on the technological characteristics of the selected industries: 08. Textile and 14. Chemical Industry. For both industries, the parameter $\beta^{TT} < 0$, representing the technological trend, is positive for main production, intra-firm R&D, and intra-firm ICT.

We can infer the following technological characteristics from the results of the trans-log price. In the 1990s, Japan began to show signs of a new era, shifting away from a growth pattern that had continued since the end of World War II—one characterized by the expansion of heavy and chemical industries through capital investment, which leveraged an increasing labor force and economies of scale to enhance efficiency and production capacity. Instead, there emerged a move toward building a new economic structure centered on the expansion of investments in intangible assets, often referred to as intellectual assets, and the development of an industrial structure that capitalizes on them. As is evident from Table 6.1, a key feature of economic growth since 1995 has been a sharp decline in the labor force, reflecting demographic changes, and a shift in capital accumulation from tangible to intangible assets. In Chap. 6, we will examine the outcomes of this expansion in intangible capital investment and consider its impact on the economy of the twenty-first century.

During this period, industries such as Textile Manufacturing, Chemical Products, Fabricated Metal Products, Heavy Electric and Transportation Machinery, Robot, and Other Precision Machinery showed significant technological progress. Notably, the Textile and Chemical industries demonstrated substantial advancements in their main production activities, intra-firm research and development, and intra-firm information and communication technology (ICT).

Both industries exhibited positive technological trends characterized by cost-saving properties for tangible fixed capital and intangible fixed capital related to R&D activities. This systematic advancement not only led to improved efficiency in main production but also fostered enhanced technological capabilities in subsidiary activities. Specifically, the technological characteristics in these industries, represented by parameters such as $\beta^{TT} < 0$ and $\beta^{KT} < 0$, revealed a consistent pattern of progress.

Such advancements in technology and cost-saving measures created a synergy that further propelled the growth and efficiency of the textile and chemical sectors. The technological bias in intra-firm R&D and ICT activities, though not identical in both industries, contributed to a harmonious development in the overall production processes. The impact of price fluctuations in research and development and labor exhibited contrasting reactions in technological efficiency changes, underscoring the dynamic nature of technological progress in these industries.

For the textile manufacturing and chemical industries, concerning main production, common characteristics include $\beta^{TT} < 0$, and $\beta^{KT} < 0$. As a technological progress trend, there is systematic advancement, and this progress exhibits cost-saving characteristics for tangible fixed capital and for intangible fixed capital related to R&D activities. Therefore, the technological property in the main activity is more progressive with the saving properties in the two subsidiary activities.

In the main production activity of the textile industry, they exhibit the common technological properties, $\beta^{KT} < 0$, $\beta^{KrT} < 0$, and $\beta^{KiT} < 0$. Labor property is $\beta^{LT} > 0$ with labor using and intermediate raw materials are cost-saving with $\beta^{MT} < 0$.

Therefore, concerning the main production activities of the textile, the impact of price fluctuations in research and development and labor would lead to opposite reactions in their technological efficiency changes.

For intra-firm R&D and intra-firm ICT activities, the technological bias is not necessarily the same for both industries of Textile and Chemical, with technological biases respectively. The increase in the price of tangible capital services and intermediate goods tends to improve the technological efficiency of this activity, while the increase in the price of labor tends to reduce technological efficiency (Figs. 6.4 and 6.5).

6.6 Technology Evaluation Sheet

6.6.1 Technology Evaluation Sheet of Textile and Chemical Industry

Tables 6.7 and 6.8 are the technology evaluation sheets of the textile manufacturing industry and chemical industry. In both the textile manufacturing industry and the chemical industry, as well as in the main production and both intra-firm activities, the parameter β^{TT} for the technological progress trend is negative ($\beta^{TT} < 0$). Both indicate a trend in technological efficiency improvement. However, the observed Total Factor Productivity (TFP) growth rates, which are shown in the Technology Evaluation Sheet, tell a different story. In the textile manufacturing industry, TFP growth rates have decreased in both infra-firm activities. In both countries, TFP growth rates are negative in both main production and intra-firm activities. Despite the small but slightly positive technological progress could have been expected the positive productivity growth indicated by the parameters of the price function's technological trend ($\beta^{TT} < 0$), the observed TFP growth rates during this period in both sheets are showing the declining trends in technological efficiency. What does this imply?

The relationship between the main production activities and intra-firm R&D and ICT activities in an industry is interdependent, but it is not necessarily contrasting or equal. As a part of their long-term cost minimization behavior, companies determine the technology choice and factor demand in main production activities. The demanding factor includes investments in tangible fixed capital and intangible fixed capital, such as R&D and information processing, for main production activities, along with labor and raw materials. Assuming that to realize these investments in intangible fixed capital, companies determine the demand for tangible fixed capital, labor, and intermediate materials in both intra-firm activities based on the cost-minimization behavior. The trends in service prices for intangible fixed assets such as R&D and ICT services are determined because of the technology and factor demands in intra-firm activities. Then, the impacts of main production activities are leading to the interdependence of the three activities. In the background of this interdependence among the three activities, there is the technology choice made for each activity, and the effectiveness of combining these choices efficiently serves as an indicator of the synergy effect efficiency of the three activities. The prices of

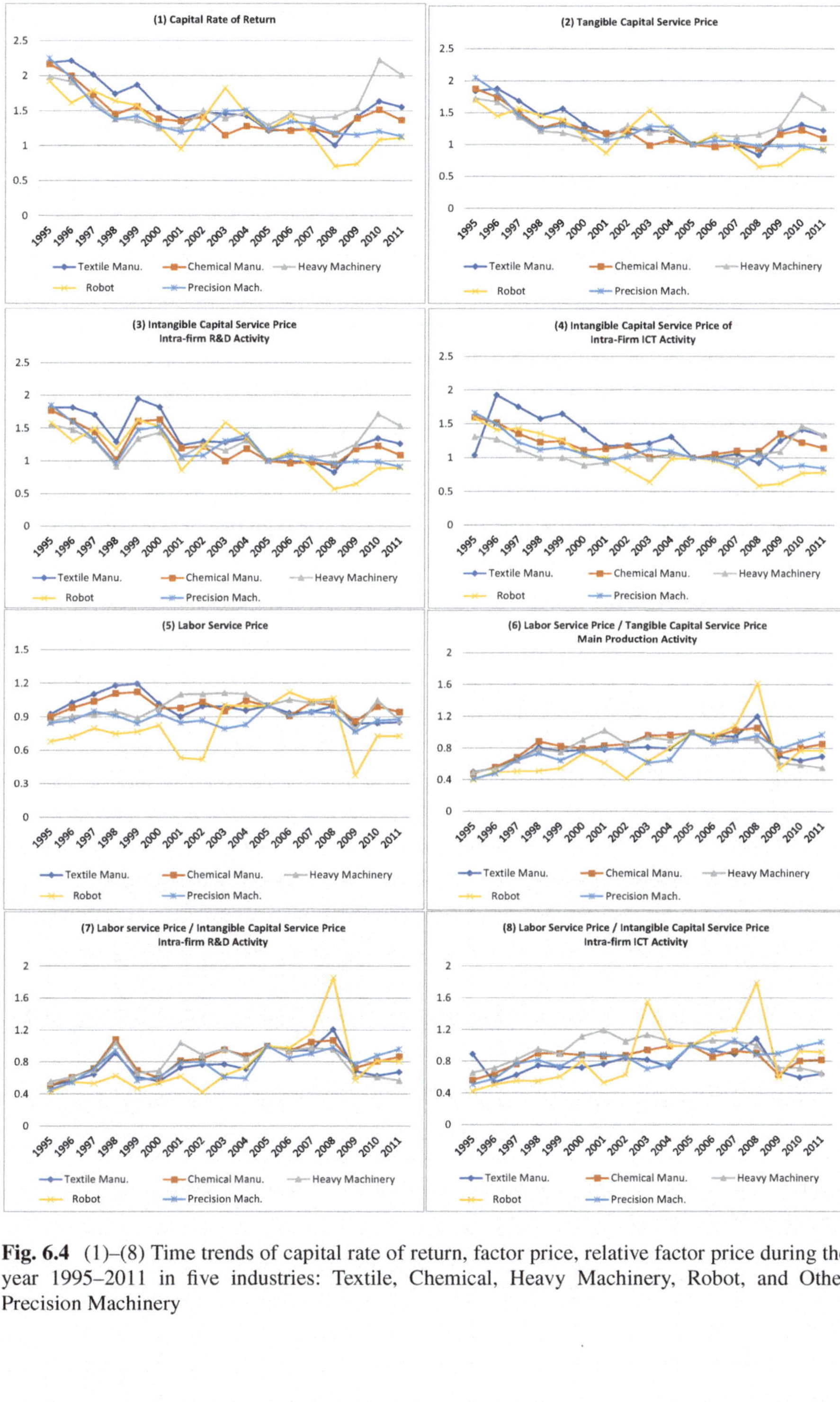

Fig. 6.4 (1)–(8) Time trends of capital rate of return, factor price, relative factor price during the year 1995–2011 in five industries: Textile, Chemical, Heavy Machinery, Robot, and Other Precision Machinery

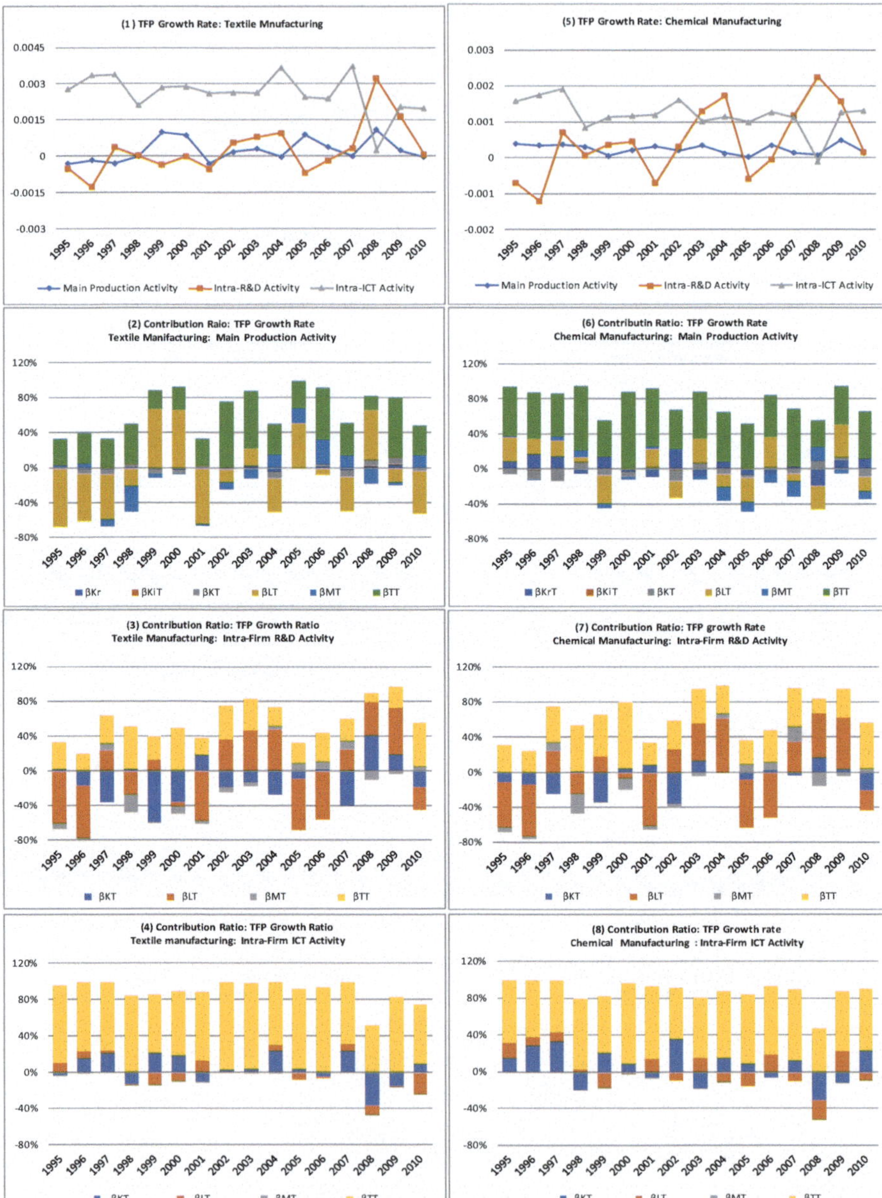

Fig. 6.5 (1)–(8) Trend of TFP growth rate and contribution ratio of the sources: Textile Industry and Chemical Products, year 1995–2011

Table 6.7 Technology evaluation sheet: Textile manufacturing industry

a-2-1 BkrT < 0	BKrT	BKiT	BKT	BLT	BMT	BTT
Main activity 08. Textile manufacturing	−0.00013	−1.50E−05	−0.00034	0.005525	−0.00517	−0.00027
Average price increase rate (annual %)	−2.2441	−2.2465	−2.52761	0.2955	0.0407	
Technology bias trends	Saving	Saving	Using	Using	Saving	
Productivity effect and TFP (%/year)	−	−	−	+	−	0.0017
Intra-firm computer R&D activities			−0.00454	0.014908	−0.00229	−0.00045
Average price increase rate (annual %)			−2.20131	0.1266	−0.24904	
Technology bias trends			Saving	Using	Saving	
Productivity effect and TFP (%/year)			−	−	−	−0.0324
Intra-firm computer ICT activities			0.005185	0.031406	−0.00175	−0.00257
Average price increase rate (annual %)			−5.34389	−2.1097	−0.37909	
Technology bias trends			Using	Saving	Saving	
Productivity effect and TFP (%/year)			+	+	−	−0.2398

Table 6.8 Technology evaluation sheet: Chemical industry

a-2-1 BkrT < 0	BKrT	BKiT	BKT	BLT	BMT	BTT
Main activity: 14. Chemical industry	0.00578	−3.00E−05	0.006095	0.008469	−0.00379	−0.00251
Average price increase rate (annual %)	−2.78401	−2.0678	4.098442	−0.00181	−2.23175	
Technology bias trends	Using	Saving	Saving	Saving	Using	
Productivity effect and TFP (%/year)	+	−	−	+	−	−0.1302
Intra-firm R&D activity: semiconductor			−0.0026	0.015853	−0.01326	−0.00057
Average price increase rate (annual %)			−2.91613	0.001266	−0.24904	
Technology bias trends			Saving	Using	Saving	
Productivity effect and TFP (%/year)			−	−	−	−0.0519
Intra-firm ICT activity: semiconductor			0.003418	0.026717	−0.03023	−0.0014
Average price increase rate (annual %)			−3.31456	−0.00367	−0.38594	
Technology bias trends			Using	Using	Saving	
Productivity effect and TFP (%/year)			+	−	−	−0.2704

factors such as tangible fixed services, labor services, and intermediate materials are determined because of supply and demand equilibrium in their respective markets and are not dependent solely on the three activities of a specific industry. On the other hand, while the two intra-firm activities of a company are influenced by the overall supply and demand situations in factor markets, they have the characteristic of mutual interdependence with the main production activity in the internal market of the company. Therefore, this internal market has different relationships depending on the industry, and its technological characteristics reflect the effectiveness of the three activities in the industry. Keeping these relationships of the three activities in mind, let's focus on the technology evaluation sheets for the textile manufacturing industry and the chemical industry, again.

In Textile and Chemical Manufacturing industries, the TFP growth rate for information processing activities was high during this period. Both industries also show significant impacts from the IT bubble in 1998 and the Lehman Shock in 2008. The TFP growth rate for R&D activities exhibits a characteristic of large fluctuations. Additionally, the TFP growth rate for main production activities, while stable, is relatively low compared to the two intra-firm activities. The growth in TFP for main production activities in the textile manufacturing industry is significantly influenced by the factor-saving technological bias of labor services. When labor service prices decrease, it enhances technological efficiency, and vice versa. The impact of tangible and intangible capital services is relatively small, and the influence of technological trends (β^{TT}) and labor services reflects the trend of technological progress rates. For intra-firm R&D and ITC activities, the contribution of efficiency improvement due to technological trends is significant, while the technological bias characteristics of tangible capital and labor services have a strong impact on efficiency variations, given the trend of their price changes. On the other hand, for intra-firm information processing activities, the effect of efficiency improvement due to technological trends is substantial, while the influence of technological bias and tangible capital services contributes to efficiency gains through the characteristic of their factor-saving bias.

In this section there are several indexes for the selected industries including Textile, Chemical, Heavy Electric and Transport Machinery, Robot and other selected industries. It includes the time-series information for wages, relative prices for labor and capital inputs.

Figures 6.6 and 6.7 show charts of the comparison of the several factors for the TFP growth rate in several industries including textile and chemical manufacturing. Almost all industries are showing significant impacts from the IT bubble in 1998 and the Lehman Shock in 2008. The TFP growth rate for R&D activities exhibits a characteristic of large fluctuations. Additionally, the TFP growth rate for main production activities, while stable, is relatively low compared to the two intra-firm activities. The growth in TFP for main production activities in the textile manufacturing industry is significantly influenced by the factor-saving technological bias of labor services. When labor service prices decrease, it enhances technological efficiency, and vice versa. The impact of tangible and intangible capital services is relatively small, and the influence of technological trends (β^{TT}) and labor services

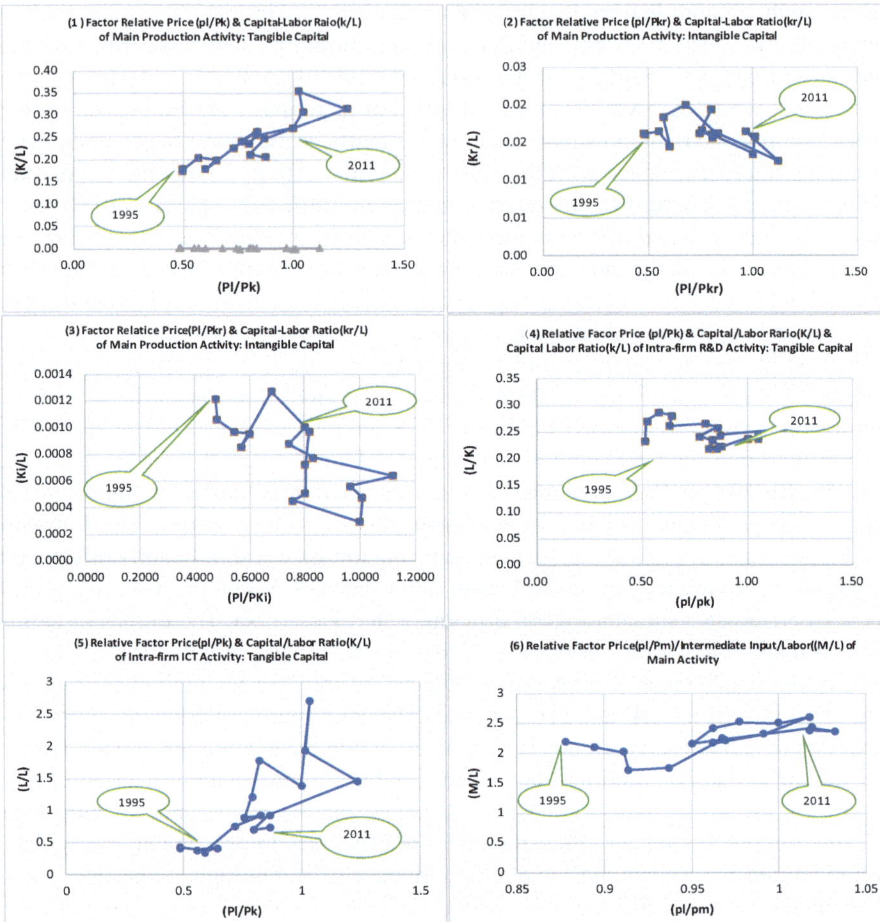

Fig. 6.6 (1)–(6) Interdependence between relative factor price (capital/labor price ratio [P^L/P^K]) and capital-labor ratio (K/L) during 1995–2011: Textile Manufacturing

reflects the trend of technological progress rates. For intra-firm information R&D activities, the contribution of efficiency improvement due to technological trends is significant, while the technological bias characteristics of tangible capital and labor services have a strong impact on efficiency variations, given the trend of their price changes. On the other hand, for intra-firm information processing activities, the effect of efficiency improvement due to technological trends is substantial, while the influence of technological bias and tangible capital services contributes to efficiency gains through the characteristic of their factor-saving biases.

For the chemical industry, it is characterized by the predominant influence of the technological progress effect due to technological trends on the overall technical efficiency trend for both main production and intra-firm activities. On the other hand, the technological bias of labor services and the fluctuation in their factor prices tend to significantly impact variations in technical efficiency. In R&D activities, the technological bias of tangible capital services shows a negative effect, whereas for information processing activities, there are multiple years with positive effects.

Based on the assumed trend in factor prices, we examine the synergistic effects between main production activities and intra-firm R&D activities as well as information processing activities using the technological assessment sheet proposed in the previous chapter. Tables 6.7 and 6.8 represent the technological assessment sheets for the textile manufacturing industry and the chemical industry.

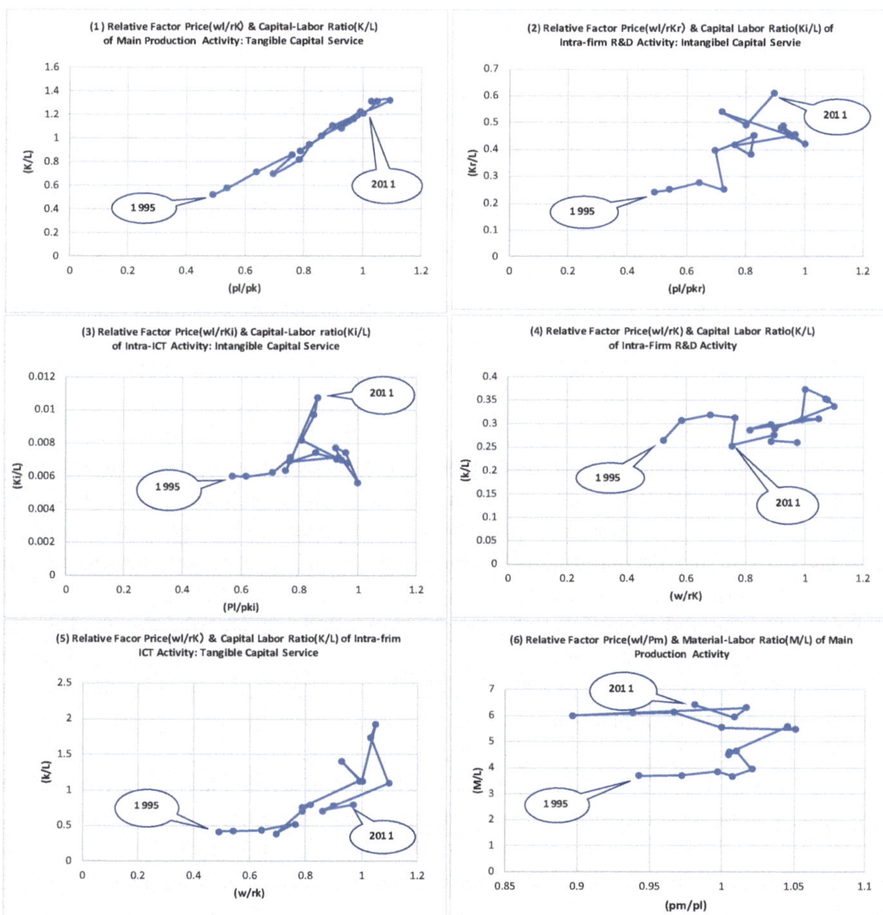

Fig. 6.7 (1)–(6) Interdependence between relative factor price (capital/labor price ratio [P^L/P^K]) and capital-labor ratio (K/L) during 1995–2011: Chemical Products Manufacturing

6.6.2 Technology Bias by Industry and General Interdependence

In the textile manufacturing industry, concerning the technological bias of main production activities, except for labor services showing a usage bias, all other factors, including intangible fixed capital services for R&D, intangible fixed capital services for information processing, tangible fixed capital services, and intermediate materials, exhibit a savings bias ($\beta^{KT} < 0, \beta^{KrT} < 0, \beta^{KiT} < 0, \beta^{LT} > 0, \beta^{MT} < 0$).

The movement of each factor's price during the observation period shows a declining trend for intangible and tangible fixed capital services and an increasing trend for labor and intermediate materials. Therefore, in correspondence with the technological bias of the factors excluding labor service, there is a tendency for technological progress to be anticipated as a technological trend. However, as concerns labor service with the nature of labor using technological bias, labor demand would be declining for this period.

As a result, overall productivity during this period has not shown significant improvement, with an average annual growth rate of 0.17%. The technological bias of intra-firm R&D activities includes factor-saving ($\beta^{KT} < 0$), usage ($\beta^{LT} > 0$), and savings ($\beta^{MT}, \beta^{LT}, \beta^{MT} < 0$) technological characteristics for tangible capital services, labor services, and intermediate goods services, respectively. The observed average fluctuation rates of factor prices during this period are −2.2013% annually for tangible capital services, 0.1266% for labor services, and −0.2490% for intermediate goods services. For all factors, the characteristics observed of technological bias and the direction of price changes during this period suggest that the average price movement does not correspond to improvements in the efficiency of R&D and ICT activities.

In both the textile manufacturing industry and the chemical industry, β^{TT} for the technological progress trends in major production activities and intra-firm activities is negative ($\beta^{TT} < 0$). This shows a trend of increasing technological efficiency. However, the observed growth rates of total factor productivity (TFP) show different results. In the textile manufacturing industry, the TFP growth rate has decreased in both intra-firm activities. In the chemical industry, the TFP growth values in major production and intra-firm activities are cancelled out. Despite the small positive technological progress indicated by the price function's technological trend parameter ($\beta^{TT} < 0$), the observed TFP growth rate during this period indicates a decline in technological efficiency.

The decrease in the price of tangible fixed capital services impedes the efficiency improvement of both factor-saving main production activities and intra-firm R&D activities. The observed decline in tangible fixed capital services prices during this period indicates a non-functional synergy effect between intra-firm R&D and main production activities. This trend is also observed in the intra-firm R&D activities of the chemical industry.

Furthermore, regarding the intra-firm ICT activities of the textile manufacturing industry, there is a factor-utilization bias for tangible fixed capital services. The decrease in the capital services price has the characteristic of enhancing the efficiency of information processing activities. However, for both labor services and intermediate goods services, there is a savings technological bias, and the declining trend in the prices of both services indicates a technological characteristic that negatively affects their technological efficiency. The characteristics of technological bias and the direction of changes in factor prices for the chemical industry also indicate that the improvement in technological efficiency of intra-firm ICT activities is not associated. The implications of these technological bias characteristics and observed changes in factor prices suggest that the average trends in technological efficiency for both intra-firm activities lead to a decline during this period.

Therefore, based on the changes in factor prices during the observation period and the direction of changes in factor prices for each activity, there is no synergy effect where both intra-firm activities contribute to the efficiency improvement of main production activities, and main production activities contribute to the efficiency improvement of both intra-firm activities.

On the other hand, the impact of intangible fixed capital services for R&D and intangible fixed capital services for information processing on main production activities in both intra-firm activities are influenced by the technological bias of both intangible fixed capital services in main production activities. In the textile manufacturing industry, both factors show a savings bias, while in the chemical industry, it is a savings bias for intangible fixed capital services for information processing. This indicates that in both industries, the decrease in the prices of both services due to the activities of intra-firm intangible fixed capital services inhibits the efficiency increase in main production activities. If both intra-firm activities are efficient and have the effect of lowering the prices of intangible capital services, it suggests that some policy guidance is needed to activate main production activities using the characteristics of intangible capital services, such as scalability and synergy effects.

The characteristics of these factor prices and technological biases are the result of technological choices reflected in the time-series changes in factor relative prices and capital intensity in the main production activities and both intra-firm activities of the industry.

Figures 6.7 and 6.8 illustrate the changes in factor relative prices and the corresponding tangible/intangible capital-labor ratio within main production activities, intra-firm R&D activity, and intra-firm ICT activity in the textile and chemical industries from 1995 to 2011.

Regarding the textile manufacturing industry, when examining the trends in the relative prices of tangible capital services and labor services, as well as the capital intensity represented by the ratio of tangible capital services to labor services (Fig. 6.7), it is observed that there is some instability between years. However, there is a positive correlation between the changes in relative prices and capital intensity. In other words, when labor services exceed the relative increase in capital service prices, a rationality is observed where tangible capital investment expands, leading to an increase in capital intensity, despite some instability over the years.

On the other hand, Fig. 6.6 (2) and (3) show the capital intensity of relative prices of intangible capital services to labor services, do not show such a correlation. In the case of the selection of intangible capital services, rationality is not clear. Additionally, Fig. 6.6 (6), depicting the substitution relationship between labor and

Fig. 6.8 (1)–(6) Time trend of the contribution ratio to the TFP growth rate year 1995–2011; Heavy and Transportation Equipment Machinery and Robot Manufacturing

raw material inputs, does not clearly demonstrate a substitution relationship between the two factors. This suggests that the technological characteristics regarding the input of these factors have a different nature compared to the capital-labor relationship.

Regarding the chemical industry, Fig. 6.7 (1)–(6) illustrate the trends in relative prices of factors and capital intensity. In the case of the chemical industry, compared to the textile manufacturing industry, there seems to be a rationality in the observed trends in relative prices of factors and capital intensity, especially concerning tangible and intangible capital services.

The difference in their technological characteristics stems from variations in the intangible fixed services' bias in R&D in main production activities, as well as differences in the technological bias characteristics of labor services and raw materials. In the case of the chemical industry, these differences appear to enhance their technological efficiency in response to fluctuations in the prices of the first two elements. Specifically, in the chemical industry, Fig. 6.7 (4) and (5) suggest that there is rationality in the substitution relationship between tangible capital services and labor services in intra-firm R&D as well as ICT activities.

Industries classified as (a-1), such as (44) Heavy Machinery and Transport Equipment (with technological characteristics $\beta^{TT} < 0$, $\beta^{KT} < 0$, $\beta^{KrT} > 0$ where $\beta^{KiT} > 0$, $\beta^{LT} < 0$, $\beta^{MT} > 0$), Robotics (47) ($\beta^{KiT} > 0$, $\beta^{LT} > 0$, $\beta^{MT} > 0$), and Other Precision Machinery (50) ($\beta^{KiT} > 0$, $\beta^{LT} > 0$, $\beta^{MT} < 0$), exhibit technological trends similar to the chemical industry ($\beta^{TT} < 0$, $\beta^{KT} < 0$, $\beta^{KrT} > 0$). However, the synergy effects between main production activities and intra-firm R&D and ICT activities may not necessarily contribute to productivity growth. In the case of (44) Heavy Machinery and Transport Equipment industry, despite all activities sharing the technological trend characteristic $\beta^{TT} < 0$, the annual average growth rate of total factor productivity for each activity during this period is sluggish. The synergy effect between the three activities is not observable. Similarly, in (47) Robotics and (50) Other Precision Machinery industries, despite the assured technological progress trend β^{TT}, the observed annual growth rates of total factor productivity for each activity are negative, indicating that the collaboration between intra-firm activities and main production activities has a negative synergy effect. Additionally, these industries classified under (a-1) exhibit a trend of decreasing capital profit rates during this period, and there is a characteristic correlation with the decreasing tangible and intangible capital service prices.

During this observation period, labor service prices are on an upward trend. Therefore, the relative prices of labor to tangible and intangible capital services (w/r) are also increasing, leading to an increase in capital intensity for both tangible and intangible capital. Although the expansion of tangible and intangible capital intensity is expected to enhance the technological efficiency of each activity and reduce factor prices, the situation arises where the savings-oriented technological characteristics of each activity's technological bias offset this effect. On the other hand, the decrease in expected returns may hinder the expansion of future planned production, reducing the effect of technological scalability.

Figure 6.8 (1)–(6) summarize TFP's contribution changes to main production, intra-firm R&D, and ICT activities for heavy electrical and transportation machinery, and robots. For two industries, it can be confirmed that the synergy effect of the three activities is not working very effectively.

When focusing on the trend of total productivity, another technological characteristic that needs to be considered is the intercept α^T of the technological progress function, as seen in Eq. (6.1). This parameter represents the trend of the price function's first-order term. As inferred from the technology progress function in Eq. (6.1), the technology progress rate is the sum of various elements such as the technological bias of each factor, factor prices, technology trends, and trends, with the addition of the intercept α^T. Therefore, the level of the technology progress rate is determined by the parameter of the intercept. Thus far, in the case of the textile manufacturing industry, chemical industry, heavy machinery and transportation machinery, robots, and other precision machinery industries, α^T is positive, and the technology progress rate is decreased by the amount of $-\alpha^T$ to reach equilibrium.

Figure 6.8 (1)–(6) depicts the changes in Total Factor Productivity (TFP) for the heavy and transport machinery and robot industries. The figures break down the TFP changes in main production activities, intra-firm R&D, and information processing activities into the contributions of technological bias for each production factor and the changes in the contribution of technological trends. It's important to note that the values of the intercept α^T in the technological progress function are universal and not included in these figures.

For heavy machinery and transportation machinery, the technological trend for main production activities and intra-firm R&D activities indicates positive technological progress with a relatively small magnitude ($\beta^{TT} < 0$). However, for intra-firm information processing activities, the technological trend suggests a negative technological progress rate. In main production activities, the technological bias for tangible capital, R&D intangible capital, and intermediate goods elements is usage oriented. Decreasing the prices of these elements would enhance technological efficiency, but the observed average annual prices of these elements have been increasing, counteracting the improvement in technological efficiency. Only the labor services element shows consistency between its technological bias and price changes, contributing to improved technological efficiency, although not significantly. In intra-firm R&D activities, technological progress due to technological trends is larger compared to main production, but the inconsistency between technological bias and element prices offsets this, leading to a decrease in the annual average TFP. In intra-firm information processing activities, technological efficiency is enhanced through the consistency of technological bias and price changes for tangible capital and labor services elements, resulting in an increase in technological efficiency beyond the effects of the technological trend. However, this increased technological efficiency is not translating into a synergy effect contributing to the improvement of technological efficiency in main production, as evidenced by the lack of a decrease in the information services prices in main production.

For the robot industry, all three activities (main production, intra-firm R&D, and intra-firm information processing) exhibit a technological trend with $\beta^{TT} < 0$, indicating technological progress. However, in main production activities, the price fluctuations of tangible capital services and labor services lack consistency with their technological bias, hindering the increase in technological efficiency. This inconsistency offsets the technological progress rate as a trend, resulting in a negative TFP growth rate. Similarly, in intra-firm R&D activities, there is a lack of alignment between the technological bias and price fluctuations for tangible capital, labor, and raw materials, preventing an increase in TFP growth rate. In intra-firm information processing activities as well, the lack of alignment between the technological bias characteristics and element price fluctuations counteracts technological progress due to the technological trend, inhibiting the rise in TFP and leading to a decreasing trend in the annual average TFP growth rate. Once again, the technological efficiency of the two intra-firm activities does not contribute to improving the efficiency of main production.

In industries belonging to the a-1 classification, the capital profit rate shows a downward trend over time, and the capital service price also exhibits a decreasing trend over time. Unlike the industries in the a-2 classification to be presented next, these industries have the characteristic of $\beta^{KT} > 0$, indicating a bias toward the use of tangible capital services. Then, the decreasing trend in the capital service price hinders technological progress.

Next, we focus on industries classified under a-2. Like the industries mentioned earlier, they are also included in the (a) category, and during the observation period from 1995 to 2011, they are industries with a time-series decline in capital rate of return. However, unlike the industries classified under a-1, these industries have a technological bias coefficient for tangible capital services in main production activities, indicated by $\beta^{KT} > 0$, meaning they exhibit an element-use bias for the tangible capital. Therefore, the decrease in tangible capital service prices has the characteristic of enhancing efficiency through technological progress.

This classification includes six industries: General machinery (20), Semiconductor devices (29), Communication equipment (32), Computers (35), Semiconductor components (38), and Construction (62). Within this classification, they are further divided based on the technological bias of intangible fixed capital services in the main production activities. The a-2-1 category, indicating an element-saving bias with $\beta^{KrT} < 0$, includes Computers (35) ($\beta^{KiT} > 0$, $\beta^{LT} > 0$, $\beta^{MT} < 0$), Semiconductor components (38) ($\beta^{KiT} > 0$, $\beta^{LT} > 0$, $\beta^{MT} < 0$), and Construction (62) ($\beta^{KiT} > 0$, $\beta^{LT} > 0$, $\beta^{MT} < 0$). The a-2-2 category, indicating an element-using bias with $\beta^{KrT} > 0$, including General machinery (20) ($\beta^{KiT} > 0$, $\beta LT > 0$, $\beta^{MT} < 0$), Semiconductor devices (29) ($\beta^{KiT} > 0$, $\beta^{LT} > 0$, $\beta^{MT} < 0$), and Communication equipment (32) ($\beta^{KiT} < 0$, $\beta^{LT} > 0$, $\beta^{MT} < 0$). The decrease in intangible capital service prices due to technological progress in corporate research and development is associated with improved skill efficiency in a-2-2 industries but inhibits technological efficiency in a-2-1 industries.

First, we focus on Computers (35) ($\beta^{KiT} > 0$, $\beta^{LT} > 0$, $\beta^{MT} < 0$), and Semiconductor components (38) ($\beta^{KiT} > 0$, $\beta^{LT} > 0$, $\beta^{MT} < 0$), which are classified under a-2-1. The technological assessment sheets for both industries are shown in the following sheet (Tables 6.9 and 6.10).

Among the six industries categorized under a-2, we hope to draw attention to two industries with specific technological characteristics in their main production activities. These industries have a technological bias coefficient for tangible capital services ($\beta^{KT} > 0$), indicating capital-using characteristics, and a technological bias coefficient for research and development's intangible fixed capital services ($\beta^{KrT} < 0$), indicating capital-saving characteristics. Both the computers industry and the semiconductor and integrated circuit industry show a technological trend coefficient (β^{TT}) indicating a trend of technological progress in both main production activities and both intra-firm activities. However, during the observed period from 1995 to 2011, the observed annual average growth rate of total factor productivity as a percentage is negative for all activities in the computers industry.

Table 6.9 Technological evaluation sheet: electric computer industry and semiconductor integrated circuit industry during 1995–2011

a-2-1 BkrT < 0	BKrT	BKiT	BKT	BLT	BMT	BTT
Main activity 35. Electronic calculator	−9.9E−05	0.00038	0.001811	0.011172	−0.01326	−0.00165
Average price increase rate (annual %)	−11.2856	−11.2316	−5.34388	−0.03021	−4.41929	
Technology bias trends	Saving	Using	Using	Using	Saving	
Productivity effect and TFP (%/year)	−	+	+	+	−	−0.00144
Intra-firm computer R&D activities			−0.00447	0.006759	−0.00229	−0.00071
Average price increase rate (annual %)			−3.96169	0.001266	−0.24904	
Technology bias trends			Saving	Using	Saving	
Productivity effect and TFP (%/year)			−	−	−	−0.07583
Intra-firm computer ICT activities			0.005185	0.031406	−0.03659	−0.0096
Average price increase rate (annual %)			−5.34389	−0.01071	−0.38706	
Technology bias trends			Using	Using	Saving	
Productivity effect and TFP (%/year)			+	+	−	−0.5079

Table 6.10 Technological evaluation sheet: Integrated circuit and semiconductor

a-2-1 BkrT < 0	BKrT	BKiT	BKT	BLT	BMT	BTT
Main activity: Semiconductor industry	−0.01088	0.000102	0.006095	0.008469	−0.00379	−0.00251
Average price increase rate (annual %)	0.668768	−0.90064	4.098442	−0.00181	−2.23175	
Technology bias trends	Saving	Using	Using	Using	Saving	
Productivity effect and TFP (%year)	+	+	−	+	−	0.0974
Intra-firm R&D activity: semiconductor			−0.01016	0.019092	−0.00894	−0.00099
Average price increase rate (annual %)			−2.20316	0.001266	−0.24904	
Technology bias trends			Saving	Using	Saving	
Productivity effect and TFP (%/year)			−	−	−	0.250573
Intra-firm ICT activity: semiconductor			0.012391	0.02301	−0.0354	−0.0043
Average price increase rate (annual %)			−2.62634	−0.00222	−0.3873	
Technology bias trends			Using	Using	Saving	
Productivity effect and TFP (%/year)			+	+	−	−0.37352

In the semiconductor industry, the capital profit rate is positive for main production and R&D activities, but negative for information processing activities. Examining the technological biases in the two industries, in the main production activities of the computers industry, prices have been decreasing for all elements during this period. However, for R&D's intangible capital services and intermediate goods inputs, there is a bias toward saving elements, and the downward trend in their prices inhibits technological efficiency improvement. Yet, for the other three production elements, there is a bias toward using elements, and the decreasing prices indicate a direction toward improving technological efficiency. Even with these technological characteristics and price changes as assumptions, the overall main production activities show a negative average annual growth rate in total factor productivity, indicating that technological efficiency has not improved during this period. Regarding intra-firm R&D activities, all three input elements do not show improvements in technological efficiency based on the values of technological bias parameters and the direction of changes in their prices. Here too, the observed average annual technological progress rate in total factor productivity is negative, offsetting the technological characteristics of the trend coefficient.

Moreover, in intra-firm ICT activities, the lack of consistency between the technological bias and price changes of intermediate goods offsets the rational movements of tangible capital services and labor services. As a result, the overall growth rate of total factor productivity is negative. Therefore, even in the computer industry, for all three activities during the observation period, there is no improvement in average annual TFP, and synergy effects between main production activities and intra-firm activities are not observed.

Figure 6.9 (1)–(2) plot the changes in the contribution rates of elements based on the technological bias in the three activities of the computer industry. The contribution of elements to TFP in main production activities is mostly explained by technological trends and the contribution of labor services. The biased use of labor services and the decrease in their prices enhance technological efficiency. For intra-firm R&D and ICT activities, the contribution is mostly explained by technological trends. In intra-firm R&D, there are periods where tangible capital and labor services negatively impact technological efficiency, and in that sense, synergy from main production activities is small. In intra-firm information processing activities, tangible capital and labor services have a nature that enhances technological efficiency, but their influence is extremely small, with most of the improvement in technological efficiency being due to the effect of technological trends. However, the extent to which this affects main production activities is very small.

In the semiconductor and integrated circuit industry, the main production activities exhibit some inconsistency in the direction of technological bias and the variation in element prices for several input elements, hindering efficiency improvement. However, overall, it is indicated that technological efficiency is increased at an average annual rate of total factor productivity. Nevertheless, it should be noted that the impact of each element's price on technological efficiency is relatively small due to the small values of technological bias, and the effect of offsetting element price variations on technological efficiency is relatively small.

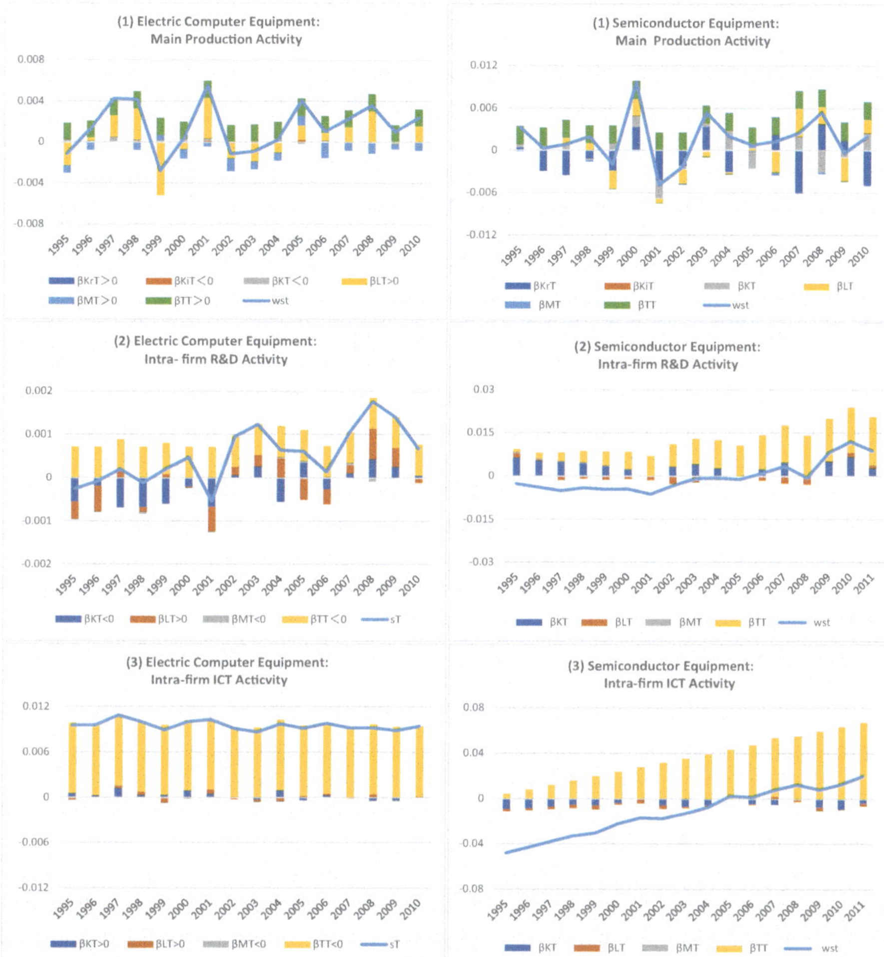

Fig. 6.9 (1)-(6) Contribution Ratio to the TFP Growth Rate in Electric Computer Equipment and Semi-conductor Equipment (1995-2011)

In intra-firm R&D activities, although there is no direction of technological bias and price variation for each element contributing to technological efficiency improvement, technological trends offset this, resulting in an efficiency improvement of around 0.2% per year. In intra-firm information processing activities, the efficiency improvement effects of tangible capital and labor services are offset by the element-saving technological bias of intermediate goods and the decrease in their prices, leading to an overall decrease in technological efficiency. Therefore, it is estimated that main production activities and the intra-firm activities plot the changes in the contribution rates of elements based on TFP in the semiconductor and integrated circuit manufacturing industry.

In main production activities, it is indicated that, along with technological trends, there is a contribution from the intangible capital services of intra-firm R&D and **information processing activities. However, the degree of this contribution over time tends to fluctuate based on the direction of element price variations. Concerning intra-firm R&D activities, the effects of tangible capital services and labor services, along with technological trends, are detected**. Again, in correspondence with technological bias and element price variations, the impact on technological efficiency tends to fluctuate. In intra-firm information processing activities, the predominance of technological trend effects is observed, and tangible capital services also contribute to improving technological efficiency. However, the effects are significantly influenced by price variations.

Figure 6.10 (1)–(4) plot the time series of capital rate of returns, tangible capital services, and intangibtle capital services (intra-firm research and intra-firm information processing) for six industries classified as a-2-1 and a-2-2. The temporal patterns of capital returns and capital service prices (tangible and intangible) are not significantly different from those of industries classified as a-1. The plots reflect the recovery pattern from the burst of the financial bubble in the early 1990s to, the recovery trend after the 2008 Lehman Shock. Tangible and intangible capital service prices also show a slight downward trend, reflecting the decline in capital returns during this period. Labor service prices, mirroring market trends, do not differ significantly from the previous industrial group classified as a-1, resulting in similar fluctuations in the relative prices of labor and capital services, as shown in the time series of capital rate of returns, tangible capital services, and intangible capital services (intra-firm research and intra-firm information processing).

Figure 6.10 (1)–(4) illustrate the variations in capital intensity and relative factor prices for tangible and intangible capital services in the electric computer-related industries during 1995–2011

For the computer industry, Fig. 6.11 (1)–(3) show that the capital intensity increases proportionally to the rise in labor-to-capital factor prices in the main production activities and intangible capital services from intra-firm R&D and ICT activities. This observation is also evident in the tangible capital intensity and factor relative prices for both intra-firm activities, as depicted in Fig. 6.11 (4)–(5). Furthermore, Fig. 6.11 (6) illustrates a relationship between intermediate raw materials and labor services in the main production activity. Therefore, in the computer industry, there is an apparent rational technological choice in the relationship between the factor relative prices of tangible and intangible capital services and labor services, and capital intensity for all three activities. However, as mentioned earlier, the impact of the technological bias and price fluctuations of each factor on technological efficiency has not yielded significant results in all three activities, and the synergy effect has not been realized.

In the semiconductor and integrated circuit industry, the relationship between the relative prices of tangible and intangible capital services and labor services, as well as the capital intensity, is illustrated in Fig. 6.12 (1)–(6). The time-series changes in the main production activities, such as the relationship between intangible capital from R&D and information processing, labor, and tangible capital intensity, as well

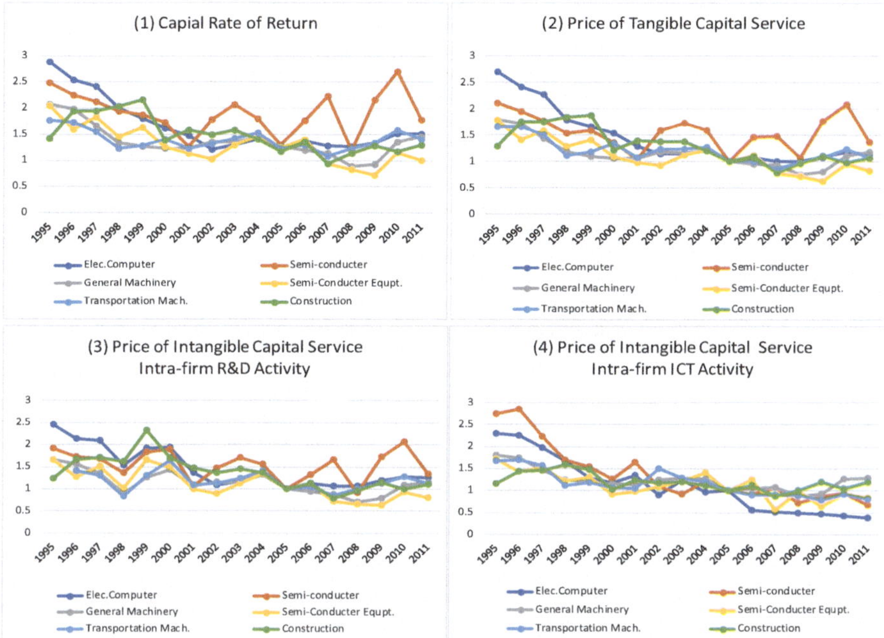

Fig. 6.10 Capital rate of return and capital service prices of tangible and intangible capital service (1995–2011)

as the relationship between tangible capital intensity and the relative prices of capital and labor in both intra-firm activities, do not exhibit economic rationality compared to the earlier computer industry. As mentioned before, one factor contributing to the lack of synergy effect in all three activities in this industry could be the absence of economic rationality in the variations of capital intensity and factor relative prices for each activity, as reflected in the observed facts here.

Figure 6.13 (1)–(3) illustrate the interdependence of the three activities in the construction industry, which is another industry classified under a-2-1. The figures depict changes in the contribution of each production factor to the TFP (Total Factor Productivity) across the three activities.

According to Fig. 6.13 (1), the TFP changes in the main production activity of the construction are generally explained by the technology bias, tangible capital services, and labor services. The influence of both intra-firm activities is almost negligible. Concerning intra-firm R&D activities, tangible capital services, labor services, and the impact of technological trends are significant. However, the effects of the technology bias in the first two factors depend heavily on their price fluctuations. For tangible capital services, which have a cost-saving technology bias, and labor services, which have a usage-based bias, the impact on TFP can vary widely. Regarding corporate information processing, technological progress due to technological trends is dominant, while the influence of other factors is minimal. Despite being in the construction industry, the patterns of technological efficiency changes

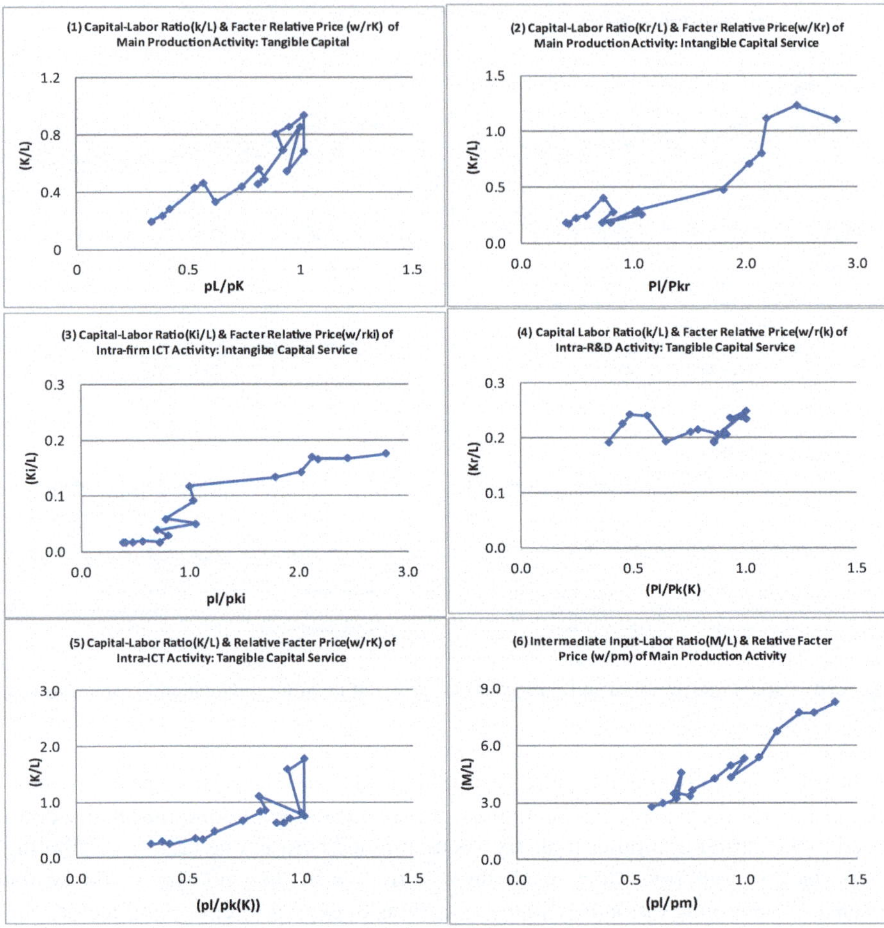

Fig. 6.11 Factor relative ratio of capital-labor (*K/L*) and factor relative price (w/r): Electric Computer Industry (1995–2011)

do not differ significantly from those observed in the computers industry and the semiconductor/integrated circuit industry mentioned earlier. The characteristic difference in the technology bias of tangible capital, where it becomes usage-based, sets the a-2-1 classified three industries apart from the a-1 classified industries mentioned earlier. As mentioned before, the impact of the technology bias of tangible capital services on the TFP changes in the three activities is significant and distinct from the a-1 classified industries. However, the impact on the main production activity of infra-firm activities is not substantial, and the synergy effect between the three activities is not clearly observed.

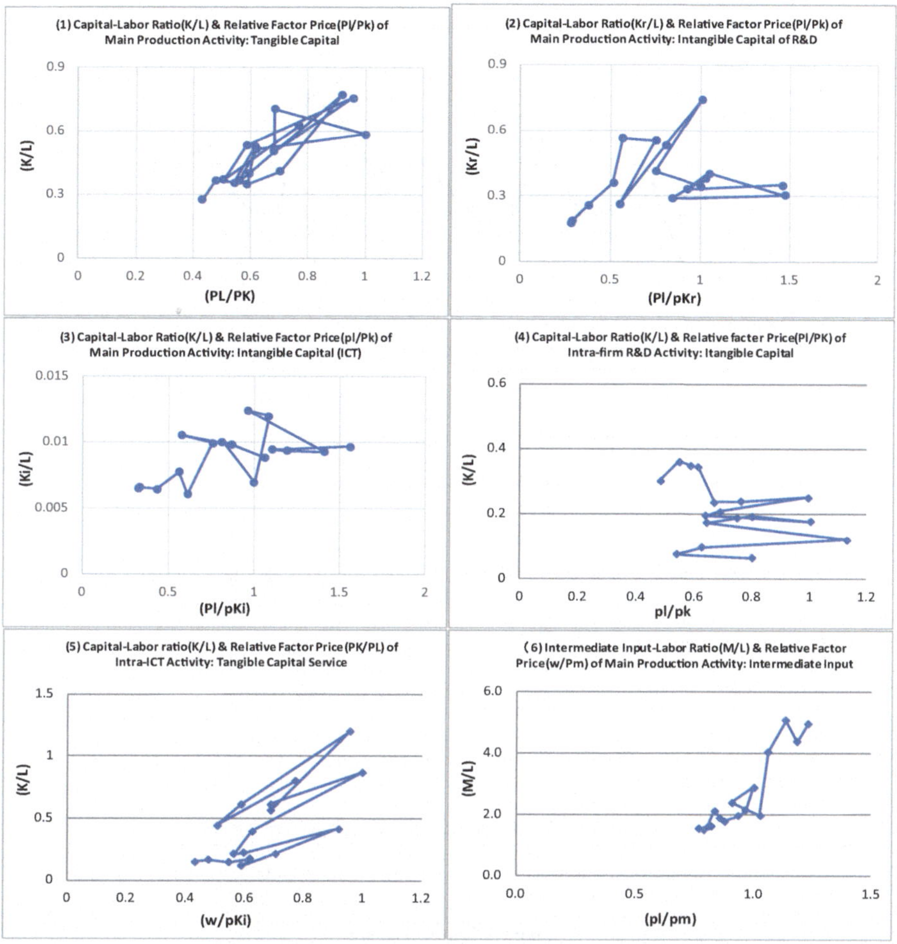

Fig. 6.12 (1)–(6) Correlation map for capital intensity (*K*/*L*) and relative factor price (w/r) in Main and Intra-firm Activities of the Semiconductor and Integrated Circuit Industry (1995–2011)

Let's focus on the three industries classified with the characteristic of a positive bias in the technical trend of R&D's intangible capital services in main production activities ($\beta^{KrT} > 0$). These industries are 20. General Machinery Manufacturing, 29. Semiconductor Device Manufacturing and 33. Communication Equipment Manufacturing. Although these three industries differ from the previous industries in terms of $\beta^{KrT} > 0$, they share the characteristics of $\beta^{KT} < 0$ and $\beta^{TT} < 0$. Additionally, in these separated three industries, there are common characteristics such as $\beta^{KT} > 0$, $\beta^{LT} > 0$, $\beta^{MT} < 0$ in main production activities, $\beta^{KT} < 0$, $\beta^{LT} > 0$, $\beta^{MT} < 0$ in intra-firm research and development, and $\beta^{KT} < 0$, $\beta^{LT} > 0$, $\beta^{MT} < 0$ in intra-firm information processing.

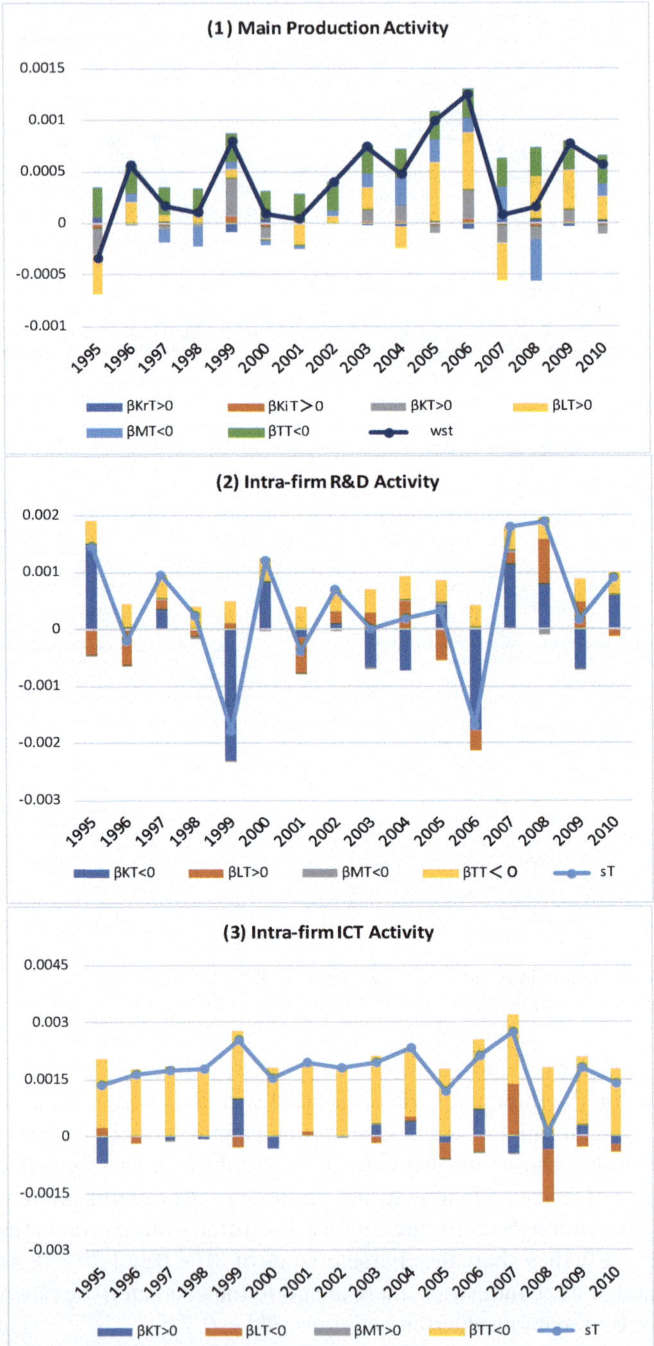

Fig. 6.13 (1)–(3) Contribution Ratio of Technology Bias to the TFP Growth in the Construction Industry (1995-2011)

We focus on the technology evaluation sheets for the selected two industries: 20. General Machinery Manufacturing and 29. Semiconductor Device Manufacturing (29) (Tables 6.11 and 6.12).

The movements in capital rate of returns, tangible and intangible capital service prices, labor service prices, and so on, in these industries do not differ significantly from those classified in the previous industries. In the general machinery industry, the technological trend indicates a trend of technological progress for both main production and intra-firm activities, with $\beta^{TT} < 0$. One feature of this industry is that the α^T value in technological progress function (Eq. 6.1) is positive and large, reducing technological biases elsewhere. This characteristic results in a small average growth rate of TFP. Despite the positive impact of the technological biases of each element on technological efficiency, which is aligned with fluctuations in element prices, the rate of TFP increase is kept small for main production activities. On the other hand, for infra-firm R&D and ICT activities, despite the technological trend having the characteristic of bringing about technological progress, the reason for the negative TFP growth rate during this period is the large value of the intercept α^T in the technological progress function. Additionally, the mismatch between the technological biases of elements and the fluctuation in their prices contributes to this negative growth rate. This mismatch is particularly significant in R&D activities, leading to a decline in technological efficiency.

Figure 6.14 (1)–(6) represents the time trend of the relationships between the relative factor price (w/r) and the capital intensity (*K*/*L*) in the general machinery manufacturing industry. The relationship between capital intensity and element relative prices for labor service input into tangible and intangible capital in the main production activities shows a nearly positive correlation, maintaining rationality. This trend is observed consistently. Even during the reversal caused by the 2008 Lehman Shock, the relationship's rationality is largely preserved for intra-firm R&D as well as information processing activities. It is interesting to note that the relationship between intermediate raw materials and labor input, as shown in Fig. 6.14 (6), maintains a stable ratio independent of the relationship with element relative prices.

We are showing the time trend of the contribution ratio to the TFP growth of each activity by the changes of the technology bias in the Figs. 6.15, 6.16, and 6.17.

The factors influencing the TFP changes in the main production activities are the technological trend, capital and labor services, and intermediate goods input. As mentioned earlier, due to the significant effect of the interception in the technological progress function, the changes in TFP caused by element biases and trends are relatively small. For intra-firm R&D activities, the effect of the technological trend is offset by the bias in labor services, leading to significant fluctuations and a small growth rate in TFP. In the case of information processing activities, the TFP growth rate is predominantly determined by the effect of the technological trend. Thus, in the general machinery manufacturing industry, the synergy effects between activities are observed to be small.

The pattern of technological efficiency changes in the semiconductor equipment industry follows a trend almost identical to that of the general machinery

Table 6.11 Technology evaluation sheet: General machinery industry

a-2-2 BKrT > 0	BKrT	BKiT	BKT	BLT	BMT	BTT
20. General machinery	0.000611	0.000283	0.000214	0.002094	−0.0032	−0.00032
Yearly average G.R.: price (%)	−2.17195	−2.0699	2.6262	−0.17	0.0941	
Technology bias and trend	Using	Using	Using	Using	Saving	
TFP effect (%/year)	+	+	+	+	+	−0.1001
22. G.M.: intrafirm R&D activity			−0.00377	0.0177	−0.01403	−0.00045
Yearly average G.R.: price (%)			−2.2112	−0.1266	−0.249	
Technology bias and trend			Saving	Using	Saving	
TFP effect (%/year)			−	−	−	0.00145
21.GM. Intra-firm ICT activity			0.00096	0.00264	0.00273	−0.00145
Yearly average G.R.: price (%)			−2.623	−0.6533	0.3856	
Technology bias and trend			Using	Using	Saving	
TFP effect (%/year)			+	+	−	−0.28121

Table 6.12 Technology evaluation sheet: Semiconductor industry

a-2-2 BKrT>0	BKrT	BKiT	BKT	BLT	BMT	BTT
29. Semiconductor	0.00048	0.00024	5.30E–06	0.00702	−0.0077	0.0048
Yearly average G.R.: price (%)	−4.5541	−4.4368	−4.691	−0.1154	0.1265	
Technology bias and trend	Using	Using	Using	Using	Saving	
TFP effect (%/year)	+	+	+	+	+	−0.00083
31. G.M.: intrafirm R&D activity			−0.0063	0.00191	−0.0128	−0.00081
Yearly average G.R.: price (%)			−4.175	0.1266	−0.249	
Technology bias and trend			Saving	Using	Saving	
30. G.M.: intrafirm R&D activity			0.00119	0.1461	0.02653	−0.00812
Yearly average G.R.: price (%)			−4.6917	0.0246	−0.3851	
Technology bias and trend			Using	Using	Saving	
TFP effect (%/year)			+	+	−	−0.2844

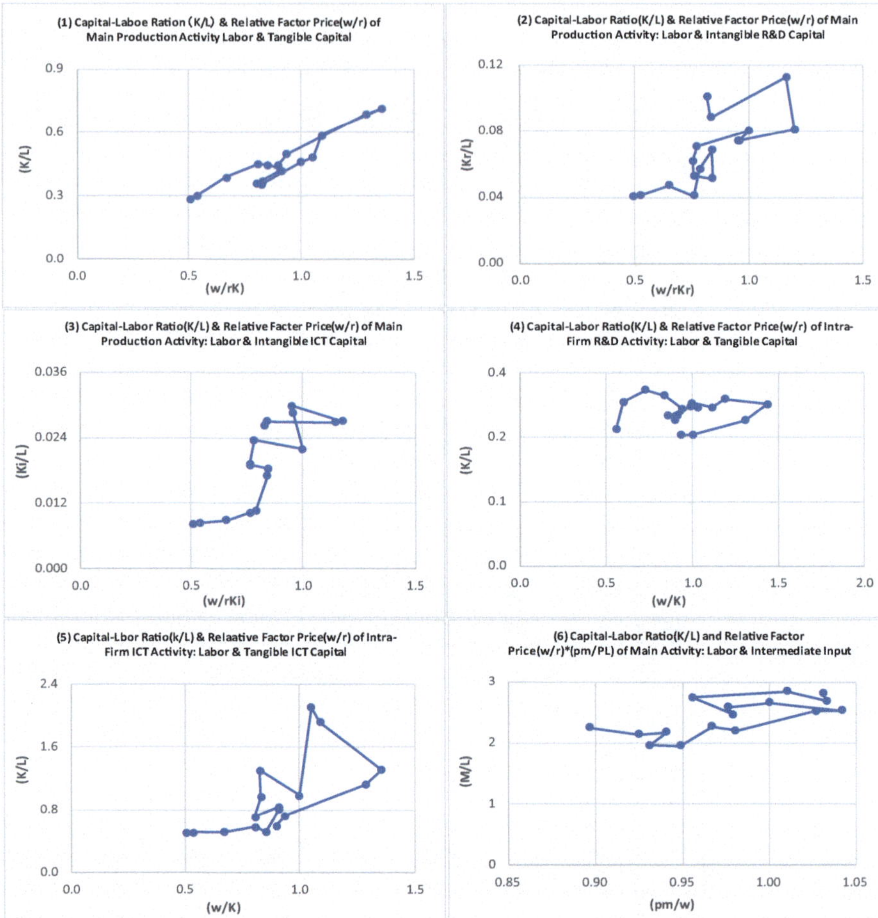

Fig. 6.14 (1)–(6) Time trend of capital-labor ration (*K*/*L*) and relative factor price (w/r)

manufacturing industry. The relationship between the technological bias of each input factor and TFP is shown on the technological evaluation sheet, and the trend is largely the same as in general machinery. As a result, despite having $\beta^{TT} < 0$, indicating a trend of technological progress, the observed average annual rate of TFP is negative.

As concerns about the semiconductor equipment industry, the contribution to the TFP growth rate of the technology bias is almost the same pattern as shown in Fig. 6.16 compared to the general machinery industry. Figure 6.17 is time series of the pattern of TFP: Semiconductor Device Manufacturing.

Regarding the Communication Equipment Industry (32) classified as a-2-2, as shown in Fig. 6.17 (1)–(3), the only difference in the technological bias is that $\beta^{KiT} < 0$, compared to the general machinery industry. However, the significant

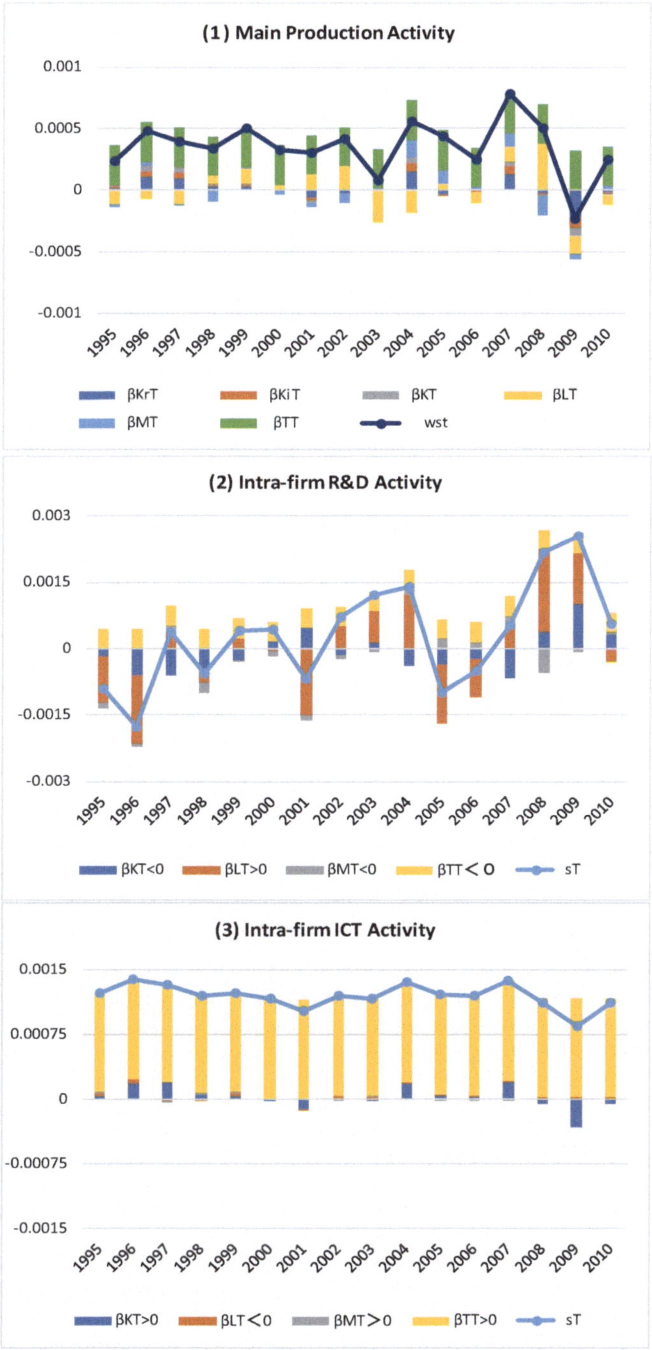

Fig. 6.15 (1)-(3) The Contribution Ratio to TFP Growth Rate by Technological Biases in General Machinery Industry (1995-2011)

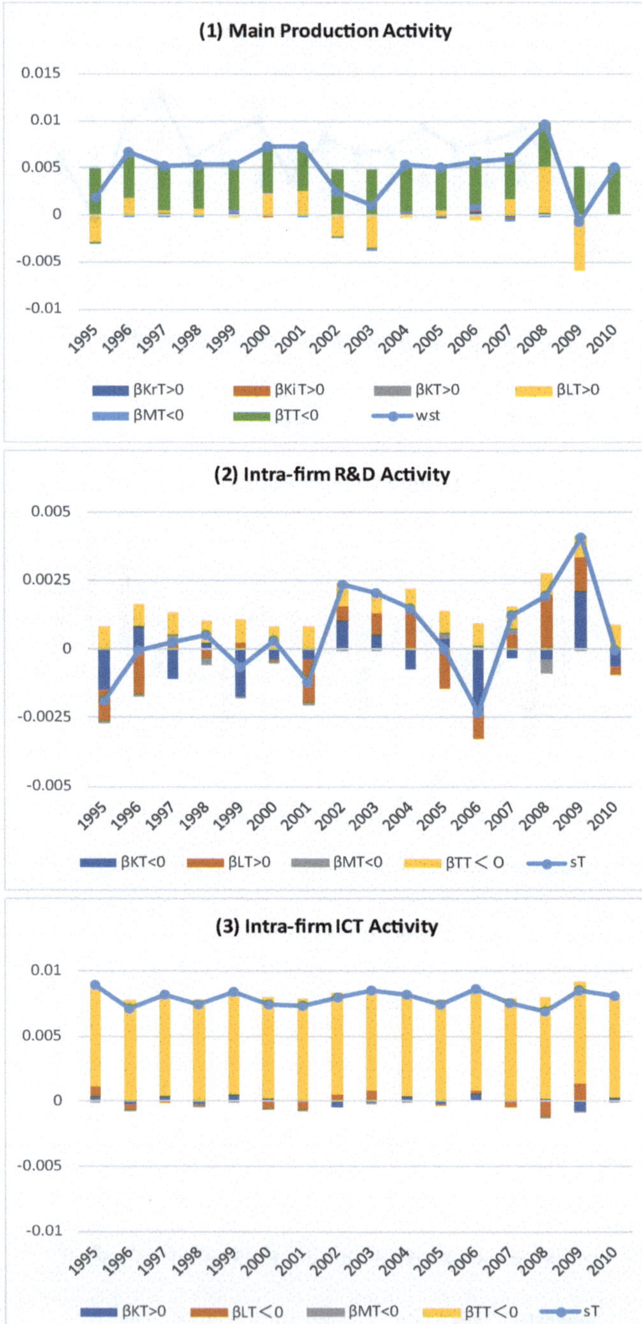

Fig. 6.16 (1)-(3)The Contribution ratio to TFP growth rate by Technological Biases in Semiconductor Equipment Industry (1995-2011)

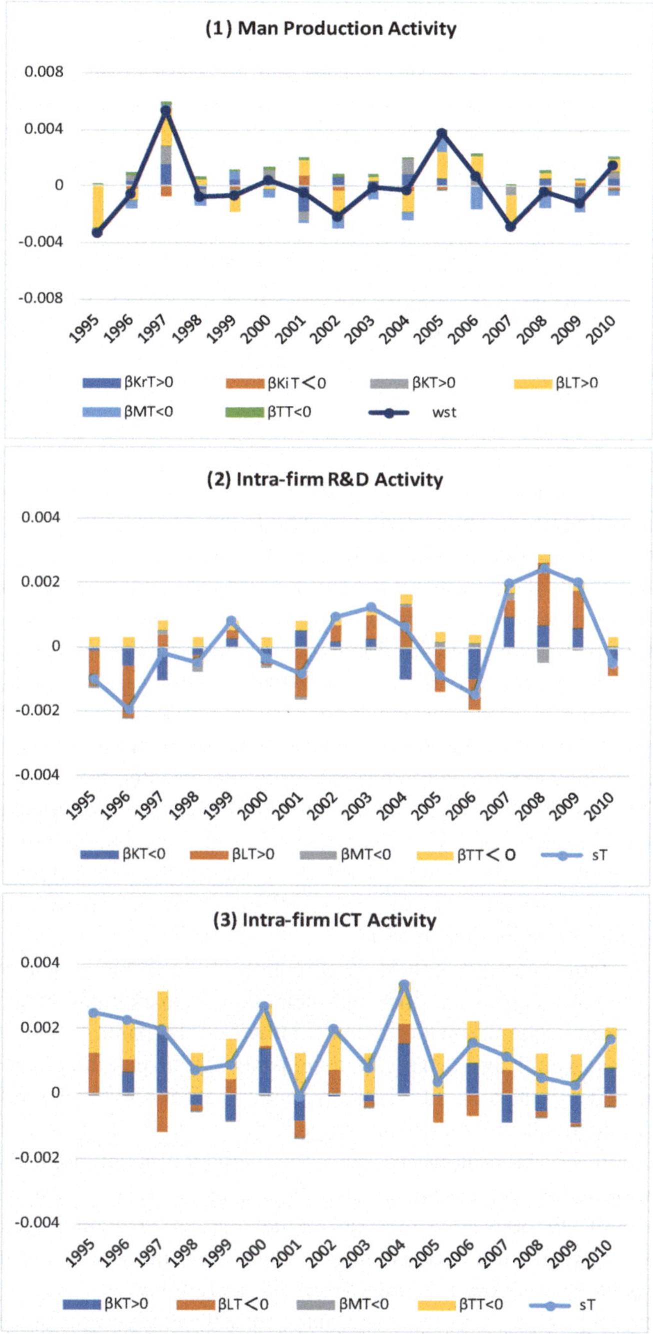

Fig. 6.17 (1)–(3) Contribution of production factors to the TFP across the three activities in Communication Equipment Industry

difference lies in the pattern of labor service price fluctuations, which is affecting the TFP variation. This is evident in the TFP fluctuation patterns of the communication equipment industry's main production and both intra-firm activities, making them distinct from the general machinery and semiconductor device industries. Nevertheless, the fluctuation in labor prices is a factor pushing down the TFP growth rate of both intra-firm activities, significantly impacting the TFP growth rate of intra-firm activities and reducing their technological efficiency. In this sense, the synergy effect between activities is not apparent here either.

Industries classified under b-1 exhibit $\beta^{TT} > 0$, indicating no signs of technological progress as a trend. Like industries classified under a-1 and a-2, during the observation period from 1995 to 2011, these industries show a declining trend in capital rate of returns and a decreasing trend in the price of tangible and intangible capital services. The industries in the b-classification include the energy industry and transportation industry. The energy industry, representing the former, has $\beta^{KT} < 0$, indicating a savings-oriented bias in tangible capital services. Meanwhile, the transportation industry, representing the latter, has $\beta^{KT} > 0$, signifying a usage-oriented bias in tangible capital services. The technological evaluation sheets for both industries can be found in Tables 6.13 and 6.14.

In both industries, the parameter β^{TT} for the technological trend in the main production activities is positive, and the TFP growth rate is, on average annually, negative, indicating a decline in technological efficiency. In the energy industry, the technological trend parameters for both intra-firm activities are negative, suggesting technological progress. However, the annual average growth rate of TFP during this period is negative. For the intangible fixed assets in R&D and ICT processing in the energy industry, they exhibit characteristics of usage-oriented bias, indicating that a decrease in the prices of these elements would improve the technological efficiency of main production activities. However, the value of these bias parameters is very small, resulting in a small savings effect. As a result, the bias effects of tangible capital, labor, raw materials, and so on, are offsetting the effects of both intangible capitals, leading to a reduction in TFP.

The transportation industry exhibits a direction in the technological bias of input elements in main production activities and their changes in element prices that has the effect of increasing technological efficiency. However, the parameter for this technological bias is very small, and as a result, there is no significant effect to counteract the characteristic of declining technological efficiency as a technological trend. Therefore, TFP does not effectively enhance technological efficiency. In the characteristics of the intra-firm activities of both industries here, the features of technological bias are not effectively utilized. Despite a positive technological progress rate due to technological trends, the overall TFP does not achieve an annual increase in technological efficiency. Hence, the synergy effect between main production and intra-firm activities in these two industries under category b is not effectively operational.

The industrial groups discussed so far have shown a tendency of decreasing capital returns during the observation period. Next, let's focus on industries where capital returns are on an upward trend during this observation period. The synergy effect between main production activities and intra-firm R&D, as well as ICT activities, is not functioning.

Table 6.13 Technological assessment sheet:energy and transportation service industry

b-1-2 BKrT > 0	BKrT	BKiT	BKT	BLT	BMT	BTT
59. Energy industry	8.52E−06	5.19E−05	−0.0071	−0.00043	0.007395	0.000145
Average price increase rate (annual %)	−5.4801	−5.1973	−4.9421	−0.1346	3.0979	
Technology bias trends	Using	Using	Saving	Saving	Saving	
Productivity effect and TFP (%/year)	+	+	−	−	−	−0.13613
61. Energy industry R&D activities			−0.00219	0.004831	−0.00265	-6.3E-05
Average price increase rate (annual %)			−4.9461	−0.12661	−0.2491	
Technology bias trends			Saving	Using	Saving	
Productivity effect and TFP (%/year)			−	−	−	−0.06712
60. Energy industry ICT activities			0.004142	0.014677	−0.01882	−0.00124
Average price increase rate (annual %)			−4.97595	−0.19811	−0.39226	
Technology bias trends			Using	Using	Saving	
Productivity effect and TFP (%/year)			+	+	−	−0.21825

Table 6.14 Technology evaluation sheet: transportation industry

b-2-1 BKrT > 0	BKrT	BKiT	BKT	BLT	BMT	BTT
65. Transportation industry	3.487E−05	2.81E−05	0.002355	0.002986	−0.0054	0.000254
Average price increase rate (annual %)	−2.8757	−2.5796	−2.3533	−0.3356	0.3677	
Technology bias trends	Using	Using	Using	Using	Saving	
Productivity effect and TFP (%/year)	+	+	+	+	+	−0.00917
67. Transportation industry R&D activities			−0.00102	0.0047	−0.00376	−0.00011
Average price increase rate (annual %)			−2.3394	−0.12662	−0.24904	
Technology bias trends			Saving	Using	Saving	
Productivity effect and TFP (%/year)			−	−	−	−0.0644
66. ICT activities in the transportation industry			0.003101	−0.00298	−0.00012	−0.00055
Average price increase rate (annual %)			−2.4387	0.043944	−0.38181	
Technology bias trends			Using	Saving	Saving	
Productivity effect and TFP (%/year)			+	+	−	−0.19057

There is no market mechanism that allows the determination of intangible asset service prices to enhance the efficiency improvement effect of intra-firm activities. In the labor market, the balance between households as suppliers and companies as demanders determines service prices. However, for intangible capital services, the demand is determined by the cost-minimization principle in the main product market, and the suppliers are influenced indirectly through the demand and supply market of main production activities. It is important to consider whether the selection of intangible services as demanders contributes to increasing the marginal productivity of intangible investments (*Kr* and *Ki*) and whether the setting of service prices based on $P = MC$ is justified. Additionally, a reevaluation of the series of intangible asset investments and the fluctuation in service prices determined by main production activities would be necessary.

It would be considered that a clear market should be established to determine the service prices related to intangible asset investments (R&D, ICT activities). The opacity of the intangible capital investment market seems to obscure the utilization of intangible capital in the Japanese economy and hinder the realization of the benefits of information accumulation. For example, the advantages of agglomeration in Silicon Valley are believed to be formed by an implicit market where suppliers and demanders of intangible capital confront each other and determine the prices of knowledge assets.

In the reconstruction of market structures after the pandemic, it may be essential to establish a market for information in order to expand the benefits of urban concentration. Concrete measures for realizing this need to be considered.

In this context, it is crucial to distinguish and consider the establishment of an intra-firm intangible asset investment market, focusing on the efficiency of synergy effects through collaboration between intra-firm activities and main production activities. Additionally, it is important to consider the creation of a market for intangible services, considering the spillover effects between industries.

How to analyze the temporal variations in inter-industry differences in prices for intangible asset services and labor services? By adjusting the series indexed to 2005 = 1.0, can we calculate intangible asset service prices per man-hour, like the wage rates, to estimate inter-industry disparities?

Here we aim to clarify the implications of industrial technology choices inferred from the parameter estimation results of the trans-log cost function in relation to the changes in factor relative prices during the observed period.

Industries 84–93 focus on activities dedicated to R&D. Industries 84–88 categorize activities related to universities and research institutions, both public and nonprofit, into five scientific fields (Life, Information and Communication, Materials, Environment and Energy, Others). Similarly, industries 89–93 classify for-profit institutions specializing in research into the same five scientific fields. This categorization distinguishes non-profit R&D activities of public entities from for-profit R&D activities conducted by industries for business purposes. Industry 94 encompasses other service industries' main production activities and their intra-firm research and development activities. In terms of industry classification, it covers 41 industries, and in terms of activity classification, it spans 95 sectors (81 and 82 are not assigned, hence effectively 93 sectors).

Chapter 7
Science for Science Technology and Innovation Policy and The Development of Economy and Society

Masahiro Kuroda and Michael C. Huang

Abstract This chapter investigates the evolving role of science and technology in shaping Japan's economy and society from the lens of "trans-science," a discipline addressing issues that science alone cannot resolve. It explores Japan's shift from tangible to intangible capital investments and the effect of R&D and information processing on industrial productivity. The chapter introduces a policy simulator to evaluate science and technology policies within economic and social contexts to assess the synergistic effects between sectors and the influence of intangible assets on economic growth. The overall findings concerning to the recent structure among the relationships between society and science-technology underscore the necessity of integrated science, technology, and industrial policies to effectively navigate the complexities of modern societal challenges.

Keywords Trans-science · Intangible assets · Total factor productivity · Economic policy · R&D investment

7.1 Development of Science and Technology and Structural Change in the Economy and Society

Over the course of Chap. 6, we attempted to grasp the structural changes in the relationship between science and technology through the overview of the Japanese economy after World War II. Additionally, entering the twenty-first century, we

M. Kuroda (✉)
Professor Emeritus, Keio University, Tokyo, Japan

M. C. Huang
Ocean Policy Research Institute, The Sasakawa Peace Foundation, Tokyo, Japan
e-mail: michael-huang@spf.or.jp

M. Kuroda (ed.), *Intangible Capital Investment and Economic Growth*, Translational Systems Sciences 45,
https://doi.org/10.1007/978-981-96-7675-0_7

explored the challenges in science and technology policy in the era of increasingly evolving trans-science, and methods to address these challenges. The term "trans-science," coined by physicist Alvin M. Weinberg, captures the modern relationship between science and society as "Scientific and technological advancements being one of the key factors causing economic and social structural changes. Yet, despite this, the various challenges occurring in society cannot be solved by science alone."

In his paper, "Science and Trans-Science," he discussed the arrival of an era of "Transcendental Science", which challenges the new issues created by the paradigms of modern science and surpasses conventional modern science.

He said in his paper,

> Many of the issues which arise in the course of the interaction between science or technology and society—e.g., the deleterious side effects of technology, or the attempts to deal with social problems through the procedures of science—hang on the answers to questions which can be asked of science and yet *which cannot be answered by science*.[1]

In his discussion, Weinberg refers to the present era as the "Age of Trans-Science," suggesting that society is confronted with problems that are too complex to be solved through modern scientific logic alone. This implies that these challenges are so significant that current scientific methodologies cannot resolve them or that the solutions require such extensive resources that funding them is infeasible. Regardless, Weinberg believes that to address and solve these fundamental problems, we must first understand the structural changes occurring in the relationship between science and society. This can be done by clearly identifying the symptoms of the issues and problems and understanding what can and cannot be resolved. See Weinberg (1972).

The concept of "Science for Science, Technology and Innovation Policy (SciREX)," proposed by the Japanese government at the 2014 Comprehensive Science and Technology Conference, aims to tackle these challenges. This initiative seeks to engage with the issues directly, emphasizing the necessity to adapt and evolve our scientific approaches to effectively address the complexities of modern societal challenges.

Following the trajectory of Japan's socio-economic reconstruction, the relationships among science, technology, and society had evolved and transformed. How far the construction of science and technology policy has progressed and the contemporary issues requiring trans-science have been the key focus throughout this period.

The following section summarizes the achievements and challenges from the immediate postwar period to the twenty-first century:

[1] Alvin M. Winberg "Science and Trans-Science" Minerva 10(2): 209–222 (1974).

7.1.1 *Post-War Reconstruction and High Economic Growth (1945–1970s)*

After World War II, the science and technology framework prioritized industrial recovery and the formation of rational markets, along with advancements in technological adaptation to large-scale production and the revitalization of the market economy.

7.1.2 *Economic Stagnation and Technological Advancement (1980s–1990s)*

From the 1980s onward, Japan faced economic challenges, such as asset bubbles and their collapse. This era was also characterized by technological breakthroughs in heavy machinery, the automotive industry, robotics, and consumer electronics. The relationship between science and society also began to address new challenges, including broader social issues such as environmental concerns and the aging population. Japan had to simultaneously deal with the difficulties of a strong yen in the fluctuating foreign exchange market.

7.1.3 *The Twenty-First Century Shift (2000s–)*

The twenty-first century shift has been characterized by further globalization and the rise of information technology. The focus has shifted toward "sustainable development," "energy efficiency" and "addressing global climate change." Science and technology policy has evolved to emphasize innovation not just in technology but also in regulatory frameworks and societal integration.

7.1.4 *Current and Emerging Challenges (Trans-Science)*

The contemporary issues facing Japan, categorized under trans-science, involve complex problems such as disaster resilience, cybersecurity, and ethical implications of technological innovations. These challenges are multidisciplinary, requiring input not just from science and technology but also from social sciences and humanities to craft comprehensive solutions.

Chapter 1 traces Japan's economic recovery and structural reforms from the end of World War II in 1945 to the publication of the 1956 Economic White Paper, which declared that Japan was no longer in the postwar period. Following Japan's unconditional surrender in August 1945, efforts began to revive the postwar economy. By 1946, Japan's per capita real gross domestic product (GDP) had fallen to less than 55% of its peak during 1934–1935, and the manufacturing production index had plummeted to one-fifth of its 1944 level. In August 1945, the Japanese government accepted the Potsdam Declaration, which called for the establishment of a peaceful and responsible government, revival and strengthening of democratic tendencies, and establishment of respect for fundamental human rights. These demands led to significant reforms of the Meiji Constitution. In April 1946, a new election law granted voting rights to women; the House of Representatives election was held, followed by the convening of the 90th Imperial Diet in May and the promulgation of the Japanese Constitution on November 3, 1946. During this period, the policies adopted by the Japanese Government Economic Stabilization Headquarters were radical institutional reforms aimed at restoring the mechanisms of the market economy. It might be deemed that these policies were extremely rational, where it might be based on Adam Smith's assertion that the principles of market economy, which the competition lies, not in laissez-faire government policies, but government interventions designed to facilitate effective market competition.

The cornerstone of market system reforms was the dissolution of the zaibatsu, which removed the prewar dominance of certain capital groups over the industrial structure, and a comprehensive land reform that transformed land ownership from landlords to farmers. This reform not only addressed the capital market but was also a revolutionary change for the labor and capital markets and helped establish a free market system. The labor law reforms carried out simultaneously were aimed at protecting the fundamental human rights of workers. Agriculture played a crucial role as a support system for more than five million soldiers and overseas Japanese nationals returning after the war. It served as a vital source of labor supply during the subsequent period of economic growth and provided essential goods to citizens suffering from food shortages.

These various reforms laid the foundation for a market economy and shaped subsequent economic policies, including an industrial policy known as the "inclined production method." Adam Smith believed that the essence of a nation's wealth lies in the expansion of industrial productivity, which is achieved through improvements in productivity brought about by the expansion of the division of labor. This division occurs through technological specialization in each production process and the expansion of inter-industrial division of labor. The postwar industrial policy, initiated with the "inclined production method," along with subsequent government financial and fiscal policies, supported Japan's high economic growth and became a critical pillar of policy.

Chapter 2 analyses the international competitiveness between Japan and the US, exploring how the industrial competitiveness between them evolved from the 1960s, a period marking Japan's recovery postwar, through the peak growth years assessed up to 1990. In this period, Japan experienced several shifts, including the high economic growth from the 1960s to the early 1970s, the transition from a fixed to a floating exchange rate system in the early 1970s, the volatile era of oil shocks, and the bubble economy phase following the 1985 Plaza Accord. These economic shifts influenced market dynamics and reflected the interplay between scientific-technological evolution, political changes, and how Japan adapted and adjusted to these factors, which influenced the international competitiveness of its industries.

The comparative analysis of industrial competitiveness between Japan and the US involves assessing the competitiveness of product prices, which was influenced by fluctuations in exchange rates, differences in factor costs (labor, capital, raw materials) internationally, and variations in production efficiency across countries. Chapter 2 presents a comparison of the total factor productivity (TFP) of Japan and the US by adjusting for factor price differences due to exchange rates and considering differences in production efficiency between the two countries. The chapter assesses how these technological efficiency changes have influenced various industries.

During the high economic growth period of 1960–1970, Japan's real GDP growth rate was exceptionally high, with an annual rate of 10.126% from 1960 to 1965 and 11.791% from 1965 to 1970. The contribution to the growth during 1960–1965 was 18% from labor, 29% from capital, and 53% from technological advancements. However, Japan was only able to surpass the US in productivity in certain industries such as agriculture, forestry, fisheries, textile manufacturing, apparel manufacturing, and other service industries in 1960, with fewer industries surpassing in 1965. Being labor-intensive and subject to lower labor costs allowed these industries to maintain a competitive edge. Meanwhile, industries centered around manufacturing showed a remarkable increase in technological advancement rates (or TFP growth rates), with annual rates of 5.402% in the first half and 6.223% in the latter half of the high-growth period. The push for heavy chemical industrialization policies facilitated capital investment growth in industries such as chemicals, steel, general machinery, and automobile manufacturing, continuing this trend until the early 1970s.

However, the global economic shifts of the 1970s, including the introduction of floating exchange rates leading to a stronger yen and the effect of soaring oil prices due to the oil shocks, had a profound effect. While the average annual real GDP growth rate was over 10% from 1960 to 1972, it dropped to 3.887% from 1972 to 1992. During the latter period, while the initial half still saw growth driven by increased capital input and technological advancements, the second half saw a decline in the contribution from technological progress, with a higher reliance on labor and capital inputs. Postwar heavy industrialization stabilized capital service prices and thus increased capital intensity through the adoption of capital-intensive technologies. Although labor service prices were still lower than those in the US, their rapid increase, coupled with expanded capital investment during the high economic growth period, drove personal income growth, and expanded effective demand through personal consumption, contributing significantly to market expansion.

Chapter 3 continues the analysis, focusing on the structural changes in labor and capital inputs that occurred after 1972 and the decline in the contribution of technological progress. This chapter explores the substantial shift from traditional tangible fixed asset investments, such as those in heavy and chemical industries, to intangible fixed asset investments, which have different characteristics and effects on economic efficiency.

During the high economic growth period of 1960–1970, tangible fixed asset investments supported significant growth in real GDP, achieving annual rates of 10.126% from 1960 to 1965 and 11.791% from 1965 to 1970. The contribution to growth during 1960–1965 was 18% from labor, 29% from capital, and 53% from technological progress. In this era, the promotion of heavy chemical industrialization boosted capital inputs into chemicals, steel, general machinery, and automobile manufacturing, facilitating a trend of productivity improvement that persisted until the early 1970s. Global economic shifts of the 1970s, such as the transition to a floating exchange rate system resulting in a stronger yen and oil price surges from the oil shocks, significantly affected the Japanese economy. From 1960 to 1972, Japan maintained an average annual real GDP growth rate exceeding 10%, but this rate declined to 3.887% from 1972 to 1992. In the latter period, while initial growth was driven by increased capital input and technological advancement, the second half saw a decline in the contribution from technological progress.

A major transformation during this period was the shift in capital input. Investment began moving away from traditional physical capital assets, such as equipment investment, and toward intangible assets, such as R&D and information processing. This shift marked a transition from supporting postwar mass production and consumption structures primarily within heavy chemical industries to fostering the development of intangible intellectual assets, such as technological knowledge and information. These intangible assets influence industrial economic efficiency in significantly separate ways compared with tangible fixed asset investments.

To evaluate the effect of intangible assets on production efficiency, Chap. 3 proposes adding an analytical framework to the industrial interconnection tables that explicitly capture activities related to R&D and information processing. It suggests incorporating the concept of the output of intangible intellectual assets, along with the traditional concept of tangible fixed assets, to measure and assess the effect of the service quantity of intellectual assets emerging from the stock of intangible fixed assets. This innovative approach provides a more comprehensive assessment of how these investments influence industrial economic efficiency and overall economic performance.

Chapter 4 builds on the discussions from previous chapters about the shift toward intangible capital assets such as R&D and information processing. It analyses the structure of corporate investments in these intellectual assets as a rational cost-minimizing behavior based on their cost functions.

This chapter proposes how we can formulate the investment behavior of the intellectual assets as their rational actions within the framework of their respective cost minimization. To achieve this, we proposed the time-series estimation of the input-output table, in which we can observe the time-series investment behavior for both tangible and intangible investment separately. This period is significant because it encompasses a range of economic conditions, including the aftermath of the Japanese asset price bubble burst in the early 1990s, economic stagnation of the "Lost Decade," and early signs of recovery in the 2000s.

The analysis in Chap. 4 delves into the specifics of how different industries have adjusted their cost structures in response to macroeconomic pressures and opportunities. By measuring the cost function parameters, the chapter seeks to elucidate the characteristics of these investments and their effects on the cost-efficiency and innovative capabilities of firms.

It addresses questions such as:

- How do firms balance the costs and expected benefits of investing in intangible assets?
- What is the industry-specific factor that influences the extent and nature of these investments?
- How do these investments affect the overall productivity and long-term strategic positioning of firms within their respective sectors?

The findings from this analysis provided us the insights into the effectiveness of corporate strategies concerning intangible asset investments and their role in shaping industry dynamics. This is crucial for understanding the broader economic implications of the shift toward a more knowledge- and information-driven economy in the developed industrial country such as Japan.

Chapter 5 continues to explore the shift from tangible asset investments toward intangible asset investments in R&D and information processing, which has significantly influenced trends in the Japanese economy. This shift reflects a broader trend observed in advanced economies, where investment in intellectual assets has increasingly become a focal point for driving economic vitality. In the context of Japan, this chapter examines the transition from the tangible fixed asset investments that characterized the high economic growth period to the growing emphasis on intangible assets. Notably, the macro-level analysis of Japan's economy from 1995 onward reveals a general stagnation in TFP, which averaged an annual rate of −0.05% from 1990 to 2011.

Despite this, the patterns of investment have shown a discernible shift: while the average annual growth rates of tangible fixed capital inputs for the every fuve years during the periods 1995–2010 were 3.40%, 2.08%, and −3.94%, the growth rates for intangible fixed capital were more robust in the earlier years, recording 7.57%, −0.29%, and −1.28% in the same periods. Growth rates in intangible capital were initially surpassing those of tangible capital, although they too have seen declines in more recent years. This recent investment trend, where intangible capital investment exceeds the growth rate of tangible capital investment, is not unique to Japan. In Europe and the US, the expansion of intangible capital has transformed the economic structure, and intellectual assets such as R&D and information services might be revitalizing the economy.

The decline in the growth rate of tangible fixed capital investment and the moderate performance of intangible asset investments post-1995 reflect the broader challenges faced by the Japanese economy following the burst of the asset bubble. However, unlike in the high economic growth era, substantial investments for the intangible infrastructure have not enhanced technological efficiency in recent decades. Similarly, the expected economic boost from investments in intellectual assets such as R&D and information processing has not been as effective as anticipated in Japan.

On the other hand, such phenomenon in Western economies suggests a global shift in economic structures, where intangible assets play a crucial role in driving economic growth. In Europe and the US, the expansion of intangible fixed assets has linked to significant structural changes in the economy, where intellectual assets such as R&D and information services are increasingly seen to be as a vital for economic revitalization.

Chapter 6 underscores that the observed post-bubble trends in Japan's investment patterns are indicative of a broader realignment toward a knowledge- and service-driven economy. We expected that the alignment was crucial for revitalizing and sustaining economic growth in the face of changing global economic dynamics and the challenges of a post-industrial economic landscape.

Despite the shift in investment from tangible capital to intangible capital following the burst of Japan's bubble economy, we did not feel an improvement in technological efficiency, well. Identifying the underlying issues behind this phenomenon has become an important challenge. Based on the measurement results of cost function parameters for each industry, the previous chapter proposed evaluating the economic effects of the expansion of intangible capital investments from two perspectives.

The first perspective focuses on whether there are effective synergistic effects between the three activities of an industry, considering their technical characteristics and interdependence within a company. The second perspective considers whether the structure of interdependence in technical characteristics between industries can exert effective synergistic effects through the interdependence of supply and demand for factors in the markets for capital (tangible and intangible), labor, and raw materials. These synergistic effects are referred to as endogenous and exogenous synergistic effects, respectively. Detailed explanations are provided in the previous chapter, but the shift from tangible asset investment to intangible fixed asset investment in Japan has not yet led to significant improvements in production efficiency as a technological characteristic of the industry.

7.2 Necessity of Policy Simulator

In an environment referred to as the era of "trans-science," where the specific challenges where cannot be resolved by modern science alone—an issue stemming from the aimlessness of modern science—what methods are necessary to discuss the pros and cons of science and technology policies? To address this question, an analytical method is needed to assess the effect of implementing science and technology policies on the real economic and social environment. This requires the creation of a "policy simulator" that can compare policy effects in a clear and comparable manner.

The policy simulator proposed here is a tool to measure the economic and societal effects by the introduction of a specific science and technology promotion policy. Its purpose is not to predict actual changes in society, but to demonstrate how the policy affects the economy and society under the condition that other factors remain constant. By inferring the merits, demerits, and effects of the policy, the experiment aims to understand how changes in these conditions affect the structure of the economy and society.

Leontief's inter-industry analysis is based on a similar idea, proposing inter-industry analysis as an experimental tool to understand the effect of implementing specific policies by setting the interdependence of various variables in the current economic and social context. This is the role of the policy simulator.

This experiment can accommodate multiple scenarios for a given policy. Even if the structure described by the policy simulator is provided, different policy scenarios will detect different effects on the economy and society, which then serve as indicators of the policy's outcomes. Therefore, the ability to accurately show the results of policy implementation based on the differences in selectable scenarios determines the quality of the policy simulator. The simulator's quality is not determined by the accuracy of its predictions of the future image of the economy and society.

Leontief's inter-industry analysis comprises a theoretical framework that captures the balance between the supply-demand structure; it reflects the production techniques of goods and services and the demand structure for these produced goods and services through the interdependent relationship of inputs and outputs in a matrix. This is complemented by concrete statistical data. The inter-industry table proposed in this book has been rearranged from published inter-industry tables, which helps us to analyze science and technology policies by quantitatively capturing the outcomes of these policies as changes in the input-output structure of industries. Specifically, to grasp the economic and social effects of R&D and information processing technologies, we have categorized the activities of each industry into the output and input of main products and the R&D and information processing activities that support their output. This way, we capture the influence of R&D and information processing activities within companies on the main production activities and understand how they change the supply-demand structure in the product market.

To observe the socio-economic effects of the national science and technology policies, we have structured the output and input activities of universities and research institutions dedicated to scientific and technological development and information processing, as well as private research and information processing institutions primarily engaged in these activities. This allows for evaluating the effect of technological advancements in R&D and information processing on other industries utilizing these outcomes. By distinguishing the output and input of intangible fixed assets, such as knowledge and information generated through R&D and information processing, from traditional tangible fixed assets such as equipment investments, we can observe the effects of intangible assets on the economy and society. This helps devise policies that reliably link these outcomes to economic and social reforms.

However, we cannot measure how and to what extent the expansion of investment in intangible fixed assets, such as knowledge and information from R&D and information processing, will advance research in each department. This depends on the research efficiency of universities and research institutions, and the direction and degree of advancement can vary significantly by research field. Thus, we do not discuss this issue here. Regarding the evolution of research fields that are expected to address the challenges of modern science and society, we aim to depict how factor prices and price changes in the goods and services market under these conditions bring about resource allocation outcomes, thereby evaluating the effect of the given policies.

By observing the inter-industry tables as time-series data (1995–2011 in this book), we capture the time-series changes in economic structure and the effects of various industrial technology policies on the economic system, based on the interdependent structure of industries.

The time series of observed values reflects not only the science and technology policies of that period but also the results of other policies at each point, such as fiscal policy, monetary policy, labor policy, and changes in various legal systems, encompassing the actual behavior of economic entities. The possibility of conducting controlled experiments, as in natural sciences, is extremely limited in economic analysis. Thus, we must adopt a methodology where we set the structure of the real economic society as a vast laboratory, theorize the outcomes of science and technology policies and other economic policies by disentangling their effects, propose an analytical framework to identify policy effects, and verify this through observed facts to elucidate the structure of science, technology, and society in the real economic world.

If the purpose of the policy simulator is not to forecast the future but to accurately capture the current interdependence between technology and society, including its structure, then the most crucial perspective in evaluating the simulator's performance is how it captures the current structure and characteristics of the "science, technology, and society" interdependence in the economy and society. This book observes the transition of Japan's economic situation from the post–World War II defeat, through the so-called high economic growth period, the era of fluctuating exchange rates and oil price shocks, the bubble economy period post-1985

Plaza Accord, and up to the present characterized by rapid evolution of information technology amid aging, declining birthrates, and a globalized society. Therefore, the policy simulator accurately captures the current relationship between science, technology, and society.

7.3 Policy Simulator in Trans-Science Age

We created the policy simulator to develop an analytical model that captures the structure of economic interdependence based upon the observed inter-industry input-output tables as time-series data. This allows us to observe the time-series changes in the output and input structures of various industries and endogenously understand the effects of various industrial and science and technology policies on the economic system within the interdependent structure of Japan's industries during this period. The measurement of the various structural parameters included in the model will be based on past observational data. These measured parameters are derived from observations reflecting past social conditions and policies at each point in time and include the effects of changes in various legal systems and policies.

In this context, analysts observe the results of actual economic behavior, including changes in various external factors and policies, considering that conducting controlled experiments, as in the natural sciences, is extremely limited in the field of economic analysis. Therefore, the analysis treats the real economic society as a vast laboratory, inferring the structure of this laboratory from observational data and creating an analytical framework that can distinguish the effects of science and technology policies from those of other economic policies. This methodology is designed to evaluate and analyze the socio-economic effects when certain science and technology policies are introduced.

The previous chapter focused on the analysis of the structure of the production, in which the input-output structures shifted from an investment strategy centered on tangible fixed capital investments, which led postwar economic growth, to an increased emphasis on intangible fixed capital investments such as R&D and information processing investments in the latter half of the twentieth century. This shift highlights the new structural changes in economic development. Specifically, a theoretical framework was constructed to estimate the capital stock of tangible and intangible fixed assets, the volume of tangible and intangible capital services generated from that stock, and service prices. Within this framework, TFP indicators for each industry were measured separately for main production, R&D, and information and communication processing activities. This approach captured the structure of how the collaboration of these three activities within industries affects their production efficiency. Additionally, the analysis identified the effects of technological efficiency in each industry on the productivity levels of related industries through the interdependence of activities across industries, thereby advancing the understanding of the structural shift from tangible fixed assets to intangible knowledge and information assets.

Changes in production efficiency in the economy are thought to arise from the selection and rationalization of technology in industrial production processes, driven by the accumulation of intellectual assets as research outcomes of science and technology at each point in time. Universities and research institutions, as research entities, are among the key players in the evolution of these intellectual assets. If these are public institutions, the intellectual assets produced are often considered as intangible public goods on a non-rivalrous basis. However, recent cases provide evidence that the use of these intellectual assets involves patent rights and is market-oriented. In the case of private research institutions, intellectual assets may be offered in the market with associated patent rights, turning them into market goods. In this case, the propagation of technology occurs through trading markets for patent rights, and establishing such markets is an important issue for science and technology policy. Although there are various routes for the distribution of scientific and technological insights from the accumulation of intellectual assets, intellectual property information is introduced into industrial sectors that utilize it to rationalize their technology according to certain standards. This introduction leads to changes in the amounts of labor, capital, and intermediate materials used in the production activities of those industries.

Therefore, when selecting technological characteristics, the quantities of production factors used are determined according to the input factor prices in the respective factor markets at that time. The choice of technology and the quantities of production factors are interlinked. In each factor market, the aggregate demand for factors by industrial sectors is compiled, and the balance of supply and demand for factors progresses, leading to the determination of factor prices as market equilibrium prices. In previous chapters, the technological characteristics were specified using a trans-log cost function, parameters were measured, and the technological characteristics of each industry were depicted.

When capturing the technological characteristics of each industry, the activities within each industry are divided into main production activities, intra-firm R&D activities, and intra-firm ICT activities. Based on the principle of cost minimization, the technological characteristics of each activity are selected, with the results of these selections reflecting the interdependencies of each activity, thereby determining the technological characteristics of the industry. The nature of the relationships between these three activities within each industry becomes an important perspective that influences technological selection.

Intra-firm R&D and information and communication technology activities, as intra-firm activities, expand intellectual knowledge and information processing technologies. The output of these activities produces intellectual knowledge for R&D and information processing, which, in turn, through its accumulation, supports the main production activities. Conversely, main production activities are thought to invest in tangible capital that supports these two intra-firm activities, carrying out actions that assist them.

Thus, changes in technological efficiency due to the selection of technological characteristics in intra-firm activities influence the expansion of intellectual capital related to R&D and information processing, which, in turn, affects the technological efficiency of main production activities. Additionally, the level of tangible capital in these two intra-firm activities, as determined by main production activities, establishes the technological efficiency between the three interdependent activities.

The structure of collaboration between main production, R&D, and information processing activities within a company is determined by intra-firm technological selection, and the resulting technological spillover effects within the industry from this relationship are called endogenous synergy effects.

The results of technological selection within each industry influence the demand level for production factors, which, in turn, affects how factor prices are determined in each production factor market. This leads to changes in the factor prices of each industry, resulting in spillover effects on production efficiency. The resulting changes in technological efficiency are called exogenous synergy effects, which are distinguished from the previously mentioned endogenous synergy effects.

The importance of understanding the effect of changes in technological efficiency due to technological selection has become increasingly significant in the twenty-first century. This is due to the rapid expansion of intangible capital investments, such as intellectual assets from R&D and advances in information processing technology, which are replacing tangible capital investments, such as equipment investments, during the period of rapid economic growth. This shift has led to the emergence of a new phenomenon where technological efficiency is being expanded, which necessitates understanding the improvement in production efficiency.

Even though intangible capital investment has shown a greater increase than tangible capital investment in Japan's economy in the 2010s, the expected improvement in industrial productivity has not progressed as anticipated since the bubble economy. The synergistic effects mentioned above have not materialized effectively, possibly contributing to the decline in Japan's economic growth and ushering in a period of stagnation (as noted in the previous chapter).

Therefore, evaluating the pros and cons of the shift in investment weight from tangible capital to intangible capital in terms of how it has affected the technological efficiency of Japan's economy and demonstrating the potential outcomes of the policies implemented in response becomes significant in the policy simulator.

7.4 Structure of Policy Simulation Model

Intra-firm R&D and information and communications technologies within each industry extend intellectual knowledge and information processing technologies. The output of these activities, producing intellectual knowledge for R&D and information processing for information and communication processing activities, supports the main production activities through accumulation. Conversely, main production activities acquire tangible capital investments that support these two intra-firm activities. Thus, the shift in technological efficiency due to the selection of technological characteristics in intra-firm activities affects the main production activities by expanding intellectual assets related to R&D and information processing. Conversely, the main production activities determine the level of tangible fixed assets supporting the two intra-firm activities. The mutual interdependence of these three activities determines their technological efficiency.

The technological choices within the company determine the structure of collaboration among main production, R&D, and information processing activities. The effect of this relationship on technology within the industry is called the endogenous synergy effect. The results of technological choices within each industry also affect the demand levels for production factors, influencing the determination of factor prices in each production factor market. This leads to changes in factor prices and affects production efficiency, resulting in exogenous synergy effects, which are different from endogenous synergy effects.

Understanding the influence of technological choices on changes in technological efficiency has become increasingly important in the twenty-first century, as rapid growth in intangible fixed asset investments, such as intellectual assets from R&D and advances in information processing technology, has begun to expand technological efficiency, replacing tangible fixed asset investments like equipment investments during the high economic growth period.

Despite intangible fixed asset investments surpassing tangible fixed asset investments in Japan's economy in the 2010s, the expected improvement in industrial productivity has not progressed as anticipated since the bubble economy. The synergies mentioned above have not materialized effectively, which is a potential factor in the decline of Japan's economic growth during this period of inflection.

Therefore, evaluating the merits and demerits of the shift in investment weight from tangible capital to intangible capital, assessing how this shift has altered Japan's technological efficiency, and indicating the outcomes of the policies deployed in this context are major roles of the policy simulator.

Considering policies that contribute to technological choices in industries, two broad approaches can be considered. One approach is to promote research in various scientific and technological fields, particularly within universities and similar institutions, to enhance the technological efficiency of the intellectual assets chosen by industries. This policy approach seeks to facilitate the activities of universities and research institutions, thereby improving the technological efficiency of each industry using intellectual property. This can be seen as a science and technology policy that promotes the advancement of scientific knowledge. The technological spillover effects on each industry will vary by the technological characteristics of the industry and the extent to which scientific research outcomes from specific fields enhance the technological efficiency of that industry.

In policy implementation, both the specific scientific fields the government will focus its policies on and the kind of policy that will be provided are crucial. The policy simulator must be equipped with a mechanism that can assess the effect of these choices. In this context, it is important to consider how policy support will be provided, depending on which scientific fields' outcomes are being targeted—whether resources will be allocated to "expanding tangible capital facilities in the scientific field," invested in "personnel costs," or provided in the form of general "operating grants." These differences in policy implementation will lead to different policy effects, and the structure of the policy simulator model must reflect these differences.

7.5 Policies to Support Technological Choices in Industries

There are two broad pathways to consider for policies that support the technological choices of industries. The first is to promote research in various scientific and technological fields at universities and other institutions, thereby enhancing the technological efficiency of the intellectual assets chosen by industries. This involves revitalizing the activities of universities and research institutions to improve the technological efficiency of industries through their intellectual property. It is essentially a science and technology policy that promotes the advancement of scientific knowledge. The extent to which technological spillover effects on each industry, based on the technological characteristics of each industry, enhance its technological efficiency varies by the scientific field and the outcomes of scientific research.

The scientific fields on which the government focuses and how it provides policy support to these selected fields are both crucial for policy implementation. Therefore, a policy simulator must be capable of evaluating these effects. In this context, it is important to consider how the policy will support the scientific fields: whether to allocate resources to "expand tangible capital facilities" in the scientific field, invest in "personnel expenses," or provide general "operational subsidies." These differences in policy implementation can lead to varying policy effects, which must be reflected in the structure of the policy simulator model.

Why does this situation arise? To connect the analysis of this structure to the technological selection behavior of industries, we measure the parameters of the industry's cost function. The results of the trans-log cost function measurements are shown in Table 5.8 of Sect. 5.3 in Chap. 5.

The evaluation of the measurement results in Chap. 4 is as follows: the characteristics of technological progress in national, public, and non-profit research institutions (national and private universities and research institutions) indicate that there is a usage bias regarding tangible capital services in all scientific fields. This means that technological progress in these scientific fields tends to use tangible capital for R&D. Therefore, an increase in the price of tangible capital services slows technological progress. Regarding labor services, a usage bias is observed in the life sciences and materials fields. For intermediate materials, life sciences, materials, and other fields exhibit factor-saving characteristics, whereas information and communication, environment, and energy fields show a usage bias. R&D at national and public institutions accumulates intellectual assets as public goods, and investments in R&D are driven by governmental policy decisions. Therefore, policy managers must focus on the biased characteristics of technological progress highlighted here when making government policy decisions.

According to Chap. 5 (Table 5.11), the parameter β^{TT}, which indicates the trend of technological efficiency, is negative in all research fields except for "other fields," suggesting that technological efficiency improves with technological progress. The parameters for the technological bias of production factors are as follows: for capital services, $\beta^{KT} > 0$ in all research fields; for labor services, $\beta^{LT} > 0$ in the life sciences and materials fields; and for intermediate materials, $\beta^{MT} > 0$ in the information and communication, and environment and energy fields. This finding indicates that an increase in the price of each factor slows technological progress.

Since whether the prices of production factors rise or fall depends on the supply-demand balance in the factor market, it may be difficult to guide the prices of factors in specific industrial sectors through policies. Implementing supplementary policies, such as providing support for certain factor prices, may guide the direction of changes in technological efficiency in these research fields. This approach could be considered a policy to enhance production efficiency by focusing on the trends in factor prices and the values of technological bias parameters in each industrial sector. Even in such cases, we must confirm in advance how any form of support might alter the supply-demand structure of each factor market and its potential effect on the overall economy.

In our input–output table, we categorize the activities of universities and research institutions into public and non-profit universities, research institutions, and private industrial research institutions. Each sector is further divided into research fields, such as life sciences, information and communication, materials, environment, and energy, to capture their activities.

For the activities of public and non-profit universities and research institutions, government policies can support the production of intellectual assets by providing subsidies for tangible capital investments in research facilities, personnel expenses, and other operating grants to various research sectors. This can presumably direct the effects of these policies to stimulate the production of intellectual property as a public good. Policy evaluations consider how these policies relate to changes in industries' technological efficiency through changes in the prices of developed intellectual assets.

Although specific research outcomes often lead to technological innovations in particular industrial sectors, predicting how such innovations might influence technological efficiency in other industrial sectors presents difficulties. In the policy simulation model, the results elucidate the extent to which technological innovations in specific scientific fields, such as the life sciences, directly or indirectly affect various industrial sectors. Additionally, using the values of the cost function parameters estimated from historical data for each industry, the magnitude of these effects and the underlying mechanisms can be inferred.

Another perspective for evaluating industrial technological choices and providing policy implications is the policy response to the interdependent structure between the main production activities, intra-firm R&D, and intra-firm information processing activities within each industry. As discussed in detail in the previous chapter, the twenty-first century is witnessing a shift in industrial investment from tangible to intangible capital investments, such as R&D and information processing. This shift has not necessarily improved industrial production efficiency, a point that has been observed and noted.

This issue can be considered a challenge for industrial policy. Given that the outcomes of these policies affect the technological efficiency of each industry, coordinating science and technology policies is essential. As mentioned in the previous section, this policy is ultimately related to endogenous synergistic effects that depend on the interdependence of the three activities within the industrial sector. A similar problem faces exogenous synergy effects, which depend on the mechanisms of interdependence between industrial sectors through the determination of factor prices in factor markets, such as labor, capital, and intermediate goods.

In this sense, long-term industrial policy must consider national industrial policies implemented by the government. The direction of the industrial structure policy, in which policymakers guide the industrial structure of the entire nation, influences the form of policy support for each industry. This, in turn, affects the fluctuations in factor prices in each production factor market, leading to changes in the exogenous synergy effects in each industry and affecting the three intra-firm activities within companies. The resulting changes alter the resource allocation between industries across the entire nation. Using a policy simulator, one can predict changes in the economic structure, observe the effects on the structure of science and technology, evaluate these effects, and assess the outcomes of policy effectiveness. This approach represents a comprehensive national, industrial, and technological policy.

In our inter-industry table, the activities of universities and research institutions are divided into public and non-profit universities, research institutions, and private sector research institutions. Each of these sectors is further divided into research fields, such as life sciences, information and communications, materials, environment, and energy. In the case of public and non-profit universities and research institutions, government policies can support the production activities of intellectual assets by providing subsidies for research equipment, personnel expenses, or other operational subsidies. These policies can pre-determine the effects and activate the production of intellectual property as a public good. Policy evaluations can include how these policies, through price changes in developed intellectual assets, lead to changes in industries' technological efficiency.

Specific research outcomes often lead to technological innovations in certain industrial sectors. However, it is extremely difficult to predict changes in technological efficiency in other industrial sectors. In the policy simulation model, the simulation results show the extent to which industrial sectors are directly and indirectly affected by innovations in specific scientific fields, such as life sciences. The values of the cost function parameters for each industry, estimated from past data, allowed us to infer the magnitude and mechanisms of these effects.

Another perspective for evaluating industrial technological choices and their policy implications is the policy response to the interdependent structure between the main production activities, intra-firm R&D, and intra-firm information-processing activities within each industry. As detailed in the previous section, in the twenty-first century, industrial investments have shifted from tangible to intangible fixed assets, such as R&D and information processing. This shift does not necessarily lead to improvements in industrial production efficiency, which highlights issues within industrial policy. Thus, policy outcomes affect the technological efficiency of each industry and necessitate coordination with science and technology policies.

As indicated in the previous section, this policy is related to the endogenous synergy effect, which depends on the interdependence of the three activities within an industrial sector. It also relates to the exogenous synergy effect, which depends on the mechanism of interdependence among industrial sectors through the determination of factor prices in the labor, capital, and intermediate goods markets. This underscores the importance of positioning such policies within a long-term industrial policy framework that considers the nation's overall industrial policy.

The direction of industrial structure policies, guided by policymakers on how to lead a nation's overall industrial structure, influences the form of policy support for each industry. This affects the fluctuations in factor prices in each factor market, leading to changes in the exogenous synergy effects of each industry and affecting the three intrafirm activities. The resulting changes in resource allocation among industries at the national level can be observed through a policy simulator that predicts the effect on the structure of science and technology. Evaluating these effects and determining the outcomes of policy effects ultimately constitutes a national-level industrial and technological policy. The TFP indicators developed in the previous sections are effective metrics for evaluating the effects of these policies.

7.5.1 Taiwan's Industrial Policy on Semiconductor: Dedicated Foundries

Michael C. Huang

As of 2025, the global market was dominated by ICT-related companies, including major firms such as GAFAM. Among them, Taiwan Semiconductor Manufacturing Company (TSMC), the world's largest contract chipmaker, serves as a notable example of a firm that has significantly benefited from industrial policy. Morris Chang, the founder of TSMC, recalled his recommendations to the Taiwanese government in the 1980s, advocating for the establishment of a government-sponsored startup fund and the implementation of tax incentives to promote capital-intensive high-tech industries. These initiatives laid the foundation for Taiwan's Science Parks, which have played a crucial role in fostering high-tech clusters through both vertical and horizontal integration.

TSMC was established in 1987 as an offshoot of a research laboratory at National Chiao Tung University, with substantial support from the government. The Industrial Technology Research Institute (ITRI), a government-backed think tank, facilitated business matching and technology transfer, while the state coordinated financial support. With government guarantees, Taiwan's semiconductor industry experienced rapid growth despite external economic shocks such as the 1985 Plaza Accord. Taiwan's IT production networks extended beyond domestic and regional markets to a global scale, evolving through successive waves of digital transformation driven by ICT products, including personal computers (1980s), laptop computers (1990s), mobile phones (2000s), smartphones (2010s), and artificial intelligence (2020s).

To meet the increasing global demand for integrated circuit manufacturing and to advance technological nodes, Taiwan pioneered the “dedicated foundry” business model. Unlike companies such as Intel, NEC, Samsung, Toshiba, and Fujitsu, which integrated both design and manufacturing, Taiwan’s dedicated foundries focused solely on semiconductor fabrication without engaging in product branding. This model facilitated the growth of fabless IC design firms, allowing them to concentrate on chip design without the burden of massive capital investment in manufacturing. As a result, TSMC has maintained stable profitability, even during industry downturns, enabling it to provide dividends as incentives for employees and reinvest in the development of next-generation semiconductor technology.

As of 2022, TSMC’s market value had grown from 3.6 billion USD in 1994 to 448.7 billion USD, with R&D expenditure reaching 8% of its total profit. This sustained investment in research and development has allowed TSMC to maintain a technological lead of two to four years over its competitors.

Beyond financial gains, TSMC’s innovations have had far-reaching spillover effects across various industries, contributing to the advancement of smart appliances and AI-driven technologies.

The success of TSMC and Taiwan’s semiconductor cluster can be attributed to early and sustained government support, underscoring the importance of well-crafted industrial policies in fostering high-tech industries. Empirical evidence suggests that government R&D investment in the ICT sector stimulates economic growth, encourages private sector investment, and increases tax revenues (Yang & Liou, 2022). Additionally, such invest-ments enhance labor productivity across industries, leading to a higher relative price of labor compared to capital. However, the finance and insurance sectors may experience a decline in labor demand due to technological advancements, implying that ongoing digital transformation could reshape their business models. A critical factor in the successful establishment of TSMC and the development of Taiwan’s semiconductor cluster lies in the effective communication and collaboration between policymakers, academia, and industry. In Tai-wan, policymaking is not confined solely to bureaucrats or politicians, particularly following the country’s democratization in the 1990s. Since then, it has become common for university professors to assume key ministerial roles in economic planning and science, technology, and innovation (STI) policy. This flexible appointment system has strengthened the interaction between the public and private sectors, as well as academia, facilitating the efficient alloca-tion of substantial resources and the enactment of legal frameworks that sup-port high-tech investment and industrial transformation.

7.6 Final Remark: Collaboration of "Science and Technology Policy" and "Industrial Policy"

I would like to consider the future direction of science and technology policy, and industrial policy based on the analysis of structural changes in the economic and social systems. These changes are driven by the accumulation of intellectual assets through research and development activities and informational assets through information processing activities, which are replacing tangible fixed assets like capital investments as intangible fixed assets.

Up to this point, I have presented a theoretical framework to capture the technological choices of producers as the main production entities in industries. Using this structural model, I have analyzed the parameters of long-term cost functions for each industry, the technical structures of main production as well as R&D and information processing activities, and the nature of technological collaboration and diffusion across industries.

It is crucial to deepen the structural understanding of the technological characteristics of cost structures in industrial activities, divided into primary production activities, intra-firm R&D activities, and intra-firm ICT activities. In each industry, these activities—main production, intra-firm R&D, and intra-firm information processing—are assumed to make rational choices of input levels and technological characteristics based on the principle of cost minimization. These choices are influenced by the prices of input factors and the anticipated output levels required to achieve planned production volumes.

The primary activity of each industrial sector, along with the two intra-firm activities, is assumed to select cost-minimizing technologies in an interdependent manner, given the input prices and planned production levels. Through the coordination of these three activities within an industry, decisions are made regarding the allocation of tangible and intangible fixed capital, labor, and intermediate raw material inputs to main production, R&D, and ICT activities.

The determination of input levels for tangible and intangible fixed capital, labor, and intermediate raw materials in each industrial activity depends on the technological characteristics of the input factors, as well as the market conditions that establish input prices. This interdependence creates synergy effects, whereby changes in production efficiency in one activity impact the efficiency of other activities. These synergy effects can have either positive or negative impacts, depending on the parameter values of the cost functions that reflect the technological characteristics of each activity and the prevailing input price levels at the time.

The prices of capital services (both tangible and intangible), labor, and intermediate raw materials that support the three activities in each industry are influenced by the behavior of other industries in the respective factor markets where these prices are determined. The structure of these factor markets dictates the price fluctuations, which, in turn, affect the choice of technological efficiency in each industry's activities. The resulting impacts, distinct from endogenous synergy effects within an industry, are referred to as exogenous synergy effects, as they arise from external market dynamics.

As detailed in Chap. 5, the parameter characteristics derived from measuring the price functions of various industries reveal that during the observation period from 1995 to 2011, there were very few industries within the Japanese economy where the endogenous synergy effects among the three intra-industry activities—main production, intra-firm R&D, and intra-firm ICT—were observed to be stable and positive. This observation could be considered one of the factors contributing to the stagnation and downturn of the Japanese economy following the bubble economy of the 1990s.

This phenomenon is closely related to the shift in corporate investment priorities from tangible fixed assets, such as equipment, to intangible fixed assets, such as knowledge and information. While the expansion of intangible fixed asset investment has significantly altered the structure and efficiency of industries, these structural changes have not yet reached a state where the utilization and application of intangible assets are effectively leveraged. This indicates a critical issue to be addressed in discussions on the future of science and technology policy and industrial policy.

A key task lies in deepening the structural understanding of the technological characteristics of the cost structures in industrial activities, categorized into main production, intra-firm R&D, and intra-firm information processing. It is assumed that in each industry, given the input prices and anticipated output levels, rational input levels and technological characteristics are chosen based on the principle of cost minimization to achieve the planned production volume.

The main activity of each industrial sector, along with the two intra -firm activities, is understood to select cost-minimizing technologies in an interdependent manner concerning input prices and planned production levels. The coordination of these three activities within an industry determines the allocation of tangible and intangible fixed capital, labor, and intermediate raw materials to main production, R&D, and information processing.

The allocation of tangible and intangible fixed capital, labor, and intermediate raw materials within each industry's activities is interdependently related to the technological characteristics of each input factor and the determination of input prices through factor markets. Changes in production efficiency within one activity influence the efficiency of other activities, forming synergy effects. These effects can be either positive or negative, depending on the parameter values of the cost functions representing the technological characteristics of each activity and the prevailing input price levels at that time.

The prices of tangible and intangible capital services, labor, and intermediate raw materials supporting the three activities in each industry are influenced by the behaviors of other industries in the factor markets where these prices are determined. The structure of these factors dictates price fluctuations, which, in turn, affect the choice of technological efficiency in the activities of each industry. The impact of these price changes, distinct from endogenous synergy effects within an industry, is referred to as exogenous synergy effects.

Let me know if you need a more formal academic phrasing or integration into a larger document.

As detailed in Chap. 5, the parameter characteristics derived from measuring the price functions of various industries reveal that during the observation period from 1995 to 2011, there were very few industries within the Japanese economy where the endogenous synergy effects among the three intra-industry activities—main production, intra-firm R&D, and intra-firm ICT—were observed to be stable and positive. This observation could be considered one of the factors contributing to the stagnation and downturn of the Japanese economy following the bubble economy of the 1990s. This matter is closely related to the shift in corporate investment focus from tangible fixed assets, such as capital investments, to intangible fixed assets, such as knowledge and information. The expansion of intangible fixed asset investments is significantly transforming the industrial structure and efficiency in Japan's economy. However, it can also be interpreted that these structural changes have not yet effectively leveraged the utilization and application of intangible assets.

This represents an extremely important issue when discussing future science and technology policies as well as industrial policies. Making technological choices more rational through policy measures is a crucial task for both science and technology policy and industrial policy.

The propagation of technological efficiency involves both endogenous and exogenous synergy effects. Changes in the factor prices of tangible and intangible fixed capital services, labor services, and intermediate raw materials, transmitted through the supply and demand markets between industries, create synergy effects of endogenous and exogenous technological efficiency through those factor markets. By tracing these relationships, it is possible to examine how main production activities, intra-firm R&D, and intra-firm ICT activities are influenced by other industries through the markets for intermediate raw materials, labor, tangible capital, and intangible capital services, for each industry.

In Chap. 3, the growth rates of Unit TFP and static Unit TFP for each activity in each industry were estimated as the average annual rate over 5-year intervals from 1995 to 2011. As a result, it was confirmed that, for all three activities, the number of industrial sectors where static Unit TFP exceeds the TFP of each activity gradually increased over time. This indicates that synergy effects across industries have grown larger in each activity.

Furthermore, the impact of advancements in information processing technology, often referred to as the "information revolution," can be detected through changes in the intermediate input structure from the information processing industry to other industries. Observations across time reveal that the inter-industry synergy effects brought about by this technological evolution have grown, and the scalability of information technology has expanded significantly.

The spillover effects of the information industry—an independent industrial sector comprising telecommunications, broadcasting, software, and information processing service providers—on other industries can be understood as the provision of intangible knowledge services in the form of information services as market goods for intermediate inputs to other industries. A time-series analysis of real intermediate input coefficients from the telecommunications, broadcasting, software, and information processing service industries to other industrial sectors shows a rapid expansion of intermediate input from the telecommunications and broadcasting sectors in industries excluding electronics-related manufacturing (e.g., consumer electronics, semiconductor manufacturing equipment, communication devices, computers, semiconductors and integrated circuits, other electronic components, heavy electrical equipment, transportation machinery, robots, and other precision machinery).

On the other hand, intermediate input from the software industry to other industries, unlike those from the telecommunications and broadcasting sectors, showed a decreasing trend in many sectors. This appears to be related to the internalization of information processing within companies, driven by the widespread adoption of packaged software. However, certain sectors, such as public and non-profit research and development institutions and industrial R&D departments, exhibit a steady trend of expanding intermediate input coefficients. This can be attributed to advancements in information and digitization in scientific and technological research, as well as the increasing sophistication of software and systems for promoting research and development and their externalization. This trend may also reflect to be platforming the information service provision.

The progress of information science, often referred to as the information revolution, is steadily advancing in Japan as well. This clearly demonstrates the role of the information industry as a platform and highlights the increase in productivity within the information industry. Moreover, the technological sophistication of the electronics industry, which supports the information industry, plays a critical role in enhancing efficiency across all industries. Collaboration with the information industry is essential for revitalizing the activities of public and private research institutions, and policy support is indispensable for expanding its effects through the market.

The synergy effects of information technology and its expansion into industrial sectors exemplify the scalability of information technology. This expansion of information technology, characterized by the broadening of intangible fixed assets across industries, differs from the tangible fixed asset-based economies of scale that defined the rapid economic growth period of the manufacturing sector. It represents a cross-sectional spread across industries and can be described as a spatially heterogeneous scalability effect.

The societal impact of government-driven development of scientific and technological knowledge as a public good exhibits entirely different ripple effects compared to the societal impact of information processing technology as a market good, even though both constitute intangible capital services. Public R&D investment is carried out as government R&D expenditure and allocated as research funding to national and public universities, public research institutions, private universities, and other non-profit research organizations. These activities create a stock of knowledge as a public good, which is then disseminated as intellectual capital to private-sector R&D and commercial research institutions, promoting efficiency improvements in the private sector.

However, the dissemination of this knowledge often occurs not through market transactions but as a form of public knowledge good, provided without cost to private entities. While university-originated patents have become more prevalent in recent years, much of this knowledge transfer takes place in a non-market context, making it difficult to explicitly capture its impact as one might with intermediate goods transactions. Nonetheless, the dissemination of such knowledge undeniably contributes to production improvements for those who utilize it.

At the same time, there has been a noticeable increase in recent years in the trade of knowledge services as market goods, in the form of patent fees or license fees. In such cases, it is increasingly important to introduce clear mechanisms for the exchange and management of intellectual property as part of science and technology policy. This would enable a more precise assessment and management of the impacts of these transactions.

In the current macroeconomic national accounts, domestic exchanges are included under property income. Within a single country, such transactions are offset in aggregate and therefore do not explicitly appear. It is desirable to explicitly capture inter-industry transactions, particularly when they exhibit characteristics of non-public goods, and to account for these transactions as cost elements that are reflected in technological structures. Furthermore, making these transactions explicit as part of market activities is also considered important.

The international trade of intellectual property has also expanded in recent years, and effective management of these transactions has become a critical issue for maintaining and further enhancing a nation's industrial competitiveness. In exploring the realities of Global Value Chain formation, it is crucial to explicitly account for such exchanges to envision the future landscape of international labor division. This represents an urgent and significant policy challenge.

In the current environment, where intangible fixed capital investment has become more significant than the expansion of tangible fixed capital investment, it is crucial to recognize the unique role of intangible capital. As a characteristic of intangible assets, such investments influence productivity across activities within industries and between different sectors, contributing to changes in industrial structures. The roles of knowledge and information, as outputs of research and development, are particularly critical. These forms of intangible capital must be valued even more than tangible fixed capital, as they possess distinct and essential attributes. Effectively leveraging these characteristics is an indispensable element for future societal development.

In Japan, however, the development of a market for trading intangible asset services to enhance the efficiency and effectiveness of intangible fixed capital usage remains insufficient. The formation of a sound market mechanism for determining prices and facilitating transactions is inadequate. For revitalizing market functions in the twenty-first century, it is an urgent challenge to establish a visible market for knowledge trading, as seen in Silicon Valley. Unlike conventional markets for goods, labor, or finance, where explicit transactions between supply and demand determine prices, it may be difficult to definitively value knowledge and information through market transactions. Nonetheless, the establishment of a market that explicitly determines service prices related to intangible asset investments (such as research and development and information processing activities) is essential for activating market functions.

The lack of clarity in the market for intangible asset investments hinders rational resource allocation through competitive market mechanisms in Japan. This, in turn, prevents the nation from fully enjoying the benefits of accumulated knowledge and information. The advantages of Silicon Valley's accumulation stem from the creation of an implicit market, where suppliers and consumers of intangible assets engage directly to determine the prices of knowledge assets. This implicit market is a cornerstone of its success, enabling the efficient utilization of intangible capital and fostering innovation and growth.

The twenty-first century has brought about significant changes compared to the twentieth century, necessitating a fundamental rethinking of the approach to science and technology policy. This shift has also led to the outflow of Japan's intellectual property and information assets to overseas markets, thereby weakening its international competitiveness. The era when science and technology policy and industrial policy could be discussed independently has come to an end.

Now, the outcomes of science and technology policy are transforming industrial structures, and any changes to these policies increasingly require a comprehensive framework to determine which areas of science and technology need to be promoted. It is becoming ever more critical to examine these policies in an integrated manner, ensuring that they align with industrial strategies and broader national objectives.

The transition from the medieval period to the seventeenth and eighteenth centuries brought significant changes to the relationship between science and society in the modern era. This development was grounded in three fundamental principles: the logical norms of science and technology through induction, the expansion of markets and the principle of market competition, and the democratic ideals that connected these two norms to form the foundation of modern society. These principles facilitated the expansion of human history and the construction of a society that brought prosperity and abundance.

However, the progress of humanity in this period came to rely on growth as an end rather than as a means to an end, which became a fundamental characteristic of the modern era. As a result, distortions arose in these three principles. Growth began to hinder equality and freedom, while the evolution of science and technology exposed and exacerbated disparities between social classes, ethnic groups, and institutions, leading to increased social instability.

At the core of the social structures that have supported this progress lies a rare evolution of knowledge in human history, which underpinned economic growth. This economic growth was driven by massive "investment" in the formation of tangible fixed capital, supported by advancements in science and technology. The expansion of infrastructure and facilities through such investments enhanced productivity, enabling sustained growth. However, this expansion also resulted in the depletion of natural resources, environmental destruction, and the social stratification into capitalists and laborers, further widening disparities between social classes and intensifying related tensions.

This book seeks to address these contemporary challenges by advocating for the expansion of the role of "intangible capital" as a new form of capital formation. Intangible capital is divided here into "knowledge" and "information," with the argument that these elements should guide reforms toward a new social structure. The formation of a society that effectively utilizes intangible fixed capital—knowledge and information—is proposed as the ideal model for the relationship between science and society in the twenty-first century. This vision represents the path that "trans-science" must aim to achieve in the future.

In the current environment, where intangible fixed capital investment has become more significant than the expansion of tangible fixed capital investment, it is crucial to recognize the unique role of intangible capital. As a characteristic of intangible assets, such investments influence productivity across activities within industries and between different sectors, contributing to changes in industrial structures. The roles of knowledge and information, as outputs of research and development, are particularly critical. These forms of intangible capital must be valued even more than tangible fixed capital, as they possess distinct and essential attributes. Effectively leveraging these characteristics is an indispensable element for future societal development.

In Japan, however, the development of a market for trading intangible asset services to enhance the efficiency and effectiveness of intangible fixed capital usage remains insufficient. The formation of a sound market mechanism for determining prices and facilitating transactions is inadequate. For revitalizing market functions in the twenty-first century, it is an urgent challenge to establish a visible market for knowledge trading, as seen in Silicon Valley. Unlike conventional markets for goods, labor, or finance, where explicit transactions between supply and demand determine prices, it may be difficult to definitively value knowledge and information through market transactions. Nonetheless, the establishment of a market that explicitly determines service prices related to intangible asset investments (such as research and development and information processing activities) is essential for activating market functions.

The core of our country's science and technology policy lies in supporting national, public, and non-profit research institutions to accumulate "intellectual assets" as intangible fixed assets. On the other hand, the goal of our country's industrial policy has been to guide the industrial structure to achieve efficient economic growth. If economic growth is considered a means rather than an end, then the two policy objectives of accumulating "intellectual assets" and achieving "economic growth" must be pursued simultaneously as policies that lead the new era.

Our analysis shows that the characteristics of technological progress in national, public, and non-profit research institutions exhibit a utilization bias in all scientific fields concerning tangible capital services. In other words, tangible fixed capital for research and development tends to be utilitarian for technological progress in these scientific fields. Put differently, an increase in the price of these tangible capital services is shown to delay technological advancements in research and development. Regarding labor services, there is a utilitarian bias in the fields of life sciences and materials, whereas the fields of information and communication, environment and energy, and others exhibit cost-saving characteristics. As for intermediate raw materials, the fields of life sciences, materials, and others tend to be cost saving, while the fields of information and communication and environment and energy exhibit a utilitarian bias.

Considering the bias characteristics of each production factor in research and development and implementing measures to guide their prices could serve as a potential policy under science and technology strategies. The accumulation of intellectual assets as public goods and investment in such research and development are expected to be significantly influenced by governmental policy decisions. However, it is essential to align these decisions with the characteristics of technological progress biases highlighted here, ensuring consistency with industrial policies.

On the other hand, the technological characteristics of research and development in private research institutions (industries) engaged in profit-oriented activities exhibit certain distinctions compared to non-profit research and development institutions. Specifically, regarding tangible fixed assets, there is little bias in technological progress, and it remains largely neutral. Conversely, labor services exhibit a clear labor-saving bias across all scientific fields, where increases in labor service prices significantly stimulate technological development. Regarding raw materials, a utilitarian bias is consistently and distinctly observed across all fields.

Chapters of the research and development sector and promoting research and development in our country fall under the role of science and technology policy, the ability to effectively promote these efforts depends on the trends in the prices of each production factor. Guiding these trends in a desirable direction is the role of industrial policy. Moreover, the scientific and technological achievements realized through the advancement of science and technology policy have a profound impact on the future formation of industrial.

Finally, the purpose of the preceding analysis has been to present a normative framework for future social development by illustrating the evolving relationship between science, technology, and society. This relationship emphasizes how research and development, as well as information processing activities, support the main production activities of traditional industries. It highlights how intangible assets such as "knowledge" and "information" invigorate economic development and promote social progress through the enhancement of spiritual and cultural civilization.

We believe we have provided a method to conceptualize, measure, and detect the economic effects of such intangible fixed assets. While there is considerable room for improvement to enhance the precision of these methods, what we wish to emphasize in concluding this document is the necessity of regularly compiling input-output tables that clearly capture the role and activation mechanisms of intangible fixed as proposed here. These tables should serve as a basis for creating a "policy simulator" that can evaluate the successes and failures of policies aimed at guiding the development of science, technology, and society.

Establishing this as a national project is indispensable for fostering a healthy relationship between science and society in the future. By doing so, we can ensure the effective integration of science and technology policy with industrial policy, paving the way for sustainable societal advancement.

Based on the analysis of how intellectual assets accumulated through R&D activities and how information assets from information processing activities, as intangible fixed assets, transform the economic and social structure in place of tangible fixed assets such as equipment investment, we draw conclusions about the future direction of science and technology policy and industrial policy.

To this end, we present a theoretical framework for capturing producers' technological choices in industrial production activities. Through this structural model, we measured the parameters of the long-term cost functions of various industries to understand the technological characteristics related to the input of production factors in the main production activities, as well as the R&D and the ICT activities within each industry. We also examine the nature of technological linkages and spillovers between industries.

As concerns the subject about "the decomposition of technological change and the accumulation of capital in Japanese economic growth", see Kuroda and Nomura (2004).

Furthermore, the historical perspectives offered by Leon Walras in *Elements of Pure Economics* laid the foundational framework for understanding social wealth theory. Wassily Leontief's studies, notably *The Dynamic Inverse* and his comprehensive examination of the American economy, underscore the significance of input-output analysis, which remains influential in contemporary economic discourse.

The OECD's "Handbook on Driving Capital Measures of Intellectual Property Products" serves as a crucial reference for understanding the intricacies of intellectual property and its impact on economic development. Meanwhile, Juro Teranishi's insights into Japan's postwar economic stabilization highlight the role of government intervention during critical periods of growth and inflation control.

These seminal works contribute to a profound understanding of economic dynamics and the pivotal role of intellectual property in shaping modern economic strategies.

References

Jorgenson, D. W., & Kuroda, M. (1992). Productivity and international competitiveness in Japan and the United States, 1960–1985. *The Economic Studies Quarterly, 43*(4), 313–325.

Jorgenson, D. W., Kuroda, M., & Nishimizu, M. (1987). Japan-U.S. industry-level productivity comparison, 1960–1979. *Journal of the Japanese and International Economies, 1*, 1–30.

Kuroda, M., & Nomura, K. (2004). Technological change and accumulated capital: A decomposition of Japanese economic growth. Chapter 15. In E. Dietzenbach & M. L. Lahr (Eds.), *Wassily Leontief and input-output economics*. Cambridge University Press.

Leontief, W. W. (1970). The dynamic inverse. In A. P. Carter & A. Brody (Eds.), *Contributions to input-output analysis*. North-Holland Publishing.

Yang, H., Liou, M. (2022) Application of Computable General Equilibrium Analysis to STI Policy Evaluation: Illustrations of R&D Investment in Tai-wan's ICT Industry, *Taipei Economic Inquiry, 58*(1), 49–106.

Author Index

M. Kuroda (ed.), *Intangible Capital Investment and Economic Growth*, Translational Systems Sciences 45, https://doi.org/10.1007/978-981-96-7675-0

Subject Index

M. Kuroda (ed.), *Intangible Capital Investment and Economic Growth*, Translational Systems Sciences 45, https://doi.org/10.1007/978-981-96-7675-0

The manufacturer's authorised representative in the EU is Springer
Nature Customer Service Centre GmbH, Europaplatz 3, 69115 Heidelberg,
Germany. If you have any concerns regarding our products, please
contact ProductSafety@springernature.com

Printed and bound by CPI Group (UK) Ltd, Croydon, CR0 4YY
21/09/2026
02220008-0006